p.132

Introductory
Computer
Methods
and
Numerical
Analysis

Introductory

Second Edition

THE MACMILLAN COMPANY

COLLIER-MACMILLAN LIMITED, LONDON

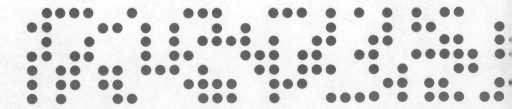

RALPH H. PENNINGTON

Computer Methods and Numerical Analysis

PRINTING   3456789   YEAR   456789

Earlier edition © copyright 1965 by Ralph H. Pennington.

Library of Congress catalog card number: 74-78965

The Macmillan Company
Collier-Macmillan Canada, Ltd.
Toronto, Ontario

Printed in the United States of America

Preface

In the five years since the publication of the first edition, the computer field has continued to develop at a phenomenal rate, in programming and application concepts as well as in hardware performance. In this same period instruction techniques and training methods in the field have also been undergoing a revolution. The computer sciences department has become accepted as a separate organizational entity in many institutions of higher learning, the beginning of a trend that seems likely to continue. The impact of computers is so broad, however, that within these same institutions the departments devoted to science, engineering, business administration, and so on, generally find it necessary to include some computer courses for their own purposes.

In spite of these developments the author is somewhat tempted to promote the immodest claim that the first edition has not really become outdated. A very large portion of the material is as applicable now as when first written. (Critics may interpret that observation as they wish.) It would seem, however, that proper revision may make the material more directly applicable in the present-day environment. The new edition is the author's attempt in this direction.

In performing the updating task the author was of course faced with a number of choices concerning what material to delete, what to add, and how to treat the various subjects selected for coverage. In rethinking the aims of the book, the author found himself drawn more toward the needs of departments of science and engineering, who view the computer as a tool for problem solving, rather than toward the computer science department, who view the computer as a subject in itself. Thus while a strong attempt was made to expand on fundamental concepts and add more rigor, these areas were allowed to suffer to some degree in order to include a sufficient number of descriptions of algorithms to leave the reader with a reasonably versatile beginner's kit of problem solving tools.

Probably the most significant single development in recent years related to computer instruction is the improvement in man-machine interaction. No longer must each computer program be punched on punch cards and carried to the computer room, there to be logged in and remain for some indeterminate period before the answers (or messages indicating that the program was wrong and no answers are forthcoming) are returned. No longer is the computer a dumb brute, slaving its way through the assigned problems one at a time. The computer is fast becoming a brilliant conversationalist, carrying on direct repartee with a score of users simultaneously, on a score of different problems. This development has such an impact on the way a user, especially the occasional user, gets his problems solved that it appeared advisable to make remote terminal usage and remote terminal language an integral part of the text. Thus in Chapter 5 a language for man-machine interaction is introduced, and in nearly all the succeeding chapters the first algorithm presented is one designed for remote terminal use. It seems to the author that this approach opens exciting possibilities for student participation and classroom exploration of interesting avenues of investigation by quick and easy modification of the remote terminal programs. By the addition of extra print statements to allow the student to observe the machine's progress through computational steps or to watch the convergence or divergence of an iteration process, or by the rerun of calculations with other input numbers, the clever instructor can provide the student with insights well beyond those available from the few examples given in the text itself.

Some hesitancy regarding the inclusion of remote terminal language was occasioned by the fact that there is as yet no standard language for man-machine interaction. Thus examples of remote terminal program may be directly applicable for only a very few of the potential users of the text. However, the author felt that (1) with the particular language adopted, the modifications from standard FORTRAN are so slight that the programs are readily converted to conventional form at installations not having interactive capability, (2) some installations may well find it easy and worthwhile to provide an interactive language matching the text, (3) the text itself may help promote a move toward standardization of an interactive language, and (4) the potential advantages of the interactive approach far outweigh any inconveniences occasioned by its adoption.

Several lesser modifications in the text may be worth mention. In Chapter 1, some of the simpler material was deleted and the converage of hexadecimal arithmetic was expanded in deference to the growing use of machines oriented toward hexadecimal notation. Descriptions of machine hardware and machine language programming were condensed into Chapter 2, to make inclusion or omission of the material in a course an easy option for the instructor. Several agonizing reappraisals were made on the subject of inclusion of a hypothetical machine language. In the end, the language was included. However, the completely artificial form of the first edition was replaced by one paralleling modern third-generation computers. The coverage, while highly condensed, is actually more complete than that of the first edition, in that it includes a somewhat more realistic treatment of input-output commands and a sufficient coverage of logical instructions to give a direct, though sketchy, introduction to assemblers, compilers, and operating systems.

Chapter 3, on accuracy considerations, has been raised to a slightly more advanced level, with a somewhat more elaborate treatment of error propagation. Chapter 4 consolidates the description of the FORTRAN language from the old Chapters V and VII into a single package, with an expanded description of subprograms and an increased number of examples. Chapter 5 is largely new, introducing flow charts from a viewpoint that seems somewhat more satisfactory to the author than that in the first edition. Chapter 5 also introduces the remote terminal language discussed earlier. Chapter 6 consolidates some of the old Chapters VII and VIII, combining the discussion of evaluation of functions with the accuracy considerations relating thereto. Chebyshev series are discussed, perhaps to a greater extent than their current usage warrants. The author was moved to include this extra coverage largely to have a vehicle for displaying some useful programming techniques for manipulation of the recursion relations involving coefficients in the series. Chapter 7 is an expanded treatment of quadrature, including the Romberg method and an expanded treatment of Gaussian quadrature. Chapter 8 is nearly the same as the old chapter on solution of algebraic and transcendental equations, the main change being the inclusion of a computer-assisted version of the method of false position. In Chapter 9, the solution of polynomial equations, the main change has been abandonment of Sturm sequences in favor of the root squaring method, now popularly preferred because of better understanding of its accuracy problems. Chapter 10 combines the old Chapters XII and XIII on linear equations and matrices. The main difference in treatment is use of the Gauss-Jordan method in place of the Gaussian elimination method, the former being slightly less efficient in running time but giving shorter and more understandable FORTRAN subroutines. To further shorten these subroutines, the operations relating to interchanging rows and columns have been removed to a separate subroutine. Chapters 11 and 12 are the old Chapters XIV and XV, with only trivial changes. Chapter 13 is a significantly revised treatment of differential equations with reduced emphasis on mechanics and increased emphasis on concepts. More introductory material on the Runge-Kutta method is included, along with a discussion of accuracy and stability. The Euler-Romberg method has been added. Milne's method has been replaced by the Adams-Moulton formulas as an example of predictor-corrector methods, on the basis of better stability properties.

Throughout the book an attempt has been made to make the problem sets more extensive and more instructive. However, the author believes that, especially with remote terminal access, the computer is its own best teacher, and that additional problems beyond those in the text will and should suggest themselves to the imaginative reader.

R. H. P.

Contents

Introductory Computer Methods and Numerical Analysis

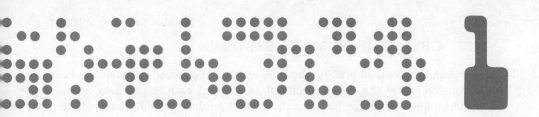

Number Systems

1.1 INTRODUCTION

In this book we intend to explain the use of the digital computer as a tool for mathematics through calculus. Specifically we are going to address ourselves to that part of mathematics having to do with calculating, that is to say, performing operations with numbers. If we wish to start at the beginning, then, we must first talk about numbers and number systems. This might sound like an altogether too elementary approach. After all, we learned about numbers and counting in grade school. Why rake over a subject so thoroughly learned so long ago? Surprising as it may seem to readers not already acquainted with digital computers, there is indeed a compelling reason for reviewing such basic topics as numbers, counting, and the basic arithmetic operations of addition, subtraction, multiplication, and division. The reason is this: In school we learned to count by tens, and we ordinarily do all our arithmetic in a number system based on ten. There are other ways of counting, other ways of doing arithmetic. While the tens system seems the simplest to us, because of our long acquaintance with it, there are certain disadvantages to it. For reasons which will appear as the reader proceeds, digital computers can often be made to perform more rapidly and more simply if they are designed to use number systems other than our familiar tens system. Thus there is a real profit to be gained by starting with a review of number systems, provided we do so from a broadened viewpoint and free ourselves of the habits of thought about numbers and counting that have developed from long association with one particular system, the tens system.

1.2   ORIGIN OF NUMBER SYSTEMS

Throughout recorded history, the development of civilized societies has been accompanied by the development of systems of counting. These number systems appear to have been the product of gradual growth and evolution, rather than the product of a single discovery.

In each primitive society, although such events cannot be documented by recorded historical evidence, there must have been a series of minor discoveries or developments in the use of numbers. The earliest stage was probably a process of comparison. One can easily visualize primitive man, prior even to the development of more than a rudimentary language, comparing two piles of rocks by matching the rocks pair by pair, and thus ascertaining that there are more rocks in one pile than in the other. Such a one-to-one system of comparison must have been essential to even the most elementary type of barter. A next logical development is comparison of objects with the fingers, and for numbers less than ten, at least, the representation of the number of objects in a group by holding up an equal number of fingers. This representation of numbers by fingers is in all probability by far the oldest means used by man. From primitive beginnings, the representation of numbers on the fingers was gradually extended to include numbers in the thousands, by the introduction of special finger combinations for units, tens, hundreds, and thousands. At some early stage, names and symbols for these numbers developed. Egyptian hieroglyphic numerals as early as 3300 B.C. were based on symbols for 1, 10, 100, 1000, and 10,000.

Although the usefulness of the fingers in number representation has led to widespread development of number systems based on multiples of ten, the number ten is by no means the only base used for number systems. The Babylonian system made combined use of ten and sixty as bases. Traces of number systems based on three, four, five, six, eight, and twenty are found among the Indians of North America. Some suggestion of a number system based on twelve is contained in our twelve-inch foot and the British twelve pence in a shilling. In modern societies, however, the number system based on ten seems to have won out: not because of any intrinsic advantage, but rather, it would appear, because of the presence of ten fingers on the hands. The science of computing has introduced a requirement of number systems based on practical considerations of a different sort. In particular, the binary system (base 2), the octal system (base 8), and the hexadecimal system (base 16) have been found to offer certain advantages for use with computers. We will need to develop some degree of familiarity with these number systems before proceeding toward an understanding of digital computers.

1.3   THE DECIMAL SYSTEM

Our present decimal system is a development of the Hindu-Arabic system which originated in India, probably in the third century B.C. A major step was the invention, sometime between the second century B.C. and the ninth century A.D., possibly by an unknown priest or scholar, of a symbol for zero. Simple as it sounds to us now, this was a real stepping stone in the development of our number system; with the use of a symbol for zero, a so-called "place value" system became feasible. It became possible to write any integer, however large, by the use of the proper combination of characters chosen from ten basic ones. The advantage of this development becomes clear when one reflects on the difficulties encountered in trying to write very large numbers in the Roman numeral system. The next development of consequence, which probably occurred in about the sixteenth century, was the introduction of the decimal point and generalization of the symbolism to include decimal fractions. Since that time, it has been possible to represent all positive whole numbers and fractions in the decimal system by the use of the zero and nine other characters. The use of the minus sign allows extension of the notation to include negative numbers. In modern notation, the basic characters of the decimal system are

$$0, \quad 1, \quad 2, \quad 3, \quad 4, \quad 5, \quad 6, \quad 7, \quad 8, \quad 9$$

The number next after 9 is represented by re-using the character 1 followed by the character 0. In general, for any number the digit first to the left of the decimal point stands for the number of units, the digit next to the left stands for the number of tens, the digit next to the left stands for the number of hundreds (ten times ten), and so forth. Likewise, the digit to the right of the decimal point stands for the number of tenths; the next, the number of hundredths; and so forth. For example, the number 963.05 stands for 9 hundreds plus 6 tens plus 3 units plus 0 tenths plus 5 hundredths. These values are indicated by the following diagram:

$$
\begin{array}{ccccc}
9 & 6 & 3 \quad . & 0 & 5 \\
\downarrow & \downarrow & \downarrow & \downarrow & \downarrow \\
9 \times 10^2 & 6 \times 10^1 & 3 \times 10^0 & 0 \times 10^{-1} & 5 \times 10^{-2}
\end{array}
$$

The operations of decimal arithmetic are well known to the reader. The grade school student learns addition and multiplication tables for the nine basic characters and zero and some rules for handling carry from one operation to the next. With these, he learns eventually to perform arithmetic operations on any finite collection of decimal numbers. In other number systems, there are analogous sets of rules for arithmetic. The sections below will describe these for some other number systems.

1.4 THE DOZEN SYSTEM

We are so accustomed to counting by tens that when we try to utilize a different number system there are many habits of thinking which are quite difficult to discard. In order to point out some of these pitfalls, we shall next discuss the dozen, or duodecimal, system. In this system let us use the symbols

$$0, \quad 1, \quad 2, \quad 3, \quad 4, \quad 5, \quad 6, \quad 7, \quad 8, \quad 9, \quad D, \quad E$$

as the basic characters. The letter D is for the decimal number ten and the letter E for decimal eleven. Let us call them dec and el to keep from confusing them with the decimal names. The number next after el is one dozen, which in this notation would be written 10. The next number, which is the decimal number thirteen, is one dozen and one, which would be written 11. In some ways it would have been better to discard the symbols 1 through 9 and use entirely new ones for our basic characters, because the use of the decimal symbols suggests decimal rules which are not true in the duodecimal system. For example, the decimal addition rule six plus five equals eleven is to be replaced by six plus five equals el:

$$6 + 5 = E$$

The decimal rule six plus seven equals thirteen is to be replaced by six plus seven equals one dozen and one, or

$$6 + 7 = 11$$

Thus care must be used not to slip back into thinking by decimal rules. A new table of addition must be learned for the basic characters of duodecimal arithmetic, and likewise a new table of multiplication. For example, five times eight equals a decimal forty, or duodecimal three dozen and four, or

$$5 \times 8 = 34$$

To write duodecimal numbers of any size, we employ the place value system, using a digit location with respect to a duodecimal point (*not* decimal point) to determine its value. Each place to the right or left of the duodecimal point differs in value from its neighboring one by a factor of a dozen. The diagram shows the decimal value of each of the digits in the duodecimal number 4E7.2D9:

4	E	7	.	2	D	9
↓	↓	↓		↓	↓	↓
4×12^2	11×12^1	7×12^0		2×12^{-1}	10×12^{-2}	9×12^{-3}

The dozen system is in some respects more convenient than the decimal system. This convenience results primarily from the fact that twelve has many more divisors than ten. Twelve is divisible by one, two, three, four, six, and twelve. Thus many hand calculations are somewhat simpler in duodecimal than in decimal notation. Several of the common fractions which are repeating in decimal form are not in duodecimal form. For instance, the fraction one-third is the same as four-twelfths, or in duodecimal form .4. Some of the simple fractions in duodecimal form are

$$1/6 = \text{decimal } 2/12 = \text{duodecimal } .2$$
$$1/4 = \text{decimal } 3/12 = \text{duodecimal } .3$$
$$1/3 = \text{decimal } 4/12 = \text{duodecimal } .4$$
$$1/2 = \text{decimal } 6/12 = \text{duodecimal } .6$$

Despite its convenience, the duodecimal system will in all probability never be adopted for hand calculations. It is not particularly advantageous for computer applications so we leave it without further discussion and turn to the systems that are most useful for digital computer applications.

In doing so, we may take along certain principles about numbering systems which carried over from the decimal system to the duodecimal system and which will carry over to other systems. First, the system must have a set of symbols to represent digits. The decimal system had ten, 0 through 9. The dozen system had twelve. The number of basic digit symbols determines the base of the number system, or radix, as it is usually called. With the digit symbols, numbers of any size can be represented by using the place value system. A period is used to make the reference point in determining value of digits. In the decimal system, this was termed the decimal point; in the duo-decimal system, the duodecimal point. In general, it is called the radix point. The first position to the left of the radix point is the units position. A digit in this position has its basic value only. If a digit is moved to the left or right of this position, its value changes. A movement of one place to the left multiplies its value by a factor equal to the radix. A movement of one place to the right divides the value by an amount equal to the radix. Arithmetic for numbers of any radix can be performed in a manner analogous to decimal arithmetic, once the addition and multiplication tables for the basic symbols are learned.

1.5 THE BINARY SYSTEM

A very useful system for digital computer work is the binary system, or the number system with base two. The only basic characters of this system are

$$0, \quad 1$$

The number next after one, or the decimal number two, is written as 10. The next number, or decimal three, is written as 11. As in any other system, we use the place value method, where the location of a digit with respect to the binary point determines its value. Each place to the right or left of the binary point differs in value from its neighbor by a factor of two. The diagram shows the decimal value of each of the digits in the binary number 101.101:

$$
\begin{array}{cccccc}
1 & 0 & 1 & . & 1 & 0 & 1 \\
\downarrow & \downarrow & \downarrow & & \downarrow & \downarrow & \downarrow \\
1 \times 2^2 & 0 \times 2^1 & 1 \times 2^0 & & 1 \times 2^{-1} & 0 \times 2^{-2} & 1 \times 2^{-3}
\end{array}
$$

A major advantage of binary notation is the fact that only two digits, 0 and 1, are required in the representation of any number. These digits, or bits, as they are frequently called, can readily be represented by any of many physical systems capable of being in either of two different states. For example, on a paper tape or card, a 1 can be represented by a hole and a 0 by the absence of a hole; on magnetic tape or other magnetizable material, a 1 can be represented by a magnetized spot and a 0 by no magnetization or by magnetization of opposite polarity; in an electric circuit, a 1 can be represented by a voltage pulse and a 0 by no pulse or by a pulse of opposite polarity.

Another advantage of binary notation is that, with only two characters, very few laws are required to cover the possible combinations of these in addition and multiplication. For example, the basic multiplication tables consist of $0 \times 0 = 0$, $0 \times 1 = 1 \times 0 = 0$, $1 \times 1 = 1$. This will be pursued further in later sections.

A major disadvantage of the binary system is the large number of bits required to express numbers of very moderate size. For example,

$$\text{binary } 1000000000000 = \text{decimal } 2^{12} = \text{decimal } 4096$$

Thus a thirteen-digit binary number may be required to represent a four-digit decimal number. In general, since

$$\log_{10} 2 = .30103$$

then

$$2^N = 10^{N \log_{10} 2} = 10^{.3N}$$

so that an N-bit binary number is roughly equivalent to a $.3N$-digit decimal number. While computers frequently use binary numbers internally to exploit the advantages mentioned above, the numbers are frequently read into and out of the machine in some other form to avoid the tedious process of handling numbers having very large number of digits.

TABLE 1
Powers of 2

2^n	n	2^{-n}
1	0	1.0
2	1	0.5
4	2	0.25
8	3	0.125
16	4	0.062 5
32	5	0.031 25
64	6	0.015 625
128	7	0.007 812 5
256	8	0.003 906 25
512	9	0.001 953 125
1 024	10	0.000 976 562 5
2 048	11	0.000 488 281 25
4 096	12	0.000 244 140 625
8 192	13	0.000 122 070 312 5
16 384	14	0.000 061 035 156 25
32 768	15	0.000 030 517 578 125
65 536	16	0.000 015 258 789 062 5
131 072	17	0.000 007 629 394 531 25
262 144	18	0.000 003 814 697 265 625
524 288	19	0.000 001 907 348 632 812 5
1 048 576	20	0.000 000 953 674 316 406 25
2 097 152	21	0.000 000 476 837 158 203 125
4 194 304	22	0.000 000 238 418 579 101 562 5
8 388 608	23	0.000 000 119 209 289 550 781 25
16 777 216	24	0.000 000 059 604 644 775 390 625
33 554 432	25	0.000 000 029 802 322 387 695 312 5
67 108 864	26	0.000 000 014 901 161 193 847 656 25
134 217 728	27	0.000 000 007 450 580 596 923 828 125
268 435 456	28	0.000 000 003 725 290 298 461 914 062 5
536 870 912	29	0.000 000 001 862 645 149 230 957 031 25
1 073 741 824	30	0.000 000 000 931 322 574 615 478 515 625
2 147 483 648	31	0.000 000 000 465 661 287 307 739 257 812 5
4 294 967 296	32	0.000 000 000 232 830 643 653 869 628 906 25
8 589 934 592	33	0.000 000 000 116 415 321 826 934 814 453 125
17 179 869 184	34	0.000 000 000 058 207 660 913 467 407 226 562 5
34 359 738 368	35	0.000 000 000 029 103 830 456 733 703 613 281 25
68 719 476 736	36	0.000 000 000 014 551 915 228 366 851 806 640 625
137 438 953 472	37	0.000 000 000 007 275 957 614 183 425 903 320 312 5
274 877 906 944	38	0.000 000 000 003 637 978 807 091 712 951 660 156 25
549 755 813 888	39	0.000 000 000 001 818 989 403 545 856 475 830 078 125

7

1.51   Binary-Decimal Conversion

A number given in binary form is easily converted to decimal form, since
each bit simply represents some power of 2. The process will be illustrated.

Example 1.   Convert the binary number 11001101.1011 to decimal form.

This number is equivalent to $2^7 + 2^6 + 2^3 + 2^2 + 2^0 + 2^{-1} + 2^{-3} + 2^{-4}$:

$$
\begin{array}{rcl}
2^7 & = & 128 \\
2^6 & = & 64 \\
2^3 & = & 8 \\
2^2 & = & 4 \\
2^0 & = & 1 \\
2^{-1} = (1/2) & = & .5 \\
2^{-3} = (1/8) & = & .125 \\
2^{-4} = (1/16) & = & .0625 \\
\hline
& & 205.6875
\end{array}
$$

1.52   Decimal-Binary Conversion

The conversion of a decimal to binary form is somewhat more difficult
than the reverse process. It is performed most easily with a table of the
powers of 2 available. Table 1 gives all powers from 2^{39} to 2^{-39}.

A decimal number is converted to binary by subtracting out the largest
power of 2 contained, then the largest contained in the remainder, and so on.
The binary number is constructed by writing a one in each bit position
corresponding to a power successfully subtracted out.

An example will serve to clarify the process.

Example 1.   Convert the decimal number 2168.33 to binary form.

$$
\begin{array}{rcl}
& & 2168.33 \\
\text{Subtracting out } 2^{11} & = & 2048 \\
\hline
& & 120.33 \\
\text{Subtracting out } 2^6 & = & 64 \\
\hline
& & 56.33 \\
\text{Subtracting out } 2^5 & = & 32 \\
\hline
& & 24.33 \\
\text{Subtracting out } 2^4 & = & 16 \\
\hline
& & 8.33
\end{array}
$$

$$
\begin{array}{ll}
\text{Subtracting out } 2^3 \;\; = & \underline{\quad 8 \quad} \\
 & .33 \\
\text{Subtracting out } 2^{-2} = & \underline{\quad .25 \quad} \\
 & .08 \\
\text{Subtracting out } 2^{-4} = & \underline{\quad .0625 \quad} \\
 & .0175 \\
\text{Subtracting out } 2^{-6} = & \underline{\quad .015625 \quad} \\
 & .001875
\end{array}
$$

The process can be continued if a smaller remainder is desired. To the accuracy so far attained, our binary number has ones in the 12th, 7th, 6th, 5th, and 4th places left of the decimal point (one higher in each case than the power of 2 removed) and in the 2nd, 4th, and 6th places to the right of the decimal point. Thus the number is

$$100001111000.010101$$

Note that the conversion from decimal to binary was not exact. The fractional part of the decimal number terminated after two digits, whereas the complete binary representation would have a nonterminating fractional part.

1.53 Binary Addition

The basic laws of binary addition are $0 + 0 = 0; 0 + 1 = 1 + 0 = 1; 1 + 1 = 10$. With these laws, and handling carry just as for decimal addition, we can add any two binary numbers.

Example 1. Find the sum of 11010.1 and 10110.0.

The sum is

$$
\begin{array}{r}
11010.1 \\
10110.0 \\
\hline
110000.1
\end{array}
$$

The steps are, starting at the right-hand end, $1 + 0 = 1; 0 + 0 = 0; 1 + 1 = 10; 1 + 0 + 1 = 10; 1 + 1 + 0 = 10; 1 + 1 + 1 = 11$.

The basic rules of binary addition are so simple that it is worthwhile to compile a summary of slightly more elaborate addition rules which include the carry digit directly. Note that, with the carry digit, for some digit positions in the above problem the combination of three bits is involved—the two binary bits of the summands and the carry from the next lower digit position. Two quantities result from this operation—the sum bit to be recorded and the carry to be taken to the next digit position. The complete process of addition for a single digit position consists of combining the summand digits and the incoming carry to obtain a sum bit and an outgoing carry. The possible combinations are given in Table 2.

TABLE 2

Binary Addition Rules

Digit Values								
First Summand	0	0	0	0	1	1	1	1
Second Summand	0	0	1	1	0	0	1	1
Incoming Carry	0	1	0	1	0	1	0	1
Result								
Sum Digit	0	1	1	0	1	0	0	1
Outgoing Carry	0	0	0	1	0	1	1	1

Binary addition is simply the process of applying the rules of Table 2 to each digit position, starting at the right and moving to the left. The incoming carry for the right-hand digit position is automatically zero. Addition of a large column of binary numbers is most readily accomplished by adding the first two, then the third to the sum, and so forth.

1.54   Binary Subtraction

As in decimal arithmetic, binary subtraction may be accomplished by applying the rules for binary addition in reverse.

Example 1.   Add algebraically 11011 and −10110.

The operation is

$$
\begin{array}{r}
11011 \\
-10110 \\
\hline
101
\end{array}
$$

The steps are $1 - 0 = 1$; $1 - 1 = 0$; $10 - 1 = 1$; $0 - 0 = 0$; $1 - 1 = 0$.

In binary arithmetic (and in decimal arithmetic as well), it is possible to avoid the requirement for a set of rules for subtraction by the use of so-called "complement" notation. In complement notation, each negative number is represented by a positive number related to it in some simple way. Then the operation of subtraction is replaced by addition using complements.

Example 2. Add algebraically -11011 and 10110.

Without using complements, we would subtract the second number from the first, since the second has the small numerical value, thus:

$$
\begin{array}{r}
11011 \\
(-)10110 \\
\hline
101
\end{array}
$$

Then because the numerically larger number was negative, we would affix a minus sign, obtaining -101 as our final answer.

Let us now use the one's complement for this problem. Since we are using five-digit numbers, we would form the one's complement of 11011 by subtracting it from 111111, giving 100100. It is interesting to note that this number differs from the original by having a one in the place of the minus sign and in each digit position having each " one " replaced by " zero " and each " zero " replaced by " one." Thus it is quite easy to form the one's complement of a binary number. With this notation, the above example becomes

$$
\begin{array}{r}
100100 \\
10110 \\
\hline
111010
\end{array}
$$

The answer is just the one's-complement representation of -101.

A slight modification of the one's-complement representation for negative numbers is two's-complement representation. In the above example, the one's complement was obtained by subtracting from 111111. The two's complement would be obtained by subtracting from a number just one greater, or 1000000. Thus the two's complement of 11011 is

$$
\begin{array}{r}
1000000 \\
-11011 \\
\hline
100101
\end{array}
$$

The two's complement can be obtained by subtraction, as we have just done, or by adding one in the last bit position of the one's complement, which we have already seen to be acquired by a very simple operation. With two's-complement notation, Example 2 becomes

$$
\begin{array}{r}
100101 \\
10110 \\
\hline
111011
\end{array}
$$

which is the two's-complement representation of -101.

1.55   Binary Multiplication

Binary multiplication can be performed in a manner analogous to decimal multiplication by use of the simple multiplication rules $0 \times 0 = 0$; $0 \times 1 = 1 \times 0 = 0$; $1 \times 1 = 1$.

Example 1.   Find the product of 1101.1 and 1.0111.

Starting this operation as in decimal arithmetic we have

$$
\begin{array}{r}
1101.1 \\
1.0111 \\
\hline
11011 \\
11011 \\
11011 \\
11011 \\
\hline
\end{array}
$$

At this point we arrive at the problem of adding several binary numbers together. As mentioned earlier, this is most readily accomplished by adding only two numbers at a time. In the following rearrangement, this is done, the addition being carried out after each step of the multiplication:

$$
\begin{array}{r}
1101.1 \\
1.0111 \\
\hline
11011 \\
11011 \\
\hline
\end{array}
$$
   first intermediate sum

$$
\begin{array}{r}
1010001 \\
11011 \\
\hline
\end{array}
$$
   second intermediate sum

$$
\begin{array}{r}
10111101 \\
11011 \\
\hline
10011.01101
\end{array}
$$
   final answer

The rule for placement of the binary point in the product is the same as for decimal arithmetic.

In ordinary decimal multiplication the practice is to start multiplying at the right-hand digit of both multiplicand and multiplier. There is some advantage to this, since one has the problem of carry from one column to the next one on the left. For example, in the decimal product

$$16$$
$$\times\ 27$$

the first step is to take 7 times 6, obtaining 42, and record the 2 and carry the 4 until the next product, 7 times 1, is formed. In binary multiplication, this problem never arises, since the only products are $0 \times 0 = 0$; $0 \times 1 = 0$; $1 \times 0 = 0$; $1 \times 1 = 1$. In no case does one of these operations give a two-digit result. Since no carry is involved in binary multiplication, the operation is just as easily performed from the left as from the right. In the binary product

$$1011.01$$
$$\times\ 1110.10$$

the first step can be taken by multiplying the multiplicand by the *left*-hand bit of the multiplier, and recording the answer beginning with the *left* bit under the left bit of the multiplier. Then the second bit of the multiplier can be used, and that product recorded starting under the second bit of the multiplier. At this stage, the work will appear as follows:

$$1011.01$$
$$1110.10$$
$$\overline{101101}$$
$$101101$$

In accordance with our practice of adding binary numbers only two at a time, the next step is addition of the two numbers so far obtained. Since binary addition *does* sometimes give a carry, it is necessary to perform the addition starting at the right-hand end of the numbers. After this is done, the calculation appears as

$$1011.01$$
$$1110.10$$
$$\overline{101101}$$
$$101101$$
$$\overline{10000111}$$

The next step is another multiplication, this time by the third digit of the multiplier. This operation, recorded in the proper location, gives

$$
\begin{array}{r}
1011.01 \\
1110.10 \\
\hline
101101 \\
101101 \\
\hline
10000111 \\
101101 \\
\hline
\end{array}
$$

Continuing this process of multiplications and additions, we finally obtain

$$
\begin{array}{r}
1011.01 \\
1110.10 \\
\hline
101101 \\
101101 \\
\hline
10000111 \\
101101 \\
\hline
100111011 \\
101101 \\
\hline
10100011.001 \\
\end{array}
$$

This procedure has the advantage that the most significant part of the answer is developed first and the less significant parts later. If it is known that only a certain number of significant digits are required in the answer, it is possible to carry the problem far enough to obtain that many significant digits and then quit.

1.56   Binary Division

In performing decimal division, we ordinarily use a trial and error process. A trial value is chosen for the first digit of the quotient and then multiplied by the divisor and the result subtracted from the appropriate digits of the dividend. Several guesses may be required to find the digit which produces a remainder which is neither negative nor larger than the quotient. The process is then repeated for each digit of the quotient.

Binary division is somewhat more direct than decimal division in that the trial and error nature of the process is removed. Each digit in the quotient will be either 1 or 0, indicating that the divisor will divide into a group of digits of the dividend either once or not at all. An example will serve to make the process clear.

Example 1.   Divide 1011 into 110.110100.

SOLUTION:

$$
\begin{array}{r}
.100111 \\
1011\overline{)110.110100} \\
101\ 1 \\
\hline
1\ 0101 \\
1011 \\
\hline
10100 \\
1011 \\
\hline
10010 \\
1011 \\
\hline
111
\end{array}
$$

EXERCISE 1

1. Write the first 64 decimal integers as binary integers.

2. Write the following binary numbers in decimal form.
    a. 101.11   b. 100111.011
    c. 111.1111   d. -100000.0001
    e. .101010101   f. -11000000001

3. Write the following decimal numbers in binary form.
    a. 128.375   b. 4.5
    c. 63.875   d. 5460.7
    e. .931842   f. 764893.9

4. Perform the following binary additions and subtractions.
    a. 1.11011   b. .110010
        1.00100   .000111

    c. 11.0111   d. .111001
        10.1101 $-.101001$

    e. .11110   f. .10001
        $-.10101$ $-.01111$

5. Perform the following binary multiplications.
    a. .110101   b. .011
        $\times .011011$ $\times .011$

    c.   .1001   d.   .1111
        $\times$.1111 $\times$.1111

    e.   .10111001   f.   .11001100101
        $\times$.11100111 $\times$.10111101111

6. Perform the following binary divisions.

    a.   .01 $\div$ .1   b.   .01101 $\div$ .101
    c.   .1011101 $\div$ .11001   d.   .11001101 $\div$ .111

1.6 OTHER NUMBER SYSTEMS

We have seen that the laws of arithmetic take on an especially simple form in the binary number system, a fact which makes the binary system a natural one for digital computers to use for their internal operation. On the other hand, it should be clear from the example problems in binary arithmetic given above that the binary notation is cumbersome and long, and that a requirement to convert all numbers to binary before using them in a computer would impose quite an inconvenience. Actually, the problems given above as examples used very short binary numbers. In computers it is quite frequently desirable to use numbers that, in decimal form, would require nine decimal digits. In binary form such numbers require roughly 30 digits, an unmanageable size for hand calculations. Consequently, while the binary system is frequently used for operation within a computer, circuits are frequently provided to allow the machine to accept or put out numbers in some other system. The decimal system is sometimes used for this purpose. However, the decimal to binary conversion is a rather difficult one, and some machines are constructed to use number systems which stand in a simpler relation to the binary system. In particular, the octal system, based on 8, and hexadecimal system, based on 16, are frequently used. We will discuss each of these systems briefly.

1.7 THE OCTAL SYSTEM

The number system based on 8 has as its basic characters

$$0, \quad 1, \quad 2, \quad 3, \quad 4, \quad 5, \quad 6, \quad 7$$

The next number after 7, or decimal 8, is formed by writing 10. The next number, decimal 9, is octal 11. If arithmetic were to be performed in octal notation, new addition and multiplication rules would have to be learned. For example, five plus six is equal to the decimal number eleven, which is eight and three, or octal 13. Thus, in octal, $5 + 6 = 13$. Since we are not particularly interested in the octal system for arithmetic purposes, we shall not pursue this point.

In an octal number, the value of a digit is determined by its position in relation to an octal point. Moving a digit to the left by a place increases its value by a factor of octal 10 (or decimal 8) and moving it to the right reduces its value by the same factor. The diagram shows an octal number, and below each digit the decimal value of that particular digit is given:

$$
\begin{array}{ccccccc}
7 & 6 & 3 & 4 \ . & 1 & 5 & 2 \\
\downarrow & \downarrow & \downarrow & \downarrow & \downarrow & \downarrow & \downarrow \\
7 \times 8^3 & 6 \times 8^2 & 3 \times 8^1 & 4 \times 8^0 & 1 \times 8^{-1} & 5 \times 8^{-2} & 2 \times 8^{-3}
\end{array}
$$

It is somewhat unfortunate that the symbols 1 through 7 are used in octal notation because through long usage we have learned to associate the decimal meanings and laws of decimal arithmetic with such numbers. There is nothing in the appearance of the above number to indicate that it is not an ordinary decimal number. When there is the possibility of confusion, as in this case, we shall indicate the number system being used by a subscript after the number which gives the decimal value of the base being used. Thus the above number would be written

$$7634.152_8$$

1.71 Octal-Decimal Conversion

Octal to decimal and decimal to octal conversion can be accomplished in a manner completely analogous to binary-decimal conversion. A detailed discussion of the process will not be given here. A different and somewhat simpler way of converting between octal and decimal is to convert first to binary and then to the other system. As will be seen shortly, octal-binary conversion is so simple as to be trivial, so that this double conversion is not at all difficult.

1.72 Octal-Binary Conversion

A clue to the process of octal-binary conversion can be obtained by writing the first several octal and corresponding binary numbers, as in Table 3 (decimal values have also been listed for reference).

If we study this table we see that it takes at most three binary bits to represent one octal digit. If we look more closely, we see that the same three binary bits always represent the same octal digit. For example, the octal

TABLE 3

Decimal	Octal	Binary
1	1	1 (or 001)
2	2	10 (or 010)
3	3	11 (or 011)
4	4	100
5	5	101
6	6	110
7	7	111
8	10	1000
9	11	1001
10	12	1010
11	13	1011
12	14	1100
13	15	1101
14	16	1110
15	17	1111
16	20	10000

number 5 is binary 101. The octal number 15 is 1101, the first 1 is the binary number standing for octal 1 and the binary 101 again stands for 5. This condition arises from the fact that eight is two cubed. In the binary system, a shift to the right of one position of the binary point corresponds to a multiplication of the number by two. A shift of three positions corresponds to a multiplication by two cubed, or eight. For an octal number, a shift of one position of the octal point is equivalent to multiplication by eight also; thus three binary positions are precisely equivalent to one octal position. This means that transformation from octal to binary can be accomplished merely by replacing each octal digit by its three-bit binary equivalent. Conversely, transformation from binary to octal can be accomplished by grouping the binary bits in threes starting from the binary point and working in both directions, and then replacing each three-bit group by its octal equivalent.

Example 1.   Convert 764.301_8 to binary.

The binary groups are used to replace individual digits, as indicated:

$$
\begin{array}{cccccc}
7 & 6 & 4 & .\ 3 & 0 & 1 \\
\downarrow & \downarrow & \downarrow & \downarrow & \downarrow & \downarrow \\
111 & 110 & 100 & .\ 011 & 000 & 001
\end{array}
$$

The resulting number is the required binary number. Some additional insight into the process can be obtained from the following diagram, which illustrates for the integral part the values of the numbers in decimal form:

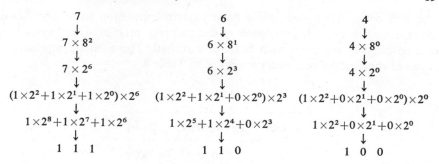

Example 2. Convert 11011.10111 to octal.

Grouping the digits by threes, starting from the binary point, we have

$$11 \quad 011 \quad . \quad 101 \quad 11$$

It is necessary to introduce an additional zero in the first group and last group in order to have three binary bits in each group, thus:

$$011 \quad 011 \quad . \quad 101 \quad 110$$

Each of these groups is now replaced by its octal equivalent, giving

$$33.56_8$$

1.8 THE HEXADECIMAL SYSTEM

The hexadecimal system, or number system based on sixteen, requires a zero and 15 other symbols for its basic characters. We shall use

$$0, \quad 1, \quad 2, \quad 3, \quad 4, \quad 5, \quad 6, \quad 7, \quad 8, \quad 9, \quad A, \quad B, \quad C, \quad D, \quad E, \quad F$$

The number next after F (or decimal 15) is 10, which stands for the decimal number 16. The decimal number 17 is represented by hexadecimal 11. As with the other number bases we have discussed, the value of a digit is determined by its location with respect to a point, in this case a hexadecimal point. Moving a digit to the left by one place increases its value by a factor of hexadecimal 10, or decimal 16, and moving it to the right reduces its value by the same factor. The diagram shows a hexadecimal number, and below each digit the decimal value of that digit:

$$
\begin{array}{ccccc}
6 & A & C & . \quad E & 9 \\
\downarrow & \downarrow & \downarrow & \downarrow & \downarrow \\
6 \times 16^2 & 10 \times 16^1 & 12 \times 16^0 & 14 \times 16^{-1} & 9 \times 16^{-2}
\end{array}
$$

As with octal, this system offers great ease in conversion to binary. Since decimal sixteen is the fourth power of decimal two, precisely four binary bits are required to represent each hexadecimal digit. These binary representations of the hexadecimal digits are given in Table 4.

TABLE 4

Decimal	Hexadecimal	Binary
0	0	0000
1	1	0001
2	2	0010
3	3	0011
4	4	0100
5	5	0101
6	6	0110
7	7	0111
8	8	1000
9	9	1001
10	A	1010
11	B	1011
12	C	1100
13	D	1101
14	E	1110
15	F	1111
16	10	10000

A hexadecimal number is converted to binary by replacing each hexadecimal digit by its four-bit binary equivalent, and a binary number is converted to hexadecimal by grouping the bits into four-bit groups, working to the right and left from the binary point, and then replacing each group by the equivalent hexadecimal digit.

Example 1. Convert $76EA.0D_{16}$ to binary.

Using the relations from Table 4,

$$
\begin{array}{cccccc}
7 & 6 & E & A & . & 0 & D \\
\downarrow & \downarrow & \downarrow & \downarrow & & \downarrow & \downarrow \\
0111 & 0110 & 1110 & 1010 & . & 0000 & 1101
\end{array}
$$

Example 2. Convert 101111.0110111 to hexadecimal.

Grouping the digits by fours, starting from the binary point, we have

$$10 \quad 1111 \quad . \quad 0110 \quad 111$$

Completing the first and last groups with zeros and using Table 4, we have

$$
\begin{array}{ccccc}
0010 & 1111 & . & 0110 & 1110 \\
\downarrow & \downarrow & . & \downarrow & \downarrow \\
2 & F & . & 6 & E_{16}
\end{array}
$$

1.81 Hexadecimal Arithmetic

Many computers in current use employ the hexadecimal system for arithmetic calculations. Consequently, we will give some introduction to the subject. It is not suggested that the reader should become skilled in performing arithmetic in the hexadecimal system. Rather, he should follow through the examples and work one or two of the exercise problems so that the notations and operations will be somewhat familiar to him instead of completely foreign.

Table 5 summarizes the rules for hexadecimal addition. Any entry in the table is easily checked by conversion to the decimal system. For example, consider the sum of A and D. A stands for decimal 10 and D for decimal 13. The sum is decimal 23, or decimal $16 + 7$. Hence in hexadecimal the sum is 17.

TABLE 5

Hexadecimal Addition Table

	1	2	3	4	5	6	7	8	9	A	B	C	D	E	F
1	2	3	4	5	6	7	8	9	A	B	C	D	E	F	10
2	3	4	5	6	7	8	9	A	B	C	D	E	F	10	11
3	4	5	6	7	8	9	A	B	C	D	E	F	10	11	12
4	5	6	7	8	9	A	B	C	D	E	F	10	11	12	13
5	6	7	8	9	A	B	C	D	E	F	10	11	12	13	14
6	7	8	9	A	B	C	D	E	F	10	11	12	13	14	15
7	8	9	A	B	C	D	E	F	10	11	12	13	14	15	16
8	9	A	B	C	D	E	F	10	11	12	13	14	15	16	17
9	A	B	C	D	E	F	10	11	12	13	14	15	16	17	18
A	B	C	D	E	F	10	11	12	13	14	15	16	17	18	19
B	C	D	E	F	10	11	12	13	14	15	16	17	18	19	1A
C	D	E	F	10	11	12	13	14	15	16	17	18	19	1A	1B
D	E	F	10	11	12	13	14	15	16	17	18	19	1A	1B	1C
E	F	10	11	12	13	14	15	16	17	18	19	1A	1B	1C	1D
F	10	11	12	13	14	15	16	17	18	19	1A	1B	1C	1D	1E

Example 1.   Add the hexadecimal numbers 1A.37 and DB.4C.

Writing the problem in the usual form we have

$$1A.37$$
$$DB.4C$$
$$\overline{}$$

The right-hand digits are added first. From Table 5, $7 + C = 13$. We record the three and carry the one to the next column, where we have $1 + 3 = 4$, thus $4 + 4 = 8$. The next column gives $A + B = 15$. We record the five and carry the one. Then in the left-hand column, we have $1 + 1 = 2$, then $2 + D = F$. The final answer is F5.83.

In later problems, we will use two's-complement notation for negative numbers, so subtraction problems per se will not occur. It will be necessary to construct the two's complement of given numbers, however. The next example demonstrates the process.

Example 2. Form the two's complement of 5B3C.

In binary form, this number is

$$0101101100111100$$

The one's complement is obtained by replacing zeros by ones and ones by zeros:

$$1010010011000011$$

The two's complement is obtained by adding one to this number, giving

$$1010010011000100$$

In hexadecimal notation this is

$$A4C4$$

The result could also be obtained directly in hexadecimal, utilizing Table 5. To do this, we subtract all but the rightmost digit from F, and subtract the rightmost digit from 10. This would give

$$F - 5 = A$$
$$F - B = 4$$
$$F - 3 = C$$
$$10 - C = F - C + 1 = 4$$

Table 6 is a multiplication table for hexadecimal numbers. The entries can easily be checked by conversion to decimal. For example, hexadecimal A times B is decimal 10 times 11, or decimal 110. This number is equal to $16 \times 6 + 14$, or in hexadecimal 6E. The next example demonstrates hexadecimal multiplication.

TABLE 6
Hexadecimal Multiplication Table

	1	2	3	4	5	6	7	8	9	A	B	C	D	E	F
1	1	2	3	4	5	6	7	8	9	A	B	C	D	E	F
2	2	4	6	8	A	C	E	10	12	14	16	18	1A	1C	1E
3	3	6	9	C	F	12	15	18	1B	1E	21	24	27	2A	2D
4	4	8	C	10	14	18	1C	20	24	28	2C	30	34	38	3C
5	5	A	F	14	19	1E	23	28	2D	32	37	3C	41	46	4B
6	6	C	12	18	1E	24	2A	30	36	3C	42	48	4E	54	5A
7	7	E	15	1C	23	2A	31	38	3F	46	4D	54	5B	62	69
8	8	10	18	20	28	30	38	40	48	50	58	60	68	70	78
9	9	12	1B	24	2D	36	3F	48	51	5A	63	6C	75	7E	87
A	A	14	1E	28	32	3C	46	50	5A	64	6E	78	82	8C	96
B	B	16	21	2C	37	42	4D	58	63	6E	79	84	8F	9A	A5
C	C	18	24	30	3C	48	54	60	6C	78	84	90	9C	A8	B4
D	D	1A	27	34	41	4E	5B	68	75	82	8F	9C	A9	B6	C3
E	E	1C	2A	38	46	54	62	70	7E	8C	9A	A8	B6	C4	D2
F	F	1E	2D	3C	4B	5A	69	78	87	96	A5	B4	C3	D2	E1

Example 3.   Find the product of the hexadecimal numbers A3.4B and C.75.

Writing the problem in the usual form,

$$A3.4B$$
$$C.75$$

The first partial product is formed by using the digit 5 as multiplier:

$5 \times B = 37$, record the 7, carry 3
$5 \times 4 = 14$, add 3, giving 17, record the 7, carry 1
$5 \times 3 = F$, add 1, giving 10, record the 0, carry 1
$5 \times A = 32$, add 1, giving 33

The first partial product is thus 33077. In like manner, the second partial product can be formed. At this stage we have

$$A3.4B$$
$$C.75$$

33077
4770D

Before forming the third partial sum, it is convenient to add these two, thus avoiding the addition of a long list of numbers in an unfamiliar number system. If this is done, and then the third partial product formed and added on, we have

$$
\begin{array}{ll}
\text{A3.4B} & \\
\text{C.75} & \\
\hline
33077 & \text{first partial product} \\
4770\text{D} & \text{second partial product} \\
\hline
4\text{AA}147 & \text{sum of first and second partial products} \\
7\text{A}784 & \text{third partial product} \\
\hline
7\text{F}22547 & \text{final result}
\end{array}
$$

The radix point is located by the same rule as in the decimal system, giving 7F2.2547.

1.82   Decimal-Hexadecimal Conversion

Decimal to hexadecimal conversion can be accomplished by converting first from decimal to binary, as described in Section 1.52, and then performing the easy binary-hexadecimal conversion. A more direct method will be demonstrated in this section.

In decimal to hexadecimal conversion, we have a number expressed in powers of 10 and we wish to express it in powers of 16. The integer part of the number represents some source of positive powers of 16. We can find these by successive division by 16. The remainder after the first division represents the zeroth power of 16, hence the units digit in hexadecimal. The remainder after the next division by 16 gives the next digit in the hexadecimal number, and so on.

The fractional part of the number represents negative powers of 16, and so successive multiplication by 16 can be used to expose the hexadecimal values. A first multiplication by 16 will make that part of the number corresponding to 16^{-1} power, that is, the first hexadecimal digit, become the integer part of the product. A second multiplication by 16 will convert the 16^{-2} part, that is, the second hexadecimal digit, to integer, and so on. The process will be clarified by an example.

Example 1.   Convert 186923.5413125 to hexadecimal.

For the integer part we have

$$186923 \div 16 = 11682, \text{ remainder 11 or hexadecimal B}$$
$$11682 \div 16 = 730, \text{ remainder 2 or hexadecimal 2}$$
$$730 \div 16 = 45, \text{ remainder 10 or hexadecimal A}$$
$$45 \div 16 = 2, \text{ remainder 13 or hexadecimal D}$$
$$2 \div 16 = 0, \text{ remainder 2 or hexadecimal 2}$$

Hence the integer part was $2 \times 16^4 + D \times 16^3 + A \times 16^2 + 2 \times 16^1 + B \times 16^0$ or $2DA2B_{16}$.

For the fractional part we have

$$.5413125 \times 16 = 8.6610000, \text{ integer part 8 or hexadecimal 8}$$
$$.6610000 \times 16 = 10.5760000, \text{ integer part 10 or hexadecimal A}$$
$$.5760000 \times 16 = 9.2160000, \text{ integer part 9 or hexadecimal 9}$$
$$.2160000 \times 16 = 3.4560000, \text{ integer part 3 or hexadecimal 3}$$
$$.4560000 \times 16 = 7.2960000, \text{ integral part 7 or hexadecimal 7}$$
$$.2960000 \times 16 = 4.7360000, \text{ integer part 4 or hexadecimal 4}$$
$$.7360000 \times 16 = 11.7760000, \text{ integer part 11 or hexadecimal B}$$

At this point, we have generated as many hexadecimal digits as there were decimal digits in the original number, so there is no point in carrying the process further. If continued, the hexadecimal digits would eventually fall into a repeating loop. The terminating decimal fraction has converted into a nonterminating hexadecimal fraction, whose value is

$$8 \times 16^{-1} + A \times 16^{-2} + 9 \times 16^{-3} + 3 \times 16^{-4} + 7 \times 16^{-5} + 4 \times 16^{-6} +$$
$$B \times 16^{-7} + \text{etc.,} \quad \text{or} \quad .8A9374B_{16}$$

Thus, the entire number in hexadecimal is approximately

$$2DA2B.8A9374B_{16}$$

1.83 Hexadecimal-Decimal Conversion

Hexadecimal to decimal conversion is performed rather easily by converting first to binary and then using Table 1. An example will demonstrate this process.

Example 1.   Convert $2DA2B.8A9_{16}$ to decimal.

Writing this number in binary we have

$$10 \quad 1101 \quad 1010 \quad 0010 \quad 1011 . \quad 1000 \quad 1010 \quad 1001$$

The highest power of 2 represented is the 17th. We have

$$
\begin{array}{rl}
2^{17} = & 131072 \\
2^{15} = & 32768 \\
2^{14} = & 16384 \\
2^{12} = & 4096 \\
2^{11} = & 2048 \\
2^{9} = & 512 \\
2^{5} = & 32 \\
2^{3} = & 8 \\
2^{1} = & 2 \\
2^{0} = & 1 \\
2^{-1} = & .5 \\
2^{-5} = & .03125 \\
2^{-7} = & .0078125 \\
2^{-9} = & .001953125 \\
2^{-12} = & .000244140625 \\
\hline
& 186923.541259765625
\end{array}
$$

The answer contains a large number of decimal places, and we might suspect that they are not truly representative of the accuracy which should be assigned to that number. It will be shown in Chapter 3 that based on the number of significant figures in the given hexadecimal number, it would be appropriate to round this number off to 186923.541.

1.9 BINARY CODING

It has been seen that, for the octal and hexadecimal systems, the digits can be represented quite easily in binary form. Thus it is extremely simple to construct devices which will automatically convert octal or hexadecimal numbers to their binary equivalent. For example, a typewriter keyboard can be wired to a tape punching unit so that pressing the key on the typewriter will cause holes to be punched in paper tape. A hole in the tape can be read as a binary digit one, and a space without a hole as a binary digit zero. If the typewriter keys are so wired that pressing the "5" key punches two holes separated by a blank space, and so forth, numbers in octal form can be typed on the keyboard and the same numbers in binary form will appear on the tape. Likewise, if four binary positions are used for each key instead of three, then the typing of a hexadecimal number on the keyboard can be made automatically to produce a binary number on the tape. There are numerous other ways in which octal or hexadecimal numbers can be automatically converted to binary. It would carry us too far from the subject of this book to attempt to enumerate in detail the various electrical or mechanical systems that can be used for this purpose. The above illustrations should serve to show that such devices are indeed feasible.

1.91   Binary Coded Decimal

Suppose a typewriter has been connected to a paper tape system in such a way that hexadecimal digits typed on the keyboard automatically produce the four-bit binary equivalents listed in Table 4 on the tape. Suppose now that instead of a hexadecimal number, a decimal number is typed on the keyboard. The result is an arrangement of holes and spaces, or binary zeros and ones, four for each decimal digit typed. For example, the decimal number 93645 would produce the combination

$$
\begin{array}{ccccc}
9 & 3 & 6 & 4 & 5 \\
\downarrow & \downarrow & \downarrow & \downarrow & \downarrow \\
1001 & 0011 & 0110 & 0100 & 0101
\end{array}
$$

Conversely, given the binary digits shown, we can reconstruct the decimal number intended. Thus the arrangement of binary digits can be used to represent the decimal number. Note carefully, however, that the array of binary digits is *not* the binary number having the same *values* as the original decimal number. In fact, considered as a number, it has no simple relation to the decimal number. For example, the decimal number 96 would appear as follows if each digit were replaced by its 4-bit binary representation:

$$
\begin{array}{cc}
9 & 6 \\
\downarrow & \downarrow \\
1001 & 0110
\end{array}
$$

On the other hand, since $96 = 2^6 + 2^5$, the binary notation for 96 is

$$
96_{10} = 1100000_2
$$

The latter number, 1100000, is a true binary number, having the same value as the decimal number 96. The former number, 10010110, which happens to be the binary number equal to the decimal number 150, can be considered as a *code symbol* for the number 96. It can be used if we are interested only in having the number 96 coded as a combination of ones and zeros, and do not intend to attempt binary arithmetic with the coded symbols. The representation of a decimal number by symbolizing each decimal digit as a group of ones and zeros is referred to as binary coded decimal.

1.92   Binary Coding of Other Characters

Just as the decimal or digits can be assigned a binary coding for purposes of representation on punched paper tape or in other systems in which only zeros and ones can be used, so also can any other symbol or character used on a typewriter keyboard or in printing be assigned a binary coding. Most computers make use of such a coding for all the letters of the alphabet and some of the other standard pointing symbols found on typewriter keyboards. From 4 to 8 bits are used for this purpose. With 4 bits, 2^4 or 16 different quantities can be represented. With 5 bits, 2^5 or 32 different quantities can be represented. This still is not sufficient to allow a different coding or representation for the 10 decimal characters plus the 26 letters of the alphabet. With 6 bits, 2^6 or 64 different quantities can be represented, so that a different representation is available for each numeral, each alphabetic character, and each of the more common printing symbols.

Six-bit codes have been frequently used in the past. However, there is now growing use of an 8-bit coding system, the extended American Standard Code for Information Interchange (ASCII). Use of this code allows compatibility with the communications world. The 256 bit combinations allowed give room for coding all alphanumeric symbols (numbers and letters), separate codings for capital and small letters, all standard typographic symbols, and still there is room for expansion.

Because these 8-bit groupings are so easily given in hexadecimal notation, we will frequently represent the 8-bit binary coding of a symbol as a pair of hexadecimal digits. For example, capital A is represented by 10100001 or $A1_{16}$, small a by 11100001 or $E1_{16}$.

Table 7 gives the complete extended ASCII code. Column labels give the first 4 ASCII bits, row labels the last 4. Note that over half of the possible entries are vacant. Some of these bit combinations have no assigned meaning, others are assigned to various special keys on teletype keyboards.

Example 1.   Using the binary code given in Table 7, write \$31.27 in binary coded form.

The solution is

    0100 0100 0101 0011 0101 0001 0100 1110 0101 0010 0101 0111

or, using the more convenient hexadecimal shorthand to avoid writing all these individual bits,

$$4453514E5257_{16}$$

Example 2.   In the binary code of Table 7, what is the meaning of the string of binary bits represented by the following?

$$59594E54544540B0F5F2E5$$

TABLE 7

Eight-bit American Standard Code for Information Interchange

Last 4 Bits Hex	Bin	First 4 Bits 0 / 0000	1 / 0001	2 / 0010	3 / 0011	4 / 0100	5 / 0101	6 / 0110	7 / 0111	8 / 1000	9 / 1001	A / 1010	B / 1011	C / 1100	D / 1101	E / 1110	F / 1111
0	0000					blank	0					@	P				p
1	0001					!	1					A	Q			a	q
2	0010					"	2					B	R			b	r
3	0011					#	3					C	S			c	s
4	0100					$	4					D	T			d	t
5	0101					%	5					E	U			e	u
6	0110					&	6					F	V			f	v
7	0111					'	7					G	W			g	w
8	1000					(	8					H	X			h	x
9	1001					)	9					I	Y			i	y
A	1010					*	:					J	Z			j	z
B	1011					+	;					K	[			k	
C	1100					,	<					L	~			l	
D	1101					-	=					M	]			m	
E	1110					.	>					N	←			n	ESC
F	1111					/	?					O	↓			o	DEL

From Table 7, the codings are

Hexadecimal Pair	Symbol Represented
59	9
59	9
4E	.
54	4
54	4
45	%
40	blank
B0	P
F5	u
F2	r
E5	e

so the message is: 99.44% Pure.

EXERCISE 2

1. Write the complete addition table for the basic characters of
   a. The octal system.   b. The hexadecimal system.

2. Write the complete multiplication table for the basic characters of
   a. The octal system.   b. The hexadecimal system.

3. Write the following numbers in binary form.
   a. 721.3_8 b. -46.25_8 c. 100_{16} d. $EF.7A_{16}$
   e. $.00613_8$ f. $.0AEC_{16}$ g. $-FF.FF_{16}$ h. 10100_8

4. Write the following binary numbers in octal form.
   a. 101.011   b. 1110.000101   c. 10000.
   d. .0000111   e. 1011.1101   f. 10111.11101

5. Write the numbers of problem 4 in hexadecimal form.

6. Write the following decimal numbers in binary, and also in a binary-coded decimal, using eight bit positions for each character as in table 7.
   a. 8   b. 24   c. 36.375
   d. 17.125   e. 4.5   f. 91.2

7. Decode the following messages, which are given in the code of Table 7.
   a. A5F440B4F54740A2F2F5F4E55F
   b. A7EF40F7E5F3F44740F9EFF5EEE740EDE1EE
   c. B4EF40E2E540EFF240EEEFF440F4EF40E2E5

8. Add the following pairs of hexadecimal numbers.
   a. 1A.B3   b. 15.26   c. FF.FF
       8 1.9F   38.47   11.11
       ‾‾‾‾‾   ‾‾‾‾‾   ‾‾‾‾‾

9. Multiply the following pairs of hexadecimal numbers.

    a. 49   b. A7   c. 21   d. B9A3

        6B   CD   67   C17D

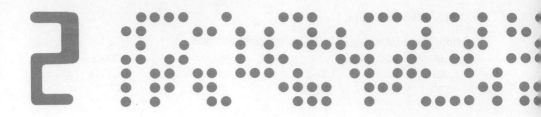

The Digital Computer

2.1 INTRODUCTION

Before entering upon the discussion of problem-solving methods which forms the major part of this text, it seems appropriate to give some description of the basic equipment involved, the digital computer. Computers are sufficiently varied in design that it would not be possible to give a detailed description of all the types of equipment now in use. However, the design of most digital computers follows the same general approach, so it is possible to give a generalized description which is applicable to most machines. Although the computer user may be able to solve his computational problems without direct recourse to this information, he may well find it useful on occasion. Consequently this chapter will be devoted to the description of the digital computer, to include both the hardware and the software operating system, from basic circuits to programming language. This is the elaborate substructure which the user ordinarily does not see but which performs highly complex feats to solve his problems for him.

2.2 COMPUTER HARDWARE ORGANIZATION

The main motivation for development of the digital computer has been the requirement to perform arithmetic very rapidly, and so historically in computer design, attention was first given to electric circuits which could add, subtract, multiply, or divide two numbers rapidly. In order to have numbers available for this arithmetic circuitry and to retain intermediate results, it became necessary to add devices which could electronically store numbers for later use. Such devices can be termed memories. In order to move the right numbers from memory to the arithmetic unit, perform the proper arithmetic operations, and store the results back in memory without waste of time, it was necessary to add more electrical circuitry. The control circuitry, combined with the arithmetic circuitry, is ordinarily termed a processing unit, or processor. In order to place numbers into the machine and to convert the answers from electric signals to printed or other readable form, input and output units were required. The basic hardware of the digital computer, then, consists of the memory, the processor, and the input-output devices. Sections 2.3 through 2.5 give further descriptions of each of these major hardware components.

2.3 MEMORY

When calculations are done by hand, or with a desk-type calculator, the person performing the calculation must frequently record the results of individual additions, subtractions, multiplications, or divisions on a work-sheet and then re-use these results at a later stage in the calculation. One of the major contributions to the speed of modern digital computers is the elimination of this cumbersome process by the use of memory units in which intermediate results are recorded, stored, and fed back into the circuits of the machine as needed, automatically and quickly. The term "memory" is usually employed to designate these functions, as well as the machine unit performing them, because they are somewhat analogous to the human mental process of remembering. The alternative term "storage" is sometimes used.

2.31 Memory Cells

A number is usually stored in memory in the form of electrical or mechanical representations of binary bits. The memory is subdivided into cells, each cell being able to store all the bits of a single number. In the ordinary computer all the memory cells, or registers as they are frequently called, are capable of storing some fixed number of binary bits. About 30 is the usual number, although some machines use as few as 12 bits or as many as 64. As was indicated in Chapter 1, 30 binary bits are equivalent to roughly $30 \times .3 = 9$ decimal digits. It might seem that four or five decimal digits would be adequate. However, the problems described in later chapters will show that rounding errors involved in calculations can easily cause the loss of several significant figures. Occasions will be found, not too infrequently, when 30 binary bits are insufficient. Because the binary bits contained in a memory cell need not actually represent a binary number, but may instead be a binary coding for some combination of letters, or other characters, the contents of a memory cell are ordinarily referred to as a "word." A "word" may be a true binary number, or binary coding for a true word in the linguistic sense, or merely some combination of letters, numbers, or other symbols in a binary coded form.

Because of the growing acceptance of the eight-bit American Standard Code for Information Interchange (see Table 7 of Chapter 1) for representation of alphanumeric and typographic characters, there is a growing tendency to use computer word lengths which are a multiple of 8 bits. This allows the coded representation of characters to be fitted neatly into computer words. In several present-day computers, such as the IBM 360 series, the word-length is 32 bits, each word being further subdivided into four 8-bit subgroups, termed bytes. Each byte is then capable of containing one character. Our later demonstration problems will assume the 32-bit word length.

2.32   Some Types of Memories

Most modern computer memories make use of the capability of iron and other so-called ferromagnetic materials, to take on and retain a magnetized state. The most common type of memory is magnetic core storage. In this memory, each storage element consists of a tiny ferrate ring less than 1 ten-thousandth of an inch in diameter. One ring, or core, can be placed in either of two different states of magnetization, thus representing a single binary bit, either zero or one. Thousands, or even millions, of such cores are strung on hair-thin wires in a fashion not unlike Indian beadwork. Electric current pulses can be passed through particular sets of wires to magnetize particular cores, and other wires can "sense" which of the cores were already magnetized and which were not.

Another widely used memory type is thin-film memory. In thin-film memories, the magnetic elements are tiny ferrite rectangles placed on a thin film of nonconducting material, and the magnetizing currents are carried by circuits printed on the other side of the film.

In memories such as these, the "writing" of a bit of information into a particular memory element or the "reading" of that bit from memory requires the passage of currents down the proper wires. This in turn requires that the wires be connected by an elaborate set of switches which can be controlled to allow switching of the currents into the proper combination of wires. Transistors are normally used to form switches, or gates, for this purpose, in a manner to be described later. A memory may contain tens or hundreds of thousands of transistors in the gating circuits.

2.33   Cell Identification

As mentioned earlier, the individual memory elements in the machine memory are ordinarily grouped into words, and numbers (or symbols) are written into memory or read from memory a word at a time. To perform such operations, it is necessary that there be some means of identifying each memory cell so that a number stored in a particular cell is not lost but can be read out when desired. For this purpose, the memory cells are normally assigned identifying numbers, starting with zero and running sequentially to the highest number required. This identifying number is referred to as the "address" of the memory cell.

2.34 Auxiliary Storage

In describing memories in the preceding sections, we have assumed that we were dealing with the storage and retrieval of numbers needed frequently in the calculation underway. Accordingly, the types of memory described were those associated with very fast random access. Any memory cell can be accessed as quickly as any other. A computer normally has such a memory as its main memory, in amounts ranging from a few thousand to over a hundred thousand words, depending on the cost of the machine. In many computer applications, it is necessary to store millions of words of data or results, and the relatively high expense of random-access storage renders it not particularly suitable for such bulk storage. Other, less expensive types of storage are normally used for such purposes.

Magnetic tape and the disk file are the most common types of auxiliary memory. On magnetic tape, information is recorded by passing the tape under a writing head, which induces magnetized spots on the tape to represent binary "ones." The recording is performed a character at a time, a single character being recorded as eight bits in a row across the tape. A typical tape can hold several million characters of information. In disk storage, magnetically coated disks rotate at high speed, and magnetic recording heads are used to read or write magnetized spots. A typical disk file can also store several million characters of information. Both the tape and disk are examples of sequential rather than random-access memory. Once written, a word cannot be reread until the unit is mechanically returned to the proper position. Whereas a word can ordinarily be retrieved from core storage in a few microseconds, retrieval from a disk ordinarily requires several milliseconds, and from a magnetic tape, milliseconds if the tape is already in the proper position, and as much as a few seconds if the tape must be repositioned.

2.35 Fixed-Point Representation

As mentioned in Chapter 1, there are two forms used in storing numbers in a computer: binary form, in which the pattern of zeros and ones is to be interpreted directly as a binary number; and coded form, in which different patterns of zeros and ones stand for different letters, digits, or typographic symbols in accordance with a prearranged code. For arithmetic the binary form is more efficient, and so numbers to be used in arithmetic are usually stored in binary form. Figure 2-1 displays a representation of a 32-bit memory

Bit Position

0   10   20   31

| 0 | 1 | 1 | 0 | 0 | 1 | 0 | 1 | 0 | 1 | 1 | 1 | 1 | 1 | 0 | 1 | 0 | 1 | 0 | 0 | 1 | 1 | 1 | 1 | 1 | 1 | 1 | 0 | 0 | 0 | 1 | 0 | 0 |

Figure 2-1: Sample representation of memory word

register with a binary word stored. Rather than write this entire string of zeros and ones each time we wish to describe a number in memory, we will use the hexadecimal equivalent as described in Chapter 1. In hexadecimal shorthand, this word can be written 657A9FC4.

There are two problems for the computer in using this binary number in a calculation. They are: What is its sign? Where is the binary point? The first problem is usually handled by making the first bit position be not part of the number itself but a separate sign bit: zero for plus, one for minus. Location of the binary point is a more difficult matter. The simplest solution is to require that numbers be stored so that the radix point is always in the same place, say at the right end of the number.

Negative numbers can be represented in two's-complement form, as described in Chapter 1.

Example 1. In a machine having 32-bit words, the number $0000000F_{16}$ is stored. What decimal number does it represent?

Since the first bit is zero, the number is positive. All the other bit positions are zero, except for hexadecimal F, or decimal 15, in the lower bit positions. Therefore, the decimal value is 15.

Example 2. In a machine having 32-bit words, the number $FFFFFFF1_{16}$ is stored. It is in twos-complement integer form, with the first bit used for sign. What is the decimal value?

The first hexadecimal digit, F, stands for the binary bits 1111. The first bit is one, so the number is negative. Hence the number is in two's-complement form. Its numerical value is obtained by performing the subtraction

$$
\begin{array}{r}
100000000 \\
FFFFFFF1 \\
\hline
F
\end{array}
$$

The numerical value is hexadecimal F, or decimal 15. Hence the decimal value is −15.

Number representation of the type just described, where the radix point is assumed to be in a certain location, is termed fixed-point or integer representation. The particular form just demonstrated is termed integer representation. It has the advantage of simplicity, but is severely limited from the standpoint of flexibility.

2.36   Floating-Point Representation

On a machine using 32-bit words, only integers having magnitude less than 2^{31}, or about 1 billion, can be handled. Because many calculations require the use of fractions, or of numbers greater than 1 billion, most computers are wired for an alternative form of number representation, termed "floating-point" representation. Floating-point representation is closely akin to the so-called scientific notation, widely used in the physical science fields. In scientific notation, a number is represented as a fractional part multiplied by 10 raised to a power. For example, the number 418.32 would be represented as $.41832 \times 10^3$. All the information contained in this number is actually contained in the fractional part, 41832, and the exponent, 3. If we were to make an agreement about notation, the decimal point and the symbol $\times 10$ could be omitted without losing any information about the number itself. We could write the number as

$$41832 \qquad 3$$

with the understanding that the first number is a fractional part, to be understood to have a decimal point in front of the first digit, and the second number is an exponent, which represents the power of 10 by which the fractional part should be multiplied. Since multiplying by a power of 10 is accomplished by merely moving the decimal point that number of places, the exponent represents the number of places the decimal point must be moved from its nominal location at the front of the fractional part to achieve its correct location. This same convention can be employed in number systems other than the decimal system. The number is given as a fractional part and an exponent. The radix point is understood to have its nominal location to the left of the leftmost digit in the fractional part. The exponent tells how many places to the right or left of this location the radix point must be moved to put it in its true location.

Example 1.   Using the convention just described, write the following hexadecimal numbers in floating-point form:

$$49.8AF$$
$$.0003D62$$

The numbers could be written

$$498AF \qquad 2$$
$$3D62 \qquad -3$$

Note that they could also be written in any of the following ways, or in many other similar ways, and still be correct according to the definitions given above.

0498AF	3	00498AF	4
03D62	-2	0003D62	0

The example just given demonstrates that floating-point notation is not unique. Leading zeros can be added to or removed from the fractional part if the exponent is adjusted to correct for them. This in turn suggests how arithmetic is done in floating-point form. Two numbers can be added or subtracted only if their exponents are equal. When two floating-point numbers are to be added, leading zeros are supplied for the one having the smaller exponent and that exponent is increased accordingly until the exponents become equal. Then the fractional parts are added in the normal fashion.

Example 2. Add the following pair of floating-point *decimal* numbers:

$$4591 \qquad 3$$
$$3712 \qquad 1$$

The second number can be rewritten

$$003712 \qquad 3$$

Now the exponents are equal and the fractional parts can be added directly:

$$\begin{array}{r} 459100 \\ 003712 \\ \hline 462812 \end{array}$$

So the answer, in floating-point form, is

$$462812 \qquad 3$$

Note that two trailing zeros had to be supplied to the first number in adding the fractional parts. The accuracy problems involved here are discussed in Chapter 3.

When a floating-point number is stored in a memory cell, the quantities which must be stored are the sign, the fractional part, and the exponent. As an example of the way this is done, the floating-point representation will be described for a 32-bit memory register of a machine in the IBM 360 series. In this machine, the first bit position is used for the sign, the next 7 for a quantity termed the "characteristic," and the last 24 for the fractional part. The fraction is expressed as six hexadecimal digits, each consisting of 4 binary bits. Instead of twos-complement form for negative numbers, as illustrated for fixed point numbers Section 2.35, the IBM 360 series uses ordinary negative numbers in the fractional part. The characteristic indicates the power of 16 by which the fractional part is to be multiplied, that is, the number of places by which the hexadecimal point is to be shifted. The seven bit positions allow storage of a number ranging from zero to 127_{10}. In order that one does not have to carry a separate plus or minus sign for the exponent, the actual exponent itself is not stored but, instead, the quantity called the characteristic, which is the exponent plus 64. Thus a characteristic of 64 (or 40_{16}) corresponds to an exponent of zero, a characteristic of zero to an exponent of -64, and a characteristic of 127_{10} (or $7F_{16}$) to an exponent of 63. Thus numbers ranging in size from 16^{-64} to 16^{63} or about 10^{-78} to 10^{75} can be represented.

Example 3.   A 32-bit register contains the following number in floating-point form:

$$41100000_{16}$$

Give the decimal value of the number.

The first two hexadecimal digits stand for the binary bits 01000001. The first bit is zero, so the sign is plus. The next 7 bits give a characteristic of 65_{10} or 41_{16}. The exponent is obtained by subtracting 64_{10} or 40_{16} from the characteristic, so the exponent is $+1$. The fractional part is $.100000_{16}$, and if the radix point is shifted to the right one place, this becomes 1.0000_{16} or simply the number 1.

Example 4.   A 32-bit register contains the following number in floating-point form:

$$C1100000_{16}$$

Give the decimal value of the number.

The first two hexadecimal digits stand for the binary bits 11000001. The first bit is one, so the sign is minus. The next 7 bits are the same as in Example 3, so the exponent is again $+1$. Since the sign is negative, the number is negative. As mentioned above, complement notation is not used in the 360 series floating point representations. The fractional part is read directly as $.100000_{16}$.

Shifting the radix point one place to the right, we have 1.00000_{16}, or affixing the sign, -1.00000_{16}, or simply -1.

2.37   Coded Representation

As mentioned earlier, the 4 bytes of a 32-bit word can contain the 8-bit codes for four alphanumeric or typographic characters. The codes given in Table 7 of Chapter 1 can be used for this purpose in a straightforward manner.

Example 1.   A set of consecutive memory cells contain the bit pattern given below. Translate these into the appropriate ASCII characters.

$$B7A9AEAE$$
$$A9A540B4$$
$$A8A540A0$$
$$AFAFA840$$

The first byte is the hexadecimal digits B7, or binary 10110111. From Table 7 of Chapter 1, this is the ASCII code for the letter W. Translation of this and the succeeding bytes is summarized below:

Byte	ACSII Character
B7	W
A9	I
AE	N
AE	N
A9	I
A5	E
40	blank
B4	T
A8	H
A5	E
40	blank
AO	P
AF	O
AF	O
A8	H
40	blank

The total message is WINNIE THE POOH.

Example 2.   Show the coded representation of the message ERROR; CHECK INPUT.

From Table 7 of Chapter 1, the coding of the characters in the message can be read directly. Grouped in fours to match the 32-bit word length, the string of characters becomes

> A5B2B2AF
> B25B40A3
> H8A5A3AB
> 40A9AEB0
> B5B44E40

EXERCISE 3

1. The following are integer binary representatives of numbers in a 32-bit machine, using two's-complement notation. Give the corresponding decimal values.

   a. 00000100_{16} b. $FFFF4FFF_{16}$ c. 00000008_{16}
   d. $00000A02_{16}$ e. $FFFFFFFF_{16}$ f. $FFFFF800_{16}$

2. Write the integers from -16 to 16 as they would appear in the notation of problem 1.

3. The following are floating-point representatives of numbers in a 32-bit machine as described in Section 2.36. Give the corresponding decimal value.

   a. 42100000_{16} b. 41200000_{16} c. $B1200000_{16}$
   d. $D2800000_{16}$ e. 44100000_{16} f. $3D100000_{16}$

4. Write the integers from -16 to 16 as they would appear in the notation of problem 3.

5. Write the powers of 2 from 2^{-16} to 2^{16} as they would appear in the notation of problem 3.

6. A series of successive words in memory contains the quantities shown below.

   > A4AF40B4
   > A8A540AE
   > A5B8B440
   > B0B2AFA2
   > ACA5AD41

   Together they form a coded message, where each byte represents an alphanumeric or typographic character as listed in Table 7 of Chapter 1. What is the message?

7. Using the coding system given by Table 7 of Chapter 1, convert the following to sets of 32-bit memory words, with four characters per word.
   a. STOP   b. GO TO 5
   c. $X = Y**2 + Z$ d. $16 \cdot 3$
   e. IF A $<$ B, THEN DO PART 1; OTHERWISE, DO PART 3

2.4 PROCESSING UNIT

The processing unit is that part of the computer which controls the entire operation of extraction of numbers from memory, performance of the required arithmetic operations, and disposition of results. It does so by means of electrical circuits, composed largely of transistors which serve as switches, in a complex network of wires. These switches can be opened or closed by voltage pulses and can thus control the paths through the network to be taken by other voltage pulses. The next two sections demonstrate that such networks can indeed be used to perform arithmetic operations and to control the operation of the computer. The succeeding sections give descriptions of the actual operations ordinarily performed by a processor.

2.41 Logic of Arithmetic Circuitry

The ability to design electrical circuits capable of performing arithmetic operations stems from the fact that all arithmetic operations can be reduced to just two logical operations, termed "negation" and "conjunction." These are operations which form part of the subject matter of the field of mathematical logic. In terms of binary bits, the operation of negation consists of replacing one by zero or zero by one. The operation of conjunction consists of considering two binary bits, and if both are one the result is one; otherwise the result is zero. These logical operations are easily realized by electrical circuits.

For example, consider a system that uses a positive voltage pulse to represent a digit "one" and a negative pulse to represent a digit "zero." Then a negation circuit has one input and one output, so connected that a positive pulse at the input produces a negative pulse at the output and a negative pulse at the input causes a positive pulse at the output. Similarly, a conjunction circuit has two inputs and one output, so connected that the presence of positive pulses at both inputs causes a positive pulse at the output, but any other combination of input pulses gives a negative output pulse. Rather than become involved in the field of electronics, let us now take for granted that negation and conjunction circuits can indeed be built, and then demonstrate how these can be used as building blocks to make a binary adder. To do this we will talk both in terms of statements from mathematical logic and in terms of the equivalent electrical circuits. Statements using the logical operations of negation and conjunction can be written in a simple symbolic form. Let us use ′ to indicate negation, & to indicate conjunction, and p, q, r, etc., to indicate binary digits either zero or one. Then the relation

$$p = q'$$

would mean p is zero if q is one, and p is one if q is zero. An electrical circuit which will produce p as an output if q is provided as an input can be represented symbolically as in Figure 2-2(a). The left-hand line represents the

(a)   (b)

Figure 2-2

input and the right-hand line represents the output. The box itself represents the particular combination of vacuum tubes, or relays, or transistors used to do the operation. A similar representation can be given for the operation of conjunction. The relation

$$p = q \,\&\, r$$

means that p is 1 if both q and r are 1, and otherwise p is zero. An electrical circuit which will produce p as an output if q and r are provided as inputs can be represented symbolically as in Figure 2-2(b). Any number of logical statements can be constructed using the operations of negation and conjunction. One of the more interesting ones is

$$p' = q' \,\&\, r'$$

which says, from the definition of conjunction, that p' is one if and only if q' and r' are one; or, stated another way, p' is zero if and only if either q' or r', or both, is zero; or, stated yet another way, p is one if and only if either q or r, or both, is one. The statement can also be written

$$p = (q' \,\&\, r')'$$

This operation can be represented by the circuitry shown in Figure 2-3. The entire part of the circuit inside the dashed line has q and r as inputs and p

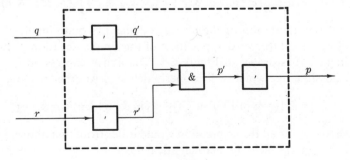

Figure 2-3

as an output. It demonstrates that the above relation can be considered as an operation on q and r to obtain p. This operation has the name "disjunction" and is sometimes represented by the symbol V. With this symbol, the above relation can be represented by

$$p = q \, V \, r$$

and the corresponding circuit by the single box shown in Figure 2-4. The conjunction operation is sometimes called the "and" operation, and disjunction the "or" operation.

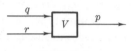

Figure 2-4

By means of a combination of "and," "or," and negation circuits, it is possible to make circuits that will obey all the rules of binary addition as given in Table 2 of Section 1.53. Let p and q represent the first and second summand bits, and r the incoming carry bit. Let s be the sum bit. From Table 2 we see that there are four conditions that will make $s = 1$. They are:

(1) p, q, and r all $= 1$.
(2) $p = 1$, q and r both $= 0$.
(3) $q = 1$, p and r both $= 0$.
(4) $r = 1$, p and q both $= 0$.

We can restate these as:

(1) p, q, and r all $= 1$.
(2) p, q', and r' all $= 1$.
(3) p', q, and r' all $= 1$.
(4) p', q', and r all $= 1$.

Finally, we can state that $s = 1$ if any of the four conditions hold and $s = 0$ otherwise by writing

$$s = (p \ \& \ q \ \& \ r) \, V \, (p \ \& \ q' \ \& \ r') \, V \, (p' \ \& \ q \ \& \ r') \, V \, (p' \ \& \ q' \ \& \ r)$$

The circuit representation of the above statement is quite involved but can be drawn by straightforward application of the simple negation, "and," and "or" circuits. It is drawn in Figure 2-5. The above expression for s is not the only possible one which will properly define s. Another is

$$s = \{p \ \& \ [(q \ \& \ r) \, V \, (q' \ \& \ r')]\} \, V \, \{p' \ \& \ [q' \ \& \ r) V (q \ \& \ r')]\}$$

The circuit to represent this expression would differ from that above. Methods are available in the field of mathematical logic to find the representation which requires the least number of circuit elements.

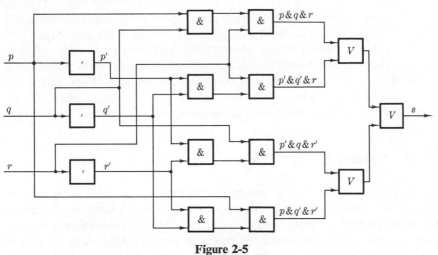

Figure 2-5

Just as we have developed an expression for the sum, an expression for the outgoing carry can be developed. If the symbol t is used for the outgoing carry bit, an expression which contains all the rules for outgoing carry contained in Table 2 of Section 1.53 is

$$t = (p \mathbin{\&} q \mathbin{\&} r) \, V \, (p' \mathbin{\&} q \mathbin{\&} r) \, V \, (p \mathbin{\&} q' \mathbin{\&} r) \, V \, (p \mathbin{\&} q \mathbin{\&} r')$$

A circuit representation can be drawn for this statement. If it is combined with that of Figure 2-5, a unit is obtained which receives three inputs, p, q, and r, and yields two outputs, s and t, obeying all the rules of Table 2. Such a circuit is a binary adder. It can be represented diagrammatically as a single unit, as in Figure 2-6. This unit represents essentially all the equipment required to perform binary addition. If the lowest order bits of two

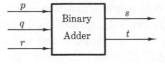

Figure 2-6

binary numbers are fed into the p and q inputs and a zero in the r input, the lowest order bit of the sum will appear on the s output, and the carry bit, if any, will appear on the t output. If then the next to lowest order bits of the summands are fed into p and q, and the t output fed into r, the next lowest order bit of the sum will appear on s, and the next carry, if any, will appear on the t output. By repetition of this process (which, incidentally, requires quite careful timing of the signals), all the digits of the sum can be generated sequentially on the s output. The process can be accelerated by use of a separate binary adder for each bit position in the numbers, connecting the "t" output of one stage directly to the "r" input of the next stage.

In the description of binary subtraction, multiplication, and division, it was pointed out that these operations involved little more than minor variations of the rules for binary addition. The statement of these operations in terms of negation and conjunction will not be given here. Suffice it to say that the circuitry for these operations is quite analogous to that for addition.

2.42   Control Logic

Section 2.41 described how arithmetic operations could be performed by electrical circuits made up of very simple basic elements. Control of selection of numbers from memory, and selection of arithmetic operation to be performed, can also be exercised by means of simple circuit elements termed "gates." In the typical computer, there are electrical connections from each of the memory cells to the arithmetic circuits. These connections are interrupted by "gates." As long as these gates are closed, there is no way for information to flow from one part of a computer to another. The process of control consists of opening the proper gates, at the proper times, to connect the desired combination of memory cells and arithmetic circuits. In actual fact, the gates are electrical circuits, which are opened or closed by voltage pulses. Each gate can be designed so that only the correct combination of pulses will cause it to open. For example, for a machine having memory cells 000 to FFF, each cell can have a gate which will be opened only by the combination of voltage pulses that corresponds to the gate number. Then binary bits corresponding, for example, to the decimal number 845, when sent to all memory cell gates, would open only that corresponding to memory cell 845, so that information could flow into or out of that cell but no other. In like manner, just as memory cell gates can be controlled by chosen combinations of binary bits (or, more accurately, the equivalent voltage pulses), so also can gates for the various arithmetic circuits be controlled by combinations of binary bits. For example, if the arithmetic unit of a machine has separate sets of "add" circuits, "subtract" circuits, "multiply" circuits, and "divide" circuits, it could be so connected that the gate for the "add" circuits is opened by the pulses which represent the letter "A," the gate for the subtractor circuits is opened by the pulses which represent the letter "S," and so forth. With such an arrangement, the proper combinations of binary bits, introduced into the control unit of the computer, can cause the computer to perform arithmetic calculations with numbers stored in the memory.

Section 2.41 described the relation between electrical circuits, some statements from mathematical logic, and the operations of arithmetic. The same considerations can be applied directly to the gating circuits used for control purposes. Conceptually, a conjunction circuit such as described in Section 2.41 can be regarded as a simple gate. The relation

$$p = q \ \& \ r$$

can be read "$p = 0$ if $r = 0$, but $p = q$ if $r = 1$." In other words, the value of r determines whether or not p is allowed to be determined by q. The value of r closes or opens the gate. If $r = 0$ the gate is closed, and whether q be zero or one, p is zero. If $r = 1$, the gate is open and p takes on the value of q. The circuit diagram in Figure 2-7 indicates r as the gate control, q as the input,

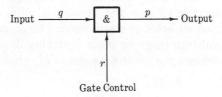

Figure 2-7

and p as the output. Gates with more complicated controls are easy to construct. For example, the relation

$$p - q \ \& \ (r \ \& \ s' \ \& \ t \ \& \ u \ \& \ v')$$

can be considered as a gate between q and p which is open only when $r = 1$, $s = 0$, $t = 1$, $u = 1$, $v = 0$. The circuitry can be drawn as in Figure 2-8.

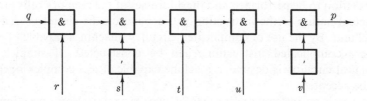

Figure 2-8

If the proper combination of ones and zeros is fed into r, s, t, u, and v, the gate opens. Otherwise it remains closed. Gates of this sort can be used to control the connections between memory cells and the arithmetic unit. To demonstrate this fact, let us assume we have a computer with only a three-cell memory, cells numbered 1, 2, and 3. For cell number 1 we will use a gate of the form

$$p = q \ \& \ (r' \ \& \ s)$$

for cell number 2 a gate of the form

$$p = q \, \& \, (r \, \& \, s')$$

and for cell number 3 a gate of the form

$$p = q \, \& \, (r \, \& \, s)$$

Then the gate for cell 1 will be opened by $r = 0$, $s = 1$; cell 2 by $r = 1$, $s = 0$; and cell 3 by $r = 1$, $s = 1$. Considering the combination r, s as a two-digit binary number, the number 01 sent to all gates will connect cell 1 only, the number 10 sent to all gates will connect cell 2 only, and the number 11 sent to all gates will connect cell 3 only. In like manner, gates can be designed that will control the access to bytes or words in very large memories. Each set of gates requires several transistors or other switching devices, so the memory control of large machines can involve hundreds of thousands of transistors.

2.43   Machine Instructions

As has been stated before, it is not our purpose to become involved in details of computer design. Enough has been said to indicate how control can be accomplished through the presentation to the gating circuits throughout the machine of binary bit combinations indicating the operation to be performed and the memory cells involved. These may be in binary-coded decimal, in octal or hexadecimal with the operation denoted by a number, or in other form. A single such combination of binary bits is termed a "command" or an "instruction." An instruction must contain those binary bits defining the operation to be performed and the addresses of the memory cells involved. The actual nature of the instruction will depend on the circuitry used in the control unit. By the use of complicated circuitry, machines capable of executing very complicated instructions can be constructed. If simpler, more economical circuitry is desired, a machine capable of less complex operations must be accepted.

   In general, an instruction consists of two basic parts: an operation code, which indicates what operation is to be performed, and one or more addresses, which indicate where the operands are to be found in the machine and where the results are to be stored. When the computer is designed, the bit patterns to represent the instructions must be selected and the machine circuits wired accordingly. Ordinarily, the number of bits used in the instruction is chosen to be equal to the word length in the machine's memory, or some even fraction thereof, so that instructions can be stored in the memory just as are the data upon which the instructions are to operate. For example, the instruction set of the IBM 360 series machines contains some instructions which are 16 bits long, some 32 bits long, and some 48 bits long. We shall not attempt to describe in detail the instruction set for that series or any other type of machine but will restrict ourselves to a brief description of how the processor handles instructions and an indication of what general classes of instructions are available on most machines.

2.44   Processor Functions

The processor performs three tasks with respect to machine instructions. They are:

(*1*)   Fetching, or bringing the instruction from memory to the processor.

(*2*)   Decoding, or determining from the instruction bit pattern what operation is to be performed, and what operands are to be used.

(*3*)   Execution, or actual performance of the task indicated by the instruction.

These three steps are performed sequentially on each instruction in a regular cycle, clocked by timing circuits within the machine. On the faster machines, the operations are overlapped, so that a machine can be executing one instruction, while decoding the next one, while fetching still the next one from memory.

To aid in instruction fetch, the processor usually contains a special register which contains the memory address of the next instruction due to be executed. Each time an instruction is needed by the processor, the contents of this register (usually termed the "location counter") are presented to the memory gating circuits. The appropriate memory location will respond, allowing the next instruction word to proceed to the processor for decoding and execution. The machine wiring ordinarily causes the contents of the location counter to be increased by one each time an instruction is executed, so that instructions will be executed in the order in which they are stored in memory. However, the location counter is just another storage register, so it is possible to have machine instructions which will store into the register, and thus it is possible for the instructions themselves to indicate what instruction should be executed next. One important application of this feature is the performance of a loop, where the machine is caused to repeat the same set of instructions over and over. Loops can be used to do the same calculation on a number of different sets of data, or to do repetitive operations within the same calculation, such as adding up a large set of numbers by repeatedly adding a new number to the previous sum.

Decoding is accomplished by presenting that portion of the instruction indicating the operation to be performed to a set of gates designed to recognize the bit patterns assigned to each instruction in the machine's repertoire. In the manner described in Section 2.42, these gates can cause the adder circuit gates to be opened if an add is called for, the divider circuit gates to be opened if a divide is called for, and so forth.

Execution is accomplished by circuits designed to perform each operation required of the machine. Section 2.41 indicated how execution of an addition, subtraction, multiplication, or division might be accomplished. It is possible to construct execution units which will perform other operations as well, such as shifting the bits in a word left or right, or combining the bits in two words by the logical "and" and "or" operations defined in Section 2.41. It is possible to wire execution units to perform square roots, or to raise to powers, or to convert numbers from binary to decimal and decimal to binary, and so on. Indeed, execution units could be wired to perform entire complex calculations. Ordinarily, the execution circuits in a general-purpose computer perform only the more basic types of operation. The more elaborate computations are accomplished by writing sets of instructions and presenting them to the machine, that is, by "programming" the machine.

2.45   General Registers

In the processing unit of a computer it is convenient to have a few special memory cells, ordinarily separate from the memory unit of the machine. They are used to hold the numbers actually being worked on by the arithmetic circuits of the machine and are somewhat analogous to the registers present in a desk calculator. In addition or subtraction, one of these registers is used ordinarily to receive the sum or difference as it is generated by the circuitry. In the simplest machines, only one register is available for this purpose, and it is usually referred to as the accumulator. The accumulator normally contains one of the addends before addition is started, and the sum after completion. For multiplication and division, a single register is not sufficient for temporary storage of the numbers involved, because the multiplier or divisor must be retained throughout the process until the answer is completely generated. In the simplest machines, a single additional register is added to fill this need. The multiplier-quotient register (MQ register) is used to contain the multiplier in a multiplication problem while the product is being built up by repeated additions in the accumulator. It is also generally used in division to contain the quotient as it is built up, while the accumulator contains the dividend as altered by repeated subtractions involved in a division problem. In more powerful machines several operand registers are available and may be used as desired for storage of operands and results. In the illustrative examples used later in the chapter, it will be assumed that we are dealing with a machine having 16 special operand registers, numbered in hexadecimal from 0_{16} to F_{16} and referred to as general registers.

2.5   INPUT AND OUTPUT UNITS

A computer consisting of a memory and a processing unit can be made to perform arithmetic and other operations in an automatic fashion. To be useful, it must also be able to accept information from external sources and transmit information to external devices. Many different types of devices may be attached to a computer to aid in its use. Some, like the magnetic tapes and disk files already described, are mainly for auxiliary storage. The latter are normally classed as input-output, or I/O, devices because they communicate with the computer in the same manner as do card readers or printers. It is perhaps more suitable to refer to all these devices as peripheral devices.

2.51   Types of Peripheral Devices

Some of the commonly used peripheral devices are described below.

Card Reader.   The card reader accepts a deck of punched cards, passes them one at a time under a light source, and produces electric signals corresponding to the holes. Typically a card has 80 columns, each column having 12 positions at which holes can be punched. Each alphanumeric character has its unique coding in terms of holes punched in the column. Thus the card reader allows characters manually punched onto cards to be transmitted electrically into the computer. A typical reading speed is 250 cards per minute, or about 300 characters per second.

Line Printer.   The line printer accepts sets of electrical signals corresponding to alphanumeric characters and activates print bars to print the appropriate characters on paper. The mechanical bars typically print an entire line of about 128 characters at each mechanical motion. Typical print rates are about 1000 lines per minute, or about 2000 characters per second.

Card Punch.   The card punch does the inverse operation of the card reader. It receives electrical signals and punches corresponding holes in punch cards. Typical speeds are 100 cards per minute, or about 130 characters per second.

Teletypewriter.   The teletypewriter has a keyboard similar to an ordinary typewriter, with a few additional, special keys. Depressing the key for any character causes the electrical pulses corresponding to the character to be transmitted, at the same time the character itself is printed by typewriter action. Conversely, an arriving signal corresponding to a keyboard character will cause that character to be printed. Thus the teletypewriter can be used for two-way communications with a computer. Typical speeds are typing speeds, 1 to 10 characters per second.

Special Keyboard Devices.   In addition to the standard teletypewriters, there are on the market a number of special keyboard devices. Some have special keyboards designed for a certain type of application, such as banking. Some have a cathode-ray scope and display characters on the scope face rather than printing the characters. These devices frequently offer more convenience than the standard teletypewriter. The operating speeds for input into the computer are in the same class. Those for output are much faster, in the hundreds of characters per second range.

Optical Character Readers.   There is a rapidly growing employment of readers which can actually read printed information and transform each alphanumeric character into the proper electrical signals. These devices offer the potential for greatly increased versatility in providing input to computers, but at this writing their specific properties are not sufficiently well defined to warrant extensive discussion.

Microfilm Recorders.   The automatic microfilmer can translate received electrical signals into characters which are recorded on film by a light beam. Such a device can produce printed material on film at rates several times those of the mechanical line printer, and can also be used to draw graphs, contour plots, engineering drawings, and so on.

Mechanical Plotters.   Graphs, engineering drawings, and so on, can also be constructed by computer-driven mechanical plotters. The incoming electrical signals are used to control motors and drive chains which mechanically position a pen on paper, and raise or lower the pen as required.

Paper Type Reader/Punch.   Holes representing alphanumeric characters can be punched in paper tape and the tape passed over an optical reader to produce electrical signals for input to the computer. Conversely, received electrical signals can be used to control the punching of holes in paper tape to record information from the computer.

Storage Devices.   As mentioned earlier, magnetic tape drives, disk files, and drums also communicate with computers in the same general fashion as do the other devices described above. Thus the input-output for the computer handles these devices exactly as it does the readers, printers, and so on.

2.52   Control of Input and Output

The communications paths between the computer and the I/O devices are called channels. Each channel is hooked into the computer control unit through a set of gates, just as are the memory cells and the arithmetic circuits. Thus the proper set of control signals sent to gates can cause a given memory cell to be open to a chosen channel for receipt or transmission of data. At the other end of the channel there may be several I/O devices, each one also connected to the channel through gates.

Just as the gates in the memory unit and the processor can be controlled by the use of machine instructions, so can the gates of the I/O unit be controlled. The instruction repertoire of the typical computer includes instructions whose purpose is to control these gates and allow transfer of selected information between the memory and peripheral devices.

2.6   FAMILIARIZATION LESSON IN MACHINE LANGUAGE PROGRAMMING

To make the ideas of Sections 2.4 and 2.5 more concrete, an excursion will be made at this point into machine language programming for a hypothetical machine. To a large extent, the machine instructions used are patterned after those possessed by the IBM 360 series computers. However, wide liberties have been taken in modifying them in order to demonstrate fundamentals without becoming overly involved in the complexities of such highly sophisticated machines.

2.61   Fixed-Point Arithmetic

It will be recalled that a machine instruction consists of two parts: an operation code portion and an address portion. For the IBM 360 series the operation code is eight bits long and occupies the first eight bit positions of the instruction. The meaning of the remaining bits depends on the type of instruction. The first instructions we will demonstrate are those for performance of fixed-point arithmetic. Of the more than 30 instructions available for fixed-point arithmetic in the 360 series, six will suffice to demonstrate the capabilities. They are shown in Table 1. There are analogous instructions for floating-point operations, but for demonstration purposes we will confine ourselves to the fixed-point set.

The load and store instructions are given completely by a set of bits which will fit into a single 32-bit word. The first 8 bits of this word are used for the operations code: 58 for load, 50 for store. The next 4 bits (or next hexadecimal digit) give the number of the general register to be used: 0 to F_{16}. The next 8 bits have a use which is beyond the scope of this description. For our purposes we may assume they are zero. The last 12 bits identify the memory address to be used. We will assume that the first memory location is 000, and that each byte is addressed sequentially, in hexadecimal. Thus the address of the first byte is 000, that of the second is 001, that of the third is 002, and so on. The address of the first full 32-bit word is 000, and it consists of four 8-bit bytes. Thus the address of the second word in memory is 004, that of the third is 008, that of the fourth is 00C, and so on.

TABLE 1
Sample IBM 360 Arithmetic Instructions

Instruction Name	Op Code	Purpose
Load	58	Bring a word from a chosen memory cell into a chosen one of the general registers.
Store	50	Take a word from a chosen one of the general registers and place it in a chosen location in memory.
Add	5A	Add the contents of a memory location to those of a chosen general register and place the result in the general register.
Subtract	5B	Subtract the contents of a memory location from those of a chosen general register and place the result in the general register.
Multiply	5C	Multiply the contents of a memory location by those of a chosen general register and place the results in the general register.
Divide	5D	Divide the contents of a general register by those of a memory location and place the results in the general register.

Example 1.   What action would the instruction

583006AC

cause to occur?

The first two digits are 58, the operations code for a load operation. The next specifies general register 3. The last three digits specify memory cell 6AC. Hence this instruction would cause the contents of memory location 6AC to be placed in general register 3.

For the add, subtract, multiply, and divide instructions, the first 8 bits again give the operation code. The next 4 specify the general register containing the first operand. The next 8 we will assume to be zero. The final 12 specify the memory location containing the second operand. It is a property of these four particular instructions that the result is placed in the register which had contained the first operand. Thus the first operand is destroyed by the operation.

Example 2.   What action would the instruction 5A700320 cause to occur?

5A is the op code for addition. Hence the instruction would cause the contents of general register 7 and memory location 320 to be added and the result to be placed in general register 7.

The six instructions described above will serve as a starting point in describing the use of the computer. At a later stage it will be found necessary to introduce some additional ones. They will be defined when needed.

It has been seen that the process of performing calculations with a digital computer consists of placing the proper numbers in the memory, having the machine execute the proper sequence of commands, and then extracting the result from the machine. The heart of the problem is the process of writing down the sequence of commands which the machine is to execute. This sequence of commands is known as the "program," and the process of preparing the program is called "programming" or "coding." A few examples of a program will be given. In these examples, it will be assumed that the number a is already stored in memory location 810, b in 820, c in 830, and d in 840.

Example 3.   Find $a + b + c + d$ and store the results in 850.

A suitable program is

```
58100810    (loads a into general register 1)
5A100820    (add a and b, puts sum in general register 1)
5A100830    (adds c to the sum of a and b, puts sum in register 1)
5A100840    (adds d to the previous sum, puts result in register 1)
50100850    (stores a + b + c + d in location 850)
```

Example 4.   Find $(ac - b)/d$ and store the result in location 870.

A suitable program is

```
58100810    (loads a into general register 1)
5C100830    (multiplies a by c, puts result in register 1)
5B100820    (subtracts b, giving ac − b in register 1)
5D100840    (divides by d, giving (ac − b)/d in register 1)
50100870    (moves answer from register 1 to location 870)
```

It is worth repeating that each instruction consists of a particular set of binary bits which when presented to the gating circuits of the machine will cause the gates to open and numbers to move into or out of particular registers or arithmetic units. Any instruction which contains an incorrect number will cause the wrong gates to be opened, with unpredictable but generally disastrous effects on the progress of the calculation. Computers are highly intolerant of even the most trivial mistakes in programming.

2.62   Input and Output

As was indicated in Section 2.52, there are gates in the computer I/O channels which permit the input and output to be controlled by machine instructions. Before information is sent out over a channel, a command must be sent which will activate the proper gates and make ready the proper I/O device. In turn, there must be a return signal to the computer to indicate whether the device is ready to send the information. The signals of this sort required to assure proper communication over a channel become rather involved, and we shall not attempt to treat them in detail. For demonstration purposes we will assume that we have a machine with just one channel, whose gates are opened by the bit combination 0001_2; and at the other end of this channel there is a card reader, whose gates we opened by the bit combination 0001_2, and a printer, which can be addressed by the bit combination 0010_2.

Several pieces of information are needed in the conduct of an I/O operation. They are:

(*1*)   Identification of the channel to be used.

(*2*)   Identification of the particular I/O device on that channel.

(*3*)   Type of operation (for example, read or write).

(*4*)   Memory location at which the information transfer is to commence.

(*5*)   Number of words (or bytes) of information to be transferred.

If the machine is to have a number of channels and peripheral devices, the information above is too much to be contained in a single instruction word. This problem can be surmounted by first storing part of the information in a preselected location in memory and then having the actual I/O instruction provide only the remaining parts of the information. When this is done, the program, in order to perform I/O, must first store the required information in these memory cells by using the normal load and store commands, and then give the actual I/O instruction. For demonstration purposes, we will assume that for our computer the information for I/O is divided between two 32-bit words; the instruction itself and a channel command word. Figure 2-9 shows the format for these words. The first 8 bits of the instruction

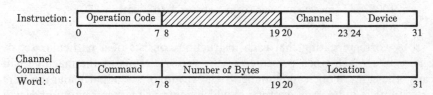

Figure 2-9: Assumed I/O Control Word Formats

word are for the operations code for I/O, which is 9C. The next 12 bits are unused, and we will fill them with zeros. The next 4 bits identify the channel. We will always use channel 1. The last 8 bits are for device number. We have two I/O devices. Number 1 is the reader and number 2 is the printer. Hence we have only two forms for the I/O instructions:

9C000101 for input
9C000102 for output

In the channel command word, the first byte is reserved for the command. We will consider two commands:

01_{16} for print
02_{16} for read

The next 12 bits specify the number of bytes of information to be transferred. Thus if five words of information are to be transferred (each word is 32 bits, or 4 bytes long), these bit positions would contain the number 20_{10}, or 014_{16}. The final 12 bits of the word specify the location in memory for the first byte of information to be transferred. If the first byte location were 400_{16}, that number would be the content of these bit positions.

Example 1. Two words are to be read from the card reader into memory starting at location 300. What instruction and channel command word would be used to accomplish this?

Because the card reader is unit 1 on channel 1, the instruction is

9C000101

The channel command word is

02008300

the 02 indicating read, 008 the number of bytes (two words) to be transferred, and the 300 indicating the starting location.

In an actual program, we must have a specified location where the channel command word is to be stored. For demonstration purposes, we will assume that the channel command word must always be stored in location 100_{16}. Any time an I/O operation is to be performed, the proper channel command word must first be placed in this location.

Example 2. Write a program which will read in three numbers, a, b, and c, compute the quantity $ab + c$, and print the answer.

The first task is to input the quantities a, b, and c in the machine. Let us store them beginning at location 200. To do so, an input instruction is needed.

Let us assume that the numbers a, b, and c are available on punch cards. To perform the input, the instruction

9C000101 (do I/O on channel 1, unit 1)

and the channel command word

0200C200 (read 12 bytes into location starting at 200)

are needed. Before issuing the input instruction we must first manage to store the channel command word in location 100. A way of doing this is to include the channel command word with the program, so as to make it available. Let us assume that we have done this, and pretend that when we load the program itself into the machine, the channel command word will end up in memory in a location which for the moment we will designate as xxx. Then we may load it into location 100 by the load and store instructions

58200xxx   (load contents of location xxx to general register 2)
50200100   (store contents of general register 2 in location 100)

Then we may perform the input, perform the calculations, and print the answer. The complete program could appear as follows:

Instruction No.	Instruction	Purpose
1	58200xxx	Load channel command word into general register 2.
2	50200100	Store channel command word in location 100.
3	9C000101	Call for input, (a, b, and c into 200, 204, and 208).
4	58100200	Load a into general register 1.
5	5C100204	Multiply a and b, place result in register 1.
6	5A100208	Add c to ab, place result in register 1.
7	50100300	Store $ab + c$ in location 300.

After these instructions are executed, the result $ab + c$ will be stored in location 300. We must now provide instructions to cause the result to be printed. The required channel command word is 01004300, the 01 indicating print, 004 indicating 4 bytes or one word, 300 indicating the starting location. To accomplish the print, we must load this channel command word into location 100, then execute an I/O instruction. For the moment, let us assume that this channel command word, like the one previously used for input, is to be placed in the machine along with the program and will appear in location yyy. Then we can continue the program started above as follows:

Instruction No.	Instruction	Purpose
8	58100yyy	Load channel command word into general register 1.
9	50100100	Store channel command word in location 100.
10	9C000102	Call for output, print the word starting at location 300).

In listing the program, we have also listed the number of the instruction word. The reason for this is that before the program will be complete we must replace the symbols xxx and yyy, which we used to indicate the storage locations of the channel control words, by the actual memory locations. We can add the channel command words on the end of the program, following the last actual instruction. If we do this, and load the program into the machine starting at location 000, then the first instruction will begin at location 000, the second at location 004, the third at location 008, the fourth at 00C, the fifth at 010, the sixth at 014, the seventh at 018, the eighth at 01C, the ninth at 020, and the tenth at 024. This means that the channel command words can be stored in locations 028 and 02C. Thus the complete program would appear as follows:

Instruction Location	Instruction	Purpose
000	58200028	
004	50200100	Input a, b, c.
008	9C000101	
00C	58100200	
010	5C100204	
014	5A100208	Form $ab + c$.
018	50100300	
01C	5810002C	
020	50100100	Print answer.
024	9C000102	
028	0200C200	Channel Command Words for input and
02C	01004300	output.

EXERCISE 4

1. In the following problems, assume that the numbers a, b, c, etc., are already stored in memory locations as indicated:

300 a	304 b	308 c	30C d
310 e	314 f	318 g	31C h
320 i	324 j		

Write programs to compute the following quantities and store the result in memory cell 328.

  a. $bd + f + h$
  b. $cdg - a$
  c. $ae + f/j$
  d. $bf - dh$
  e. $b/j - af + d/h$
  f. $a + ab + abc + abcd + abcde$
  g. $a + b(c + d(e + f(g + h)))$
  h. $(ac - fj)/(bd + gh)$

  i. $\begin{vmatrix} a & b & c \\ d & e & f \\ g & h & i \end{vmatrix}$

2. Write programs to input the necessary constants, to perform the calculations indicated, and print the results.

  a. $ab + cd$
  b. $(ab + c)/d$
  c. $ab - c$
  d. $a + 2b + 3c$
  e. $4(a + b)^2$
  f. $a + 3(b + c)$
  g. $a/b + c/d$
  h. $a^3 + a^2b + b^3$

2.63 Control Instructions

It was said earlier that the computer proceeds through a program stored in memory by taking instructions in sequential order from succeeding memory locations. This is a true statement as long as nothing interferes with the normal updating of the location counter. If, however, the contents of the location counter are modified in some fashion, this stepwise process will be altered. The machine can be made to take its next instruction from any location in memory, merely by inserting the address of that memory location into the location counter. We might say that the location counter points to the location of the next instruction to be executed.

There are two classes of operations which involve alteration of the contents of the location counter. These are branching operations and interruptions. These classes of operations will be discussed in the succeeding sections.

2.64 The Unconditional Branch

The simplest operation involving alteration of the contents of the location counter is the unconditional branch instruction. An instruction 4 bytes long will be used to exemplify the unconditional branch instruction. The first byte of this instruction is the operation code, 47. The next 4 bits are all ones, or hexadecimal F. The next 8 bits will be taken as all zeros. The final 12 bits are the location in memory from which the next instruction is to be taken. For example, the instruction

47F00320

will place the quantity 320 in the location counter and thus cause the next instruction to be taken from location 320.

Example 1. Compute $y = ax + b$ for a set of given values of x.

In order to perform this calculation, we need to read in the values of a and b, then read the first value of x, compute and print y, read the next value of x, compute and print y, and so on, until all values of x have been used. The part of the operation which inputs x and computes and prints y is done over and over, the steps being the same for each journey through the calculation, the only difference being the numerical value of x involved. The unconditional branch instruction enables us to place the instructions for these steps in the program only once and then loop back through them again and again for different values of x. A suitable program is

Memory Location	Instruction	Remarks
000	58100038	Load channel command word for input of a and b.
004	50100100	Store CCW in location 100.
008	9C000101	Read a and b into locations 200 and 204.
00C	5810003C	Load channel command word for input of x.
010	50100100	Store CCW in location 100.
014	9C000101	Read a value of x into location 300.
018	58100300	Load x into general register 1.
01C	5C100200	Multiply by a, giving ax in register 1.
020	5A100204	Add b, giving $ax + b$ in register 1.
024	50100400	Place $y = ax + b$ in location 400.
028	58100040	Load channel command word for output of y.
02C	50100100	Store CCW in location 100.
030	9C000102	Print y.
034	47F0000C	Go to location 00C for next instruction.
038	02008200	Channel command word for input of a and b.
03C	02004300	Channel command word for input of x.
040	01004400	Channel command word for output of y.

In following through the program we see that the instructions at locations 000 through 008 load the values of a and b into locations 200 and 204, where they remain for the rest of the calculation. The instructions at locations 00C through 034 will be executed over and over, as long as there are values of x available from the card reader. These instructions are said to form a loop. The ability to form loops in programs is one of the factors which give computers such great power for mathematical calculations. The program given above will perform for one value of x, or a hundred, or a thousand, without further modification. The eleven instructions which actually form the loop can cause thousands of calculations to be performed. If the above computation were required for only one value of x, hand computation would be faster than writing the program and cheaper than running the computer. For large numbers of values of x, however, the computer has the advantage.

2.65   The Conditional Branch

A somewhat more complicated, but much more powerful instruction than the unconditional branch is the branch on condition. This instruction causes the contents of the location counter to be altered sometimes and not altered at other times, depending upon some other condition within the machine. The condition which governs this selection is the quantity stored in another register within the control unit. We refer to this register as the "condition code register." It is a very short register, only two bits long, so the condition code is binary 00, 01, 10, or 11 (decimal 0, 1, 2, or 3). Certain of the instructions in the machine's repertoire will cause the condition code to be set automatically. For example, when the add instruction, 5A, or the subtract instruction, 5B, is performed, the condition code is automatically set to zero if the result was zero, to one if the result was negative, and to two if the result was positive. The multiply, divide, load, and store instructions do not affect the condition code.

For the branch on condition instruction we will use the same operations code as for the unconditional branch instruction, 47. The next 4 bit positions will specify the condition code which will cause the transfer to take place. A one in the first of these positions will cause a transfer if the condition code is zero. A one in the second position will do so if the condition code is one, a one in the third will do so if the condition code is two, a one in the fourth position will do so if the condition code is three. For example, if these 4 bits are 1010, or hexadecimal A, the transfer will be made if the condition code is either zero or two.

The next 8 bits of the instruction are zeros, and the final 12 bits specify the memory location from which the next instruction is to be taken if the transfer is actually made. For example, the instruction

$$47A00214$$

will cause the next instruction to be taken from location 214 if the condition code is either zero or two. If the condition code is either one or three, the location counter will be stepped normally and the next instruction taken from the memory location next after the one currently being used.

Example 1.   Write a program to compute and print x^3 for $x = 1, 2, 3, \ldots, 500$.

Since the values of x are equally spaced, we do not need to read all 500 values of x into the machine but instead can compute them as we need them. Further, it is quite easy to make the program more general than the problem statement requires. We can read in as inputs the initial value of x, x_0; the increment or step size to be used, Δx; and the final value of x, x_f. The problem as stated would require that we input $x = 1$ for initial value, 1 for step size, and 500 for final value. However, the program once written will work for any other set of inputs just as well. A suitable program is

Memory Location	Instruction	Remarks
000	58100044	Load channel command word for input of x_i, Δx, x_f.
004	50100100	Store CCW in location 100
008	9C000101	Read x_i, Δx, x_f into locations 200, 204, 208.
00C	58300200	Load x_i into general register 3.
010	58500208	Load x_f into general register 5.
014	5B500200	$x_f - x_i$ in general register 5.
018	58100048	Load channel command word for printing y.
01C	50100100	Store CCW in location 100
020	5C300200	Square x, giving x^2 in register 3.
024	5C300200	$y = x^3$ in general register 3.
028	50300300	Store y in location 300 for printing.
02C	9C000102	Print y.
030	58300200	Load x into general register 3.
034	5A300204	$x + \Delta x$ in general register 3.
038	50300200	$x + \Delta x$ replaces x in location 200.
03C	5B500204	Subtract Δx from register 5.
040	47A00020	If contents of register 5 are positive, go to location 020 and do another value of x.
044	0200C200	Channel command word for input of x_i, Δx, x_f.
048	01004300	Channel command word for output of y.

In this program there is a loop from locations 020 to 040, which is repeated until all the values of x less than or equal to x_f have been used. General register 5 was used to determine whether more trips through the loop were required. Initially x_f was stored in this location. Then x_i was subtracted from it, and then each time through the loop, Δx was subtracted from it. Thus register 5 always contained the quantity $x_f - x$, the difference between the final value and the current value of x. As long as this quantity remains positive or zero, the branch instruction at location 040 will change the contents of the location counter to read 020, and the loop will be repeated. When the quantity $x_f - x$ becomes negative, the loop has been performed enough times, and the branch instruction will not reset the location counter, but instead the location counter will be increased in normal fashion to read 044. At this point, with the program the way it is written above, a strange thing would happen. The quantity stored in location 044 would be fetched to the control unit and submitted to the decoder to decide what kind of instruction it is. Since it is not an instruction at all, the machine might be expected to do something odd at this point. It would seem that there should be more satisfactory ways of terminating a program, rather than letting the instruction decoder run into some garbage, and there are. These will be discussed in Section 2.66.

Before leaving this example, note how the branch on condition instruction was used. It forms the exit point for the loop of instructions that run from location 020 to 040. In example 1 of Section 2.64 the loop had no exit but only terminated on an external condition, that is, no input cards in the card reader. If a loop is to be used internally within a program, it must have at least one exit point, a point at which the computer can either proceed through the loop again or exit to other parts of the program. The conditional branch instruction is frequently used in this role.

Example 2.   Write a program which will input given values of x, compute and print $y = ax + b$ if $x < a + b$, compute and print $y = cx + d$ if $x \geq a + b$.

A suitable program is

Memory Location	Instruction	Remarks
000	58100060	Load channel command word for input of a, b, c, d.
004	50100100	Store CCW in location 100.
008	9C000101	Read a, b, c, d into location 200, 204, 208, 20C.
00C	58100200	Load a into register 1.
010	5A100204	Add b, giving $a + b$ in register 1.
014	50100210	Store $a + b$ in location 210.
018	58100064	Load channel command word for input of x.
01C	50100100	Store CCW in location 100.
020	9C000101	Read a value of x into location 300.
024	58300300	Load x into register 3.
028	5B300210	Subtract $a + b$, giving $x - (a + b)$ in register 3.
02C	47A00040	Jump to location 040 if $x \geq a + b$.
030	58100300	Load x into register 1.
034	5C100200	ax in register 1.
038	5A100204	$y = ax + b$ in register 1.
03C	47F0004C	Jump to location 04C.
040	58100300	Load x into register 1.
044	5C100208	cx in register 1.
048	5A10020C	$y = cx + d$ in register 1.
04C	50100400	Store y in location 400.
050	58100068	Load channel command word for output of y.
054	50100100	Store CCW in location 100.
058	9C000102	Print y.
05C	47F00018	Jump to location 018.
060	02010200	Channel command word for input of a, b, c, d.
064	02004300	Channel command word for input of x.
068	01004400	Channel command word for output of y.

In this program the branch on condition is used in location 02C to select which of the calculations is performed, $y = ax + b$ or $y = cx + d$. Branch instructions are quite commonly used in such a role.

2.66   Interruptions

Interruptions form the second major class of operations which involve alteration of the contents of the location counter. Some interruptions can be caused by program instructions, while others are hard-wired into the machine. This second type of interruption usually is associated with errors in the program or malfunctions of some part of the computer. For example, an attempt to divide by zero, or the use of an operation code which the decoder in the control unit cannot recognize, will cause an interruption. When the interruption occurs, the contents of the location counter are replaced by the contents of some particular memory cell in the machine. What happens next depends on what instructions have been stored at the new location where the location counter now points. For example, on a divide by zero, or an illegal instruction, we will probably want the machine to print out a message and terminate the problem. It is possible to have stored in the machine a set of instructions which will make this happen.

At this point it is necessary to introduce a new concept regarding computer programs. The programs described so far have been concerned with obtaining numerical answers to mathematical problems. Such programs are customarily termed "user programs" because their purpose is to use the machine as a tool. However, when one of these programs is running and there occurs an interruption such as the ones mentioned above, the next actions of saying what went wrong and getting the problem terminated are actions which tend to be the same for all user programs. The instructions for this purpose can be stored in the machine and left there, as user programs come in and out. On an error of the types mentioned above these instructions will be called into use and will cause the proper actions to be accomplished. These sets of instructions which are not directly a part of the user program but instead are kept continuously available in the machine are sometimes referred to as "service routines."

The way in which service routines can be used to handle user programs when a malfunction occurs has just been described. This concept can readily be extended to a host of other functions which can be handled by service routines. For example, the loading of user programs into the machine, if done manually, one at a time, is a time-consuming process, wasteful of machine time. To avoid this waste it is customary to have one master program, which is placed in the machine when it is started up, and this program causes the first program to be read into the machine and transfers control to it so that the user program will run. At the conclusion of that user program, this master program, or supervisor, causes the next user program to be read into the machine and run, and so on. If no user programs are available at the machine input, the supervisor waits and watches by executing repeatedly a small loop of instructions. If an interruption occurs, the supervisor calls on the proper service routine to take action.

The above discussions have been in preparation for introducing a machine instruction which has the purpose of causing an interruption. It was mentioned in connection with Example 1 of Section 2.65 that the program given was not really satisfactory because upon exit from the loop at memory location 040, the computer would next try to interpret the channel control word at location 044 as an instruction. From the discussions above it would appear that this would cause an interruption, and the supervisor would take over and terminate the program. Rather than depend on this somewhat chancy way of terminating a program, it is best to have an instruction which will create an interrupt directly, thus telling the supervisor that the program has come to normal completion. For this purpose we will introduce the supervisor call instruction, a 2-byte instruction with operation code 0A, and all remaining bits zero. Thus the supervisor call instruction is 0A00. The program in Example 1 of Section 2.65 could be corrected by inserting this supervisor call instruction at location 040 and adjusting the rest of the program accordingly.

EXERCISE 5

1. Rewrite the program of Example 1, Section 2.65, to include a supervisor call instruction at location 044.

2. Write a program for the computation of y according to each of the following formulas, where a, b, and c are fixed constants, and x takes on different values which are to be read in from cards.

   a. $y = (a + x)^3$ b. $y = a + bx + cx^2$
   c. $y = a + bx^2 + cx^4$ d. $y = a^2 + b^4 + ac + x$

   e. $y = \dfrac{ax + b}{cx + d}$ f. $y = a^2 + 2abx + b^2x^2$

3. Do problem 2 for the case where the values of x are to be computed starting with a value x_i, and going by equal steps Δx to a value x_f.

4. Do problem 2 for the case where formula a is to be used whenever $x \leqslant a$ and formula b whenever $x > a$.

2.67   Logical Instructions

Just as the introduction of control instructions afforded a great increase in the capability of a computer to perform useful calculations, so the introduction of yet another class of instructions will provide the capability for a much wider class of useful functions. These new instructions are termed "logical instructions." These instructions treat numbers stored in memory not as numbers but merely as sequences of bits, and apply on a bit by bit basis the operations of mathematical logic such as the "and" and "or" operations described in Section 2.41. For our purposes it will be sufficient to discuss three instructions in this class: a "compare logical" instruction, an "and" instruction, and an "or" instruction.

As an example of the compare logical instruction we will use a 4-byte instruction. The first byte is the operation code, 55. The next 4 bits are the address of one of the general registers. The next 8 bits are zero. The final 12 bits are the address of some location in memory. The instruction will cause the contents of the general register to be compared to those of the memory location, bit by bit, starting at the leftmost bit position. If each pair of bits matches, the condition code is set to zero. If a mismatch is found, the comparison is terminated and the condition code set to one if the general register contained the one bit and to two if the memory location contained the one bit.

Example 1.   General register 1 contains the quantity 47E85A4 and memory location 244 contains the quantity 47E93D7. What will be the condition code after the instruction 55100244 is executed?

The bit patterns are

$$0100011111101000010110100100$$
$$0100011111101001001111010111$$

Beginning at the left, all pairs of bits match until the sixteenth bit position is reached. In this position the second operand has a one while the first has a zero. Consequently, the condition code will be set to two.

Example 2.   We are keeping track of charge accounts for three men, John, Bill, and Joe, in the following way. In memory location 500 we have stored the name John (the ASCII coding AAEFE8EE) and in location 504 the account John owes. In location 508 is the name Bill and in 50C the amount Bill owes. In location 510 is the name Joe and in 514 the amount Joe owes. New charges are added by giving two inputs on a card, the name of the person and the amount of the charge. Write a program which will add new charges onto the proper account.

A suitable program is

Memory Location	Instruction	Remarks
000	58100050	Load channel command word for input of name and charge.
004	50100100	Store CCW in location 100.

Memory Location	Instruction	Remarks
008	9C000101	Read name and charge into locations 200 and 204.
00C	58200200	Load name into general register 2.
010	58400204	Load charge into general register 4.
014	55200500	Condition code 0 if name is John.
018	47700028	Jump to 028 if name is not John.
01C	5A400504	Add charge to John's account.
020	50400504	Store John's new debt.
024	47F00008	Jump to 008 for new input.
028	55200508	Condition code 0 if name is Bill.
02C	4770003C	Jump to 03C if name is not Bill.
030	5A40050C	Add charge to Bill's account.
034	5040050C	Store Bill's new debt.
038	47F00008	Jump to 008 for new input.
03C	55200510	Condition code 0 if name is Joe.
040	47700008	Jump to 008 for new input if name is not Joe.
044	5A400514	Add charge to Joe's account.
048	50400514	Store Joe's new debt.
04C	47F00008	Jump to 008 for new input.
050	02008200	Channel command word for input of name and charge.

If instead of three names on the list of accounts there had been several thousand, it would have been necessary to do this process differently. Suppose, for example, the computer had two tape drives. A list of the accounts could be kept on magnetic tape. This tape could be mounted on one tape drive and a blank tape on the other drive. The accounts could be read in one or a few at a time, updated if necessary as in the above example, and written out onto the other tape. A list such as the accounts list in this example is normally termed a file. The process of making changes in the file to create a new file is termed file updating. A subpart of a file which is treated as a unit, read in or out of the machine as one operation, is termed a record. Individual items in the record are termed entries. File updating is an important function for computers, not only in business applications as demonstrated in the above simple example, but also in scientific applications. We shall see later that entire computer programs are conveniently handled by treating them as files, to be updated, combined, or otherwise handled as may be convenient.

To exemplify the "and" instruction we will use a 4-byte instruction. The first byte is the operation code, 54. The next 4 bits are the address of a general register. The next 8 bits are zeros. The last 12 bits are the address of a memory location. The instruction will cause the corresponding bits from these two sources to be combined in pairs to form the logical product,

$$0 \ \& \ 0 = 0$$
$$0 \ \& \ 1 = 0$$
$$1 \ \& \ 0 = 0$$
$$1 \ \& \ 1 = 1$$

The result is placed in the general register from which the first operand came.

Example 3. General register 2 contains the quantity FFFF0000 and memory location 466 contains the quantity 7F391845. What will be contained in the two locations after execution of the instruction 54200466?

The bit patterns for the two words are

11111111111111110000000000000000
01111111001110010001100001000101

After the operation, register 2 will contain the logical product; that is, it will contain ones in each position where both of the above quantities contain ones, and zeros everywhere else. Thus register 2 will contain the bit pattern

01111111001110010000000000000000

or, in hexadecimal, 7F390000. Memory location 466 will still contain its original contents, 7F391845. It is seen that the effect of the operation is to transfer part of the quantity in location 466 into general register 2. Wherever a one appeared in register 2, the contents of 466 were accepted. Wherever a zero appeared in register 2, the zero remained. This type of operation is frequently called a masking operation. The quantity in register 2 is the mask. It determines which parts of the word in location 466 are transferred and which are not. The masking operation is a very powerful one, in that it allows, by proper choice of the mask, to separate out any character, bit, or combination of bits in a word. Then the compare logical instruction can be used to perform tests to determine what the character or bit or combination of bits is, and transfers can be made to different sets of instruction depending on the results of these tests. This means that programs can be written which will examine essentially any kind of material inserted into computer memory and act on it in accordance with any set of rules desired. Some of the major applications of this capability will be described in Section 2.7.

The logical " or " can also be exemplified by a 4-byte instruction. The first byte is the operation code, 56. The next 4 bits are the address of a general register. The next 8 bits are zeros. The last 12 bits are the address of a memory location. The instruction will cause the formation of the logical sum of corresponding bits from these two sources,

$$0 \ V \ 0 = 0$$
$$0 \ V \ 1 = 1$$
$$1 \ V \ 0 = 1$$
$$1 \ V \ 1 = 1$$

The result is placed in the general register from which the first operand came.

Example 4. General register 3 contains the op code for a load instruction into register 1 from location 0, that is, 58100000. Memory location 200 contains the number 00000536. What will be contained in register 3 after execution of the instruction 5630200?

The bit patterns for the two words are

0101 1000 0001 0000 0000 0000 0000 0000
0000 0000 0000 0000 0000 0101 0011 0110

After the operation, register 3 will contain the logical sum

0101 1000 0001 0000 0000 0101 0011 0110

or, in hexadecimal,

58100536

If the number 536 were the address of some quantity x, then the above operation would have created an instruction which says to load x into register 1. The capability to create instructions in this manner has far-reaching consequences, some of which are indicated in Section 2.7.

EXERCISE 6

1. Write a program which will read four-character words into the computer one at a time and stop when it reads the word STOP.

2. Write a set of instructions which will compare two words which are already stored at locations 200 and 300, print the word stored at location 400 if the two are the same, otherwise print the word stored at 500.

3. Write a set of instructions which will check the first character of the word stored at location 200, print the word stored at 300 if that character was a G, otherwise print the word stored at 400.

4. Given the two formulas

   a. $y = ax + b$
   b. $y = a + bx^2$

Write a program which will

(1) Read in the values of a and b.

(2) Read succeeding cards, each read instruction bringing in two words: an alphanumeric word, and a value of x. If the first character of the alphanumeric word is an A, use formula a and compute and print y. If the first character is a B, use formula b to compute and print y. If the first character is any other quantity, ignore this input and read the next word.

5. You wish to be able to code some arithmetic problems for a computer in a special language consisting of the following instructions and operands:

Instructions		Operands	
Symbol	Meaning	Symbol	Memory Location to Use
L	load	X	200
A	add	Y	204
S	subtract	Z	208
M	multiply	W	20C
D	divide		
ST	store		

you wish to be able to write in this language by giving instructions of the form

```
L   X
A   Y
ST  Z
```

where the first symbol in each pair identifies the instruction to be used and the second identifies the operand. Thus the above three instructions would mean to load X from memory into a general register, add Y, and store the result in a memory location reserved for the value of Z. Write a machine language program that will read pairs of words; check to see if the first word is L, A, S, M, D, or ST, and check the second word to see whether it is X, Y, Z, or W; and then construct and print a corresponding machine language instruction. For example, it should construct the instruction 58100200 if the pair of words L X is encountered. If a symbol is encountered which is not one of those defined above, have the program print the word "ERROR" and stop. (*Hint:* Restudy Examples 2 and 4 of Section 2.67.)

6. In the language described in problem 5, write instructions for performing the following calculations.

a. $X = Y + Z$ b. $W = X^2 + Y^2$

c. $W = X(X + Y(Y + Z))$ d. $Y = 2X + W^2$

2.7 COMPUTER SOFTWARE

It has been seen that computer hardware is capable of performing operations prescribed to it by means of instructions, which are themselves capable of being stored within the computer memory. Types of instruction available for most computers are the following.

Load instructions, which allow information to be transferred to special registers in the processing unit in preparation for further use.

Fixed-point arithmetic instructions, which allow two numbers represented in integer form to be added, subtracted, multiplied, or divided.

Floating-point arithmetic instructions, which allow two numbers represented in floating-point form to be added, subtracted, multiplied, or divided.

Logical instructions, which allow two quantities, whether they be numbers or binary-coded representations of some type, to be combined by performing logical " and " or " or," or similar operations on each pair of corresponding bits within the quantities.

Branching instructions, which cause the contents of the location counter to be modified depending on some condition within the machine, thus modifying the sequence in which instructions are executed. For example, a branching instruction can cause the machine to proceed to one set of instructions if the result of a subtraction operation is negative and to another set if the result is positive. Thus it can provide a capability for the machine, in effect, to make decisions during the execution of a program, a capability of great importance in most calculations.

Input-output instructions, which allow information or programs to be placed in the machine and answers to be received back.

Instructions of the above types form the vocabulary, or machine language, of the computer. The machine can be made to perform desired calculations by preparing a list of the proper instructions in the proper order, inserting these into the computer memory by some means, and then storing the location of the first of these instructions into the location counter and starting the machine. The process of preparing the required list of machine language instructions is termed programming.

When one attempts to write a computer program in machine language, he quickly learns two lessons. First, the instructions available in a modern computer make it an extremely versatile device. It is possible to write sequences of instructions which will examine essentially any bit patterns inserted into computer memory and act on them in accordance with any prescribed set of rules. Second, the instructions themselves are hard to remember and awkward to use in writing programs. The operations codes for ADD, MULTIPLY, LOAD, STORE, and so on, are merely bit patterns to be memorized, and the memory address of every number used in the calculation must be carefully recorded and properly used whenever the number is required. These circumstances have led to the condition that user programs are seldom written in machine language. Instead, a few service routines are constructed in machine language which are capable of analyzing user programs written in a more convenient form, and converting these user programs to the necessary machine language instructions for actual execution. Then the users are able to write their programs in one of these higher languages without being required to know any machine language. These service routines are themselves computer programs, usually very complex and difficult ones, which are normally available from the computer manufacturer along with the computer hardware. Collectively these programs are termed the computer software. The handling of user programs on a computer depends fully as much on the software as on the hardware, and many computer manufacturers are finding the software as expensive to produce as the hardware.

2.71   Assemblers

One of the first service routines usually written for a new computer is an assembly routine, or assembler. The assembler usually eases the task of the programmer in several ways:

(*1*)   It allows the programmer to use easily remembered mnemonics for the operation codes rather than the numerical codes which the machine itself requires. For example, a machine instruction might use the bit combination 01011100 for the operation code for ADD, whereas the assembler would allow the programmer to use the letters ADD directly.

(*2*)   Data to be processed can be given symbolic names by the programmer rather than actually assigned memory locations. For example, in writing in machine language, a programmer might assign a quantity X in his equations to memory cell 645. Then every time he wishes to use X in a calculation, he writes an instruction containing the number 645 for the address of X. The assembler allows him to simply write the operand as X, and wherever he uses X, it will substitute the actual number of the memory location.

(*3*)   Instructions are not assigned memory locations by the programmer. Instead, any instruction which needs to be referred to by another instruction is given a name, or symbolic tag or label, by which it can be referenced. In machine language, if the programmer wishes to branch to a set of instructions stored in memory, he must write into the branch instruction the actual number of the memory cell to which the branch is to be made. The assembler allows him to give a name to the first instruction in the new set, and then simply say branch to that name in this branch instruction.

(*4*)   The instruction vocabulary of the machine is increased by the use of "macro" instructions. Tasks which are frequently encountered by the programmers are programmed once and for all and a name given to the task similar to the mnemonics titles assigned to the actual machine instructions. Whenever the programmer wants this task accomplished, he simply uses the mnemonic for this operation as if it were another instruction built into the wiring of the machine. Operations of this sort, built from software rather than hardware, are termed pseudo operations.

The programmer writes his program in the assembly language, using the assigned mnemonics for operation codes and any symbolic names he chooses for variable names and statement labels. This program is then placed in the computer as a set of data for the assembler, which translates the mnemonic codes into the machine language instructions, assigns memory location addresses to the instructions and to the symbolic data references, and assembles a new program written in actual machine language. This assembled program, when run, will then accomplish the calculations actually desired by the programmer. Examples 2, 3, and 4 of Section 2.67 give some insight into the way in which the assembler accomplishes its tasks, and problem 5 of Exercise 6 actually calls for the construction of a simple assembler. The assembly language instructions shown in problem 5 are not unlike those in a real assembly language.

In the simplified assembly language, each instruction consisted of two symbols, one for an operations code and one for an operand. For each instruction the assembler would first load the mnemonic representing the operations code into a general register, go down a list of all the allowed op codes, already stored in the assembler, and make comparisons like those made for the names of John, Bill, and Joe in Example 2 of Section 2.67. When a match is found the assembler stores away a machine language instruction containing the actual op code for the desired operation, with zeros for the address portion. If no match is found, a prestored error message will be printed. Next the assembler will bring the symbolic name representing the operand into a general register. Since the choice of names for operands is up to the user, the assembler will have no prestored table matching operand names to addresses. It will have to construct such a table as it goes along. Each time an operand name is encountered the assembler will compare the name with each of those already appearing in this table. If a match is found, the assembler will use the memory location already assigned for this operand. If no match is found, the assembler will add this name to the table and assign the next unused memory location to it. Then, by use of the logical " or " instruction of Section 2.67, the address can be combined with the operations code already determined to form a complete machine language instruction.

The above discussion should serve to indicate how an assembler performs its tasks. Because we will not have need of an assembly language in later parts of the book, we shall not pursue the subject further, but will close this discussion with a few general observations concerning assemblers.

(1) The assembly language is much easier on the programmer than is machine language. The machine bears the burden of doing many of the tasks which are quite time consuming and burdensome for the programmer.

(2) Writing a program in assembly language is just as exacting a task as is writing in machine language, in that the assembler can only carry out a precisely defined set of operations. It is unforgiving of errors, and will not produce the desired machine language program unless presented with a program written exactly in accordance with the rules for the particular assembler being used.

(3) The actual form of the assembly language follows very closely the form of the built-in machine instructions. For this reason, assembly languages are frequently classed as "machine-oriented languages." They are closely related to the particular machine for which they are written and, consequently, vary considerably from one type of computer to another. This makes the assembly language not a particularly useful vehicle for the discussion of numerical methods applicable to all scientific computers and is the chief reason for dropping the description of the assembler at this point. Other languages, termed "procedure-oriented languages," tend to be more or less independent of the machine used and thus form a better basis for discussion of numerical analysis problems.

2.72 Compilers

Like the assembler, the compiler is a routine which accepts as input a program written in a form convenient for programmers and generates from it a set of machine language instructions which when executed will perform the calculations requested by the programmer. The original program is referred to as the source program and the resultant machine language program as the object program. The compiler allows the problem solver much more freedom of action than does the assembler, with the predictable result that the compiler is a much more difficult program to prepare than is the assembler. The assembly language might require the programmer to write the instructions for adding x and y to obtain z as

$$
\begin{array}{ll}
\text{L} \quad \text{X} & \text{for load X} \\
\text{A} \quad \text{Y} & \text{for add Y} \\
\text{ST} \quad \text{Z} & \text{for store in Z}
\end{array}
$$

whereas the compiler would allow the same operation to be represented by the statement

$$Z = X + Y$$

The difference in these two forms of representation indicates rather clearly the difference in effort required by the assembler and the compiler. The assembler receives instructions already broken down into op codes and addresses, and it merely identifies the symbols used and replaces them by machine language symbols. The compiler is confronted with mixtures of operands and operations symbols which must first be restructured into a properly arranged set of individual operations on individual operands. This operation, known as "parsing," is essentially equivalent to reducing the complex statement to the form of a tree representation, which is described in another context in Section 3.62. The methods of parsing complex statements comprise an entire new field of endeavor in computer science, a field having impact on language translation methods and having potential future impact on human language structures and hence on human forms of thought and mental capabilities.

In addition to constructing a machine language program out of some higher language source program, the compiler ordinarily performs a number of other services for the programmer. It usually provides a listing of the source language program, along with a "memory map" that gives the actual memory locations assigned to each variable named by the programmer. It also provides error messages when it finds things in the source program that it cannot interpret or that do not obey the language rules. Unfortunately, it is difficult to include in the compiler a sufficient set of prestored messages to match every kind of error that programmers will manage to make. Thus the messages provided are sometimes cryptic, and aid from a skilled programmer with knowledge of the inner workings of the compiler is occasionally required to pinpoint an error.

Several different compilers exist for most commercial computers. For scientific calculations the most widely used ones are those that handle source programs written in FORTRAN, ALGOL, or PL/I. For business applications, the commonly used higher language is COBOL. A host of other higher languages have appeared over the past several years, tailored to various fields of application for computers. A large and burgeoning software industry has developed over the past few years engaged in developing such higher languages and writing compilers for them, giving empirical support to the statement made in Section 2.7 that programs can be written which will examine essentially any kind of material inserted into computer memory and act on it in accordance with any set of rules desired. For our specific interest in this book, it is appropriate to choose one higher language suited to the problems of scientific computation and concentrate our attention on that one. We shall use the FORTRAN language because of its very widespread use in applied scientific research. An entire chapter (Chapter 4) is devoted to a detailed description of this language.

2.73 Interpretive Routines

A somewhat different class of service routine worthy of some mention is the interpretive routine. The interpretive routine is similar to the assembler and compiler in that it takes as input material written in some higher language and constructs the machine language equivalent. It differs in that each subpart of the coding is executed as soon as constructed. For example, if a FORTRAN compiler were to encounter the expression

$$Z = X + Y$$

it would construct the machine language instructions to load X into a register in the processing unit, add Y, and store the results in the memory location reserved for Z. Then it would proceed to construct the machine language for the next part of the source language program. The actual computation of Z would not take place until the programmer actually called for execution of the machine language program at some later time. The interpretive routine, on the other hand, would construct the machine language instructions to load X, add Y, and store in the location reserved for Z, and would then immediately perform these instructions, actually computing the value of Z and storing it. Historically, the interpretive routine came into being before either the compiler or the assembler. Its main purpose initially was to extend the instruction repertoire of early computers. For example, most early machines had hardware for fixed-point arithmetic only. Floating-point arithmetic could be performed by using two memory locations to store each number, one for the characteristic and one for the fractional part. To add two numbers a whole series of instructions would be required, to compare the characteristics, add leading zeros to one of the fractional parts if necessary, actually add the fractional parts, and store the final characteristic and fractional part. Using an interpretive routine, a new instruction could be introduced, whose operation code was to mean do "floating add." When the interpretive routine encountered this operation code, it would immediately perform the entire set of operations described above, with the result that the floating-point addition would be performed.

Most modern computers contain built-in hardware for the performance of floating-point arithmetic, so that the use of interpretive routines for the purpose just described is no longer prevalent. However, a new and rapidly growing use for the interpretive routine has developed in connection with usage of computers through remote terminals. In this application, which will be discussed in more detail in Chapter 5, each of many users, at teletype keyboards or other terminal devices, sends inputs through telephone lines or special high-capacity circuits to a computer located elsewhere. When a user types in an entry he obtains an almost immediate response from the computer. Because of its great speed relative to that of the human at each terminal device, the computer can jump from one user's program to another, interpret and execute the latest entry, and transmit any results back quickly enough, in many cases, to give each user the feeling that he has sole use of the machine. The higher languages used for this interactive man-machine communication are not as well developed and standardized as are the procedure-oriented languages such as FORTRAN and ALGOL. For illustrative purposes in this book, we will introduce one similar to some of the commercial time-shared computer services. Chapter 5 contains the description of that language. It is to be hoped that the future will bring a trend toward standardization of a language for man-machine communication.

2.74   Other Service Routines

The standard software of a commercial computer includes not only routines of the type described above, the so-called language translators, but it includes as well a variety of other types of service routines. One of the most important is the executive routine, sometimes called the supervisor or monitor. The executive routine is a master program which is placed in the machine when it is started up, and this program supervises the activities of the machine from then on unless interrupted by a human operator. It causes the first user program to be read into the machine and transfers control to it so that the user program will run. At the conclusion of that user program, this master program causes the next user program to be read into the machine and run, and so on. If no user programs are available at the machine input, the supervisor waits and watches by executing repeatedly a small loop of instructions. An error in a user program or a malfunction in some part of the computer will ordinarily cause an interruption signal which can be acted upon by the executive routine. For example, an attempt to divide by zero, or the use of an operation code which the decoder in the control unit cannot recognize, will cause an interruption. When the interruption occurs, the contents of the location counter are replaced by the contents of some particular memory cell in the machine. This memory cell will be the beginning of some set of instructions which are part of the executive routine. What happens next depends on what instructions have been stored at the new location where the instruction counter now points. For example, on a divide by zero, or an illegal instruction, the executive routine will probably cause the machine to print out a message, terminate work on this user program, read another user program into the machine, and start on it.

As far as the user is concerned, the executive routine, and the other service routines, are as much a part of the computer as the hardware itself. The hardware and software together form a unified whole.

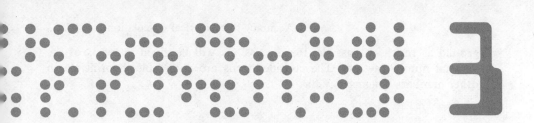

Accuracy in Numerical Calculations

3.1 INTRODUCTION

We have just learned the basic facts about computers and the first steps of computer programming. At this point there is a strong temptation to plunge into the art of programming, to learn to write programs which will perform calculations of interest. It is probably better, however, to take time out at this very early stage to face up to one very serious problem associated with performing calculations on computers, one which is all too often neglected in everyday computer operations—the problem of accuracy in numerical calculations. When we set out to perform a calculation for some practical purpose, we usually start with numbers representing such quantities as length, weight, and so on. These numbers are not ordinarily exact values, but are only approximations true to two, three, or more figures. It is possible in a complex machine program to do hundreds or even thousands of additions, subtractions, multiplications, and divisions with these inexact numbers, and it is also possible for the results to become succeedingly less and less exact, until the final results may be completely meaningless. Examples will be given later which will show that, even in a computer in which all numbers are stored with seven significant digits, it is possible to obtain completely meaningless results after a relatively short and simple calculation. Such extreme cases are not common; however, the person who trusts the result of a computer calculation without some careful check for accuracy is running a grave risk. Therefore, we shall cover the fundamental principles associated with accuracy in this chapter, and the problem of accuracy will be one of the major concerns throughout the remainder of this book. In order to remain on familiar ground as much as possible, the discussions will be given in terms of the decimal number system. The considerations are applicable in principle to other numbers systems as well.

3.2   APPROXIMATE NUMBERS

In the discussion of accuracy in calculations it is useful to make distinction between numbers that are exact and those that are only approximate values. Exact numbers are numbers whose values are completely represented by a symbol. All the integers, for example, 1, 5, 179, are of this type. Fractions can be expressed exactly as the ratio of two integers, for example, 1/2 or 2/3. Many other numbers can also be expressed completely by a symbol, for example, π, $\sqrt{2}$, and e. All the numbers mentioned have an exact meaning. Not all, however, can be exactly represented in the decimal system (or in any other place value number system, be it binary, octal, or any other, because we cannot write the exact number as a decimal or binary number containing a given number of places). For example, the decimal representation of 2/3 is .666666..., a decimal point followed by an infinite number of sixes. Since it is impossible to write an infinite number of sixes, it is impossible to write the number 2/3 exactly in the decimal system. We may write a number which is *approximately* 2/3, such as .6667, or, if we desire a closer approximation, .66666667. Likewise, we may write an approximate value for π, for example, 3.1416, but we cannot write the exact value. We shall term a decimal (or binary, or other number system) number an approximate number if it is used as a representation of a number it does not quite equal.

3.3   ABSOLUTE, RELATIVE, AND PERCENTAGE ERROR

The numerical difference between the true value of a quantity and the approximate number used to represent it is called the absolute error. The absolute error will ordinarily be regarded as positive if the approximate number is greater than the exact, and negative if the reverse is true. The relative error is the absolute error divided by the true value of the quantity. The percentage error is 100 times the relative error. For example, if Q is the true value of a number, and Q_1 the approximate number used to represent it, then

$$Q_1 - Q = \text{absolute error of the approximate number}$$
$$(Q_1 - Q)/Q = \text{relative error of the approximate number}$$
$$100(Q_1 - Q)/Q = \text{percentage error of the approximate number}$$

In nearly all cases, we will find that we use an approximate number Q_1 either (*1*) because we cannot write the exact number as a decimal number (as for π, 2/3, and so on), or (2) because we simply do not know the value of the exact number (as when we have measured something with a rule or scale). In such cases, the error $Q_1 - Q$ cannot be written exactly in decimal form either; for if it could we could simply subtract it from the approximate number Q_1 to obtain the exact value Q. Our primary concern, however, will be how well or how poorly the approximate number describes the exact number. For this purpose, we would like to know the maximum possible size of the error. For practical purposes, then, it is better to speak not of the absolute error but of the upper limit on the magnitude of the absolute error.

Let ΔQ be a number such that

$$|Q_1 - Q| \leqslant \Delta Q$$

Then ΔQ is an upper limit on the magnitude of the absolute error. In the future we will frequently somewhat loosely refer to such an *upper limit* on the *magnitude* of the absolute error as *the* absolute error.

If we have an approximate number Q_1 and a value ΔQ for the absolute error, we would like to be able to find an upper limit for the relative error. By the definition above, we can say that

$$\text{magnitude of relative error} = |(Q_1 - Q)/Q|$$
$$\leqslant |\Delta Q/Q| = \Delta Q/|Q|$$

Thus $|\Delta Q/Q|$ is an upper limit for the relative error. However, since we do not know the value of Q, the exact number, this formula is not of much use. We need a formula in terms of ΔQ and Q_1. We can write

$$|Q| = |Q_1 - (Q_1 - Q)| \geqslant |Q_1| - |Q_1 - Q| \geqslant |Q_1| - \Delta Q$$

Hence

$$|\Delta Q/Q| \leqslant \Delta Q/(|Q_1| - \Delta Q)$$

Thus, if Q_1 is known and an upper limit ΔQ on the absolute error is known, the quantity $\Delta Q/(|Q_1| - \Delta Q)$ can be calculated to give an upper limit on the magnitude of the relative error. In the future, we will frequently somewhat loosely refer to such an *upper limit* on the *magnitude* of the relative error as *the* relative error. If ΔQ is known to be small compared to $|Q_1|$, then the limit given above for the relative error is approximately $\Delta Q/|Q_1|$.

3.4 SIGNIFICANT FIGURES

In the decimal representation of a number, a significant figure is any one of the digits 1, 2, 3, ..., 9; and 0 is a significant figure except when it is used to fix the decimal point or fill the place of unknown or discarded digits. Thus in the number 0.0158, the significant figures are 1, 5, and 8. The zeros do nothing but fix the decimal point and hence are not significant. The number could just as well be represented as 1.58×10^{-2}. The number 4076 has four significant figures, the zero not being used in this number to fix the decimal point. In the number 38500, there is nothing to show whether the zeros are significant or whether they are simply for the purpose of fixing the decimal point. This ambiguity can be removed by writing the number as 3.85×10^4 if none of the zeros is significant, or 3.850×10^4 if one is significant, or 3.8500×10^4 if both are significant.

3.41   Rounding of Numbers

If we attempt to represent the fraction 4/7 in decimal form, we obtain .57014285 ..., a nonterminating decimal. For purposes of practical computation this must be represented by an approximate number, such as .57, or .5701, or .570143. The approximate number is obtained by dropping certain of the lower-order digits in the number, a process known as rounding. An error is of course introduced in rounding, and it is desirable to make this error as small as possible. By using a proper method of rounding, it is always possible to make the absolute error no greater than half a unit in the last place retained. This is done by increasing by one the last digit kept if the discarded part is greater than half a unit in that digit position. If the discarded part is exactly one-half a unit in the last place kept, it is best to increase the last kept digit by one sometimes, and not other times, so that if several round-offs are made during the problem, all will not introduce errors in the same direction. A simple way to do this is always to round the last kept digit to an even number, when the discarded part is exactly half a unit in the last kept place.

Example 1.   Round the following numbers to the two decimal places:

	The answers are
46.12416	46.12
31.34792	31.35
52.27500	52.28
2.38500	2.38
2.38501	2.39

In the majority of calculations, the relative error is of more direct interest than the absolute error. It will be shown shortly that the relative error is closely associated with the number of correct significant figures. Hence it is frequently of interest to round off to some particular number of significant figures. The rule for doing this can be stated as follows:

To round off a number to n significant digits, discard all digits to the right of the nth significant digit. If the number discarded is less than half a unit in the nth place, leave the nth digit unchanged; if the number discarded is greater than half a unit in the nth place, increase the nth digit by one; if the number discarded is exactly half a unit in the nth place, leave the nth digit unaltered if it is an even number, but increase it by one if it is an odd number.

Example 2. Round off the following numbers to four significant figures:

	The answers are
316.8972	316.9
4.167500	4.168
.0011118	.001112
19.265001	19.27

3.42   Correct Significant Digits

When an approximate number is used to represent an exact one, it is of interest to know how many of the digits in the approximate number have meaning in the description of the exact number. For example, one number which is approximately equal to 2/3 is .66667. Another is .66699842593. Although the second number has many more significant digits, it is actually a worse representation than the first. Many of the digits in the second number contribute nothing toward helping that number represent the fraction 2/3. Only the first few digits are of assistance in this representation. We will call a significant digit in an approximate number " correct " if rounding the approximate number off to just after that digit position will cause the absolute error to be no more than half a unit in that digit position. In the first representation of the fraction 2/3 above, all five of the significant digits are correct. In the second, only the first three are correct. In hand calculations, we usually try to write only the correct significant digits in writing an approximate number. The computer, on the other hand, will customarily carry all numbers to many significant digits, whether these are correct or not.

3.43   Relative Error and Correct Significant Digits

It was mentioned earlier that the relative error of an approximate number was closely related to the number of correct significant digits. If we know the number of correct significant digits in a given, approximate number we can calculate an upper limit for the relative error; vice versa, if we have an upper limit on the relative error for a given, approximate number we can determine something about the number of correct significant digits. We shall consider each of these cases separately.

First consider the problem of estimating the relative error when the number of correct significant digits is known. An example will make the process clear.

Example 1.   The number 34.152 is an approximate number, known to be correct to five significant digits. Determine an upper limit for the relative error.

From the definition of correct significant digit, we know that the absolute error is no more than half a unit in the digit position of the lowest-order correct significant digit, so in this case

$$\Delta Q \leqslant .0005$$

Hence, from Section 3.3,

$$\text{relative error} = \frac{\Delta Q}{|Q_1| - \Delta Q} = \frac{.0005}{34.152 - .0005}$$

$$= \frac{.5}{34152 - .5} = \frac{1}{68304 - 1} = \frac{1}{68,303}$$

The process just demonstrated, although correct, is somewhat cumbersome. In addition, since the result is only an estimate of the relative error anyhow, there is some question as to the value of having such a careful determination. For example, for practical purposes we would probably be just as well off if, instead of knowing the relative error were less than 1 part in 68,303, we knew only that it were less than 1 part in 68,000, or even say, only that it were less than 1 part in 10,000. To be sure,

$$\frac{1}{68,303} < \frac{1}{10,000}$$

so the first bit of knowledge is more complete than the last, but the extra knowledge may be of little practical value. Consequently, it is worth asking ourselves whether or not there are easier methods of estimating the relative error, methods which may give cruder but still useful estimates. One such method is given by the following theorem:

Theorem 1.   *If a number is correct to n significant digits, then the relative error is less than* 5×10^{-n}, *except in the case where the correct digits are a one followed by* $n - 1$ *zeros.*

PROOF: Consider first the case in which the decimal point is just after the last correct significant digit, and assume that the number is not a one followed by $n - 1$ zeros. Let Q_1 be the given approximate number, Q the exact value, and ΔQ the absolute error. Then

$$\Delta Q \leqslant .5$$

Also, since Q_1 is not allowed to be a one followed by $n - 1$ zeros, we have

$$|Q_1| \geqslant 1 \times 10^{n-1} + 1$$

so that the relative error

$$
\begin{aligned}
\Delta Q/|Q| \leqslant \Delta Q/(|Q_1| - \Delta Q) &= .5/(1 \times 10^{n-1} + 1 - .5) \\
&= .5/(1 \times 10^{n-1} + .5) \\
&< 5 \times 10^{-n}
\end{aligned}
$$

We have proved the theorem for the case in which the number has the decimal point following the last correct significant digit. To complete the proof, we note that the value of the relative error is unaffected by location of the decimal point. If we move the decimal point by m places (consider a positive value of m to indicate moving the point to the right and a negative value to indicate moving to the left), the effect is to multiply both $|Q|$ and ΔQ by 10^m and leave the value of the relative error unchanged.

For the special case where Q_1 is a one followed by $n - 1$ zeros, the relative errors are as given in the following table:

n	Q_1	Relative Error	Theorem 1 Prediction
1	1.	$.5/.5 = 1$	$1/2$
2	10.	$.5/9.5 = 1/(1.9 \times 10^1)$	$1/(2 \times 10^1)$
3	100.	$.5/99.5 = 1/(1.99 \times 10^2)$	$1/(2 \times 10^2)$
4	1000.	$.5/999.5 = 1/(1.999 \times 10^3)$	$1/(2 \times 10^3)$

The last column shows the values which would be inferred if we attempted to apply Theorem 1 to these cases. It is seen that for $n > 2$, Theorem 1 still gives a very good approximation of the upper bound, being about 5% off for $n \geqslant 2$ and much less for higher values of n. The sole major discrepancy is for the case $Q_1 = 1$, where the error inferred from the theorem is only half as large as the possible upper limit on relative error. Since this case is of no real interest to us and since we are interested in estimates of the relative error rather than a rigorous upper bound, we will henceforth ignore the exceptional case in applying Theorem 1 and use it as if it applied in all cases.

Example 2. Work Example 1 by applying Theorem 1.

There are five correct significant digits, so by Theorem 1 the relative error is less than 5×10^{-5}, or

$$\frac{1}{20,000}$$

Let us now concern ourselves with the converse to the question we have just been studying: If we have a number and an upper limit on its relative error, what can we say about the number of correct significant digits in its decimal representation? Again, the basic method for answering such a question can be illustrated by an example.

Example 3. The number 31.546824 is known to have a relative error no worse than 1 part in 100,000. How many of the digits are known to be correct?

In this case we know that

$$\frac{\Delta Q}{|Q|} \leqslant \frac{1}{100,000}$$

or

$$\Delta Q \leqslant \frac{1}{100,000} |Q|$$

We would like to know ΔQ. We do not know Q, but we do know Q_1, and we know that

$$|Q| \leqslant |Q_1| + \Delta Q$$

Hence we can say

$$\Delta Q \leqslant \frac{1}{100,000} (|Q_1| + \Delta Q)$$

or

$$\Delta Q \leqslant .00031546824 + \frac{1}{100,000} \Delta Q$$

or

$$\frac{99,999}{100,000} \Delta Q \leqslant .00031546824$$

or

$$\Delta Q \leqslant .00032$$

Since ΔQ is less than half a unit in the thousandths place, the significant digit in the thousandths place (the digit 6) is correct, so the number has at least five correct significant digits.

As in the converse problem discussed previously, it would be helpful to have a less cumbersome method of estimating the number of correct significant digits when the relative error is known. One such method is given by the following theorem:

Theorem 2. *If the relative error in an approximate number is less than or equal to* $.5 \times 10^{-n}$, *then the number is correct to n significant figures.*

The proof follows directly from the definition of correct significant digit.

EXERCISE 7

1. Round the following numbers to five significant digits.

   a. 762.186   b. 9.23196   c. .0189955
   d. .9999999   e. 192765001.   f. 3461.4612

2. Assuming all significant digits are correct, find the upper limit on the relative error by means of Theorem 1.

   a. 3728.14   b. 614.2   c. 1.000   d. 70.0

3. The following numbers are listed with known limits on relative error. Round them so as to keep only digits known to be correct, using Theorem 2.

   a. 6.17926 10^{-4}
   b. 576.8901 $1/(2 \times 10^6)$
   c. .4912634 $1/(2 \times 10^4)$
   d. .0499999 $1/(2 \times 10^3)$
   e. .0006814269 10^{-4}

3.44 Computer Storage of Numbers

It was pointed out in Chapter 2 that numbers are usually stored in a computer memory in words having a fixed number of bits and are stored in either fixed-point or floating-point form. For fixed-point form, the position of the binary point is fixed, usually at the right end of the number. This means that no fractional part can be carried, so that any number so stored must be rounded, and the error can be as large as half a unit. Actually, many machines do not round off numbers in this manner, but instead merely chop off the fractional part. In this case, the error can be as much as a whole unit.

Most calculations of large size are performed using floating-point representation for the numbers involved. In this form, a number is stored as a fractional part and an exponent (actually a characteristic representing the exponent). The fractional part occupies a fixed number of bit positions and the exponential part of the remainder. In the 32-bit storage described in Chapter 2, the fractional part occupied 24 bit positions, which are enough to allow for storage of about seven decimal digits. To simplify the later examples we will represent floating-point numbers in decimal form, that is, use radix 10 rather than radix 16 or radix 2, as actually used in most computers. Numerical results will be slightly different from those which would be obtained using radix 2 or radix 16, but the numerical differences are minor and the principles are the same. In radix 10 notation, the floating-point number can be written as

$$f \times 10^e$$

where f is the fractional part and e the exponent. The number of digits used to represent f is fixed by the word length used in storing the number. Let t stand for the number of digit positions allowed. Then every floating-point number stored in the machine will contain t significant digits. The number t is termed the "precision" of the floating-point numbers. As seen in the preceding sections, these t significant digits are not necessarily correct in any particular calculation, but we can have at most t correct significant digits. By Theorem 1 of Section 3.43, if all the digits are known to be correct, then the relative error due to roundoff is less than 5×10^{-t}. In some machines, extra digits are not rounded off but are merely chopped off. The error in this case can be a whole unit in the last digit position and the relative error will be correspondingly greater.

Computers usually also offer the option of doing arithmetic in double precision, using two memory locations to store each number, giving twice as many bit positions for the fractional part and thus much reduced roundoff errors. Operations in double precision are usually quite slow compared to single precision, and thus more expensive, but they are required in many cases to retain sufficient accuracy.

3.5 ACCUMULATION OF ERRORS IN ARITHMETIC

If two approximate numbers are added, multiplied, or otherwise combined, the result is again an approximate number. If estimates of the errors in the original numbers are available, it is possible to make an estimate of the error in the result. We shall consider this problem for both the absolute error and the relative error, for each of the fundamental operations of arithmetic. For these discussions, the numbers involved will be positive. The slight additional complication introduced by the use of signed numbers is left for the reader's investigation.

3.51 Error Accumulation in Addition

Let u_1 and u_2 be two approximate numbers and Δu_1 and Δu_2 be the absolute values of their errors. Then the true values are no larger than $u_1 + \Delta u_1$,

$u_2 + \Delta u_2$, and no smaller than $u_1 - \Delta u_1$, and $u_2 - \Delta u_2$. Hence the true sum is no larger than

$$u_1 + \Delta u_1 + u_2 + \Delta u_2$$

and no smaller than

$$u_1 - \Delta u_1 + u_2 - \Delta u_2$$

The sum of the approximate numbers is $u_1 + u_2$. Hence the absolute error of the sum is no greater than $\Delta u_1 + \Delta u_2$. To find the relative error, let r_1 stand for the relative error in u_1, r_2 for the relative error in u_2, and r for the relative error in the sum. From Section 3.3 we can say that approximately

$$r_1 = \frac{\Delta u_1}{u_1}, \qquad r_2 = \frac{\Delta u_2}{u_2}, \qquad r = \frac{\Delta u_1 + \Delta u_2}{u_1 + u_2}$$

The expression for r can be rewritten

$$r = \frac{u_1}{u_1 + u_2} \left(\frac{\Delta u_1}{u_1}\right) + \frac{u_2}{u_1 + u_2} \left(\frac{\Delta u_2}{u_2}\right)$$

or

$$r = \frac{u_1}{u_1 + u_2} r_1 + \frac{u_2}{u_1 + u_2} r_2 \qquad (3\text{-}1)$$

This relationship has a form familiar to mathematicians. If we let

$$\theta = \frac{u_1}{u_1 + u_2} \qquad (3\text{-}2)$$

then (3-1) can be written

$$r = \theta r_1 + (1 - \theta) r_2 \qquad (3\text{-}3)$$

This is the familiar formula for linear interpolation. When $\theta = 0$, $r = r_2$, and when $\theta = 1$, $r = r_1$. For values of θ between zero and one, r takes on values between r_1 and r_2. From (3-2), θ may have any value from zero to one, depending on the values of u_1 and u_2. Thus the relative error of the sum is intermediate between the relative errors of the two addends.

Example 1.   Find the sum of the approximate numbers 476.6 and 3.11918, each being correct to its last figure, but no further.

Consider first how the arithmetic would be performed in floating-point form in a computer. As described in Section 3.44, let us represent the computer arithmetic by writing the numbers in decimal form, always keeping a seven-digit fractional part. This should correspond roughly to the floating-point representation described in Chapter 2 for a computer word length of 32 bits. In this notation, which is termed seven-place floating decimal notation, the numbers are

$$.476600 \times 10^3$$
$$.3119180 \times 10^1$$

Before the number can be added, the second number must be written with an exponential part of 10^3,

$$.0031191 \times 10^3$$

We have lost the last digit in this operation, and did not round, but merely chopped it off. The practice within computers varies, but chopping as we have done here is not uncommon.

When the second number is added to the first, the result is

$$.4797191 \times 10^3$$

or, in ordinary notation,

$$479.7191$$

Actually, since one of the original numbers was accurate to only one decimal place, the result is only accurate to one decimal place. The shifting and chopping of the second number did not really affect the accuracy of the final result. Note, however, that the shifting did increase the relative error in the second number. If we were to use expression (3-1) to estimate the relative error based on the given numbers in Example 1, we would obtain a smaller value than would be obtained using the numbers as added within the computer. The difference is trivial in this case but can be important in other cases.

As indicated in the above example, there is a source of error in floating-point addition not specifically accounted for in relation (3-1). When the smaller of the two numbers is shifted to match up the exponents, its relative error may be increased. Assume that the operation is being done on a machine in which floating-point numbers have a fractional part of t decimal digits. Then the effect of the shift is to round off the sum to t decimal digits, that is, introduce an arbitrary change in the sum of as much as $5 \times 10^{-t}(u_1 + u_2)$. Thus the relative error, instead of being that given by (3-1), is

$$r = \frac{u_1}{u_1 + u_2} r_1 + \frac{u_2}{u_1 + u_2} r_2 + 5 \times 10^{-t} \qquad \textbf{(3-4)}$$

3.52 Error Accumulation in Subtraction

Again, let u_1 and u_2 be two approximate numbers and Δu_1 and Δu_2 be the absolute values of their absolute errors. Let us assume $u_1 > u_2$, and we desire the value of $u_1 - u_2$. The true value of the first number may be as large as $u_1 + \Delta u_1$, and that of the small one may be as small as $u_2 - \Delta u_2$. In this case the true difference could be as large as

$$u_1 + \Delta u_1 - (u_2 - \Delta u_2) \qquad \text{or} \qquad u_1 - u_2 + \Delta u_1 + \Delta u_2$$

On the other hand, the true value of the first number could be as small as $u_1 - \Delta u_1$ and the second as large as $u_2 + \Delta u_2$. In this case the true difference might be as small as

$$u_1 - \Delta u_1 - (u_2 + \Delta u_2) \qquad \text{or} \qquad u_1 - u_2 - (\Delta u_1 + \Delta u_2)$$

Thus we see that the true difference may be higher or lower than the approximate difference $u_1 - u_2$ by as much as $\Delta u_1 + \Delta u_2$. Thus the absolute error of the difference must be taken as $\Delta u_1 + \Delta u_2$. From this absolute error the relative error can be estimated as for addition. Let r_1 stand for the relative error in u_1, r_2 for the relative error in u_2, and r for the relative error in the difference. As long as the relative errors are small, we can say that approximately

$$r_1 = \frac{\Delta u_1}{u_1}, \qquad r_2 = \frac{\Delta u_2}{u_2}, \qquad r = \frac{\Delta u_1 + \Delta u_2}{u_1 - u_2} \tag{3-5}$$

The expression for r can be rewritten as

$$r = \frac{u_1}{u_1 - u_2} r_1 + \frac{u_2}{u_1 - u_2} r_2 \tag{3-6}$$

or

$$r = \frac{u_1 + u_2}{u_1 - u_2} \left(\frac{u_1}{u_1 + u_2} r_1 + \frac{u_2}{u_1 + u_2} r_2 \right) \tag{3-7}$$

The expression within the parentheses is just the relative error expression (3-1) obtained for the addition of two numbers. Its value is intermediate between r_1 and r_2. The factor out front is a factor which is clearly greater than one. It can be extremely large if u_1 and u_2 are near to each other. Thus r may turn out to be much greater than either r_1 or r_2. In the subtraction of nearly equal numbers, the relative error can be magnified to a value much higher than that of the original numbers. The problem of magnification of the relative error in subtraction is of great importance in numerical calculations, as will be demonstrated in later examples and discussions.

As a first example of subtraction, we take one in which the errors do not become excessive.

Example 1. Subtract 37.68151 from 754.8, where each number is approximate, correct to the last digit given.

In the seven-place floating-point notation used in Example 1 of Section 3.51, the numbers are

$$.7548000 \times 10^3$$
$$.3768151 \times 10^2$$

Before subtraction, the exponent of the second number must be adjusted by rewriting as

$$.0376815 \times 10^3$$

Now the subtraction gives the result

$$.7171185 \times 10^3$$

or, in ordinary decimal form,

$$717.1185$$

As for addition, shifting of the number having smaller absolute value causes a further roundoff error to be introduced. To account for this, expression (3-6) should be replaced by

$$r = \frac{u_1}{u_1 - u_2} r_1 + \frac{u_2}{u_1 - u_2} r_2 + 5 \times 10^{-t} \qquad \textbf{(3-8)}$$

Example 2. Subtract 938.67804 from 938.67827, each number being accurate to the number of places given.

The numbers given have eight correct significant digits. In our seven-place notation one of these will be lost and the numbers will have the form

$$.9386782 \times 10^3$$
$$.9386780 \times 10^3$$

Upon subtraction, we obtain

$$.0000002 \times 10^3$$

When numbers are handled in floating-point form within a computer, they are ordinarily adjusted so that the fractional part has no leading zeros. This answer above would immediately be readjusted to the form

$$.2000000 \times 10^{-3}$$

The original numbers had eight correct significant digits. The final answer has only one. Six significant digits were lost in the subtraction. One was lost because the word length used was not sufficient to hold the entire number.

In some machines, instead of filling zeros in the lower six positions, contents from some other register within the machine are shifted into these positions, so that numbers other than zero might appear there. The result might appear to have meaningful digits in all positions, but only the first digit is correct. If this result is then to be combined with other numbers, the fact that it is accurate to only one figure may be a major influence in determination of the accuracy of the final result. This problem of accuracy in subtraction is the most important accuracy consideration in many problems.

The loss of the leading significant figures in the subtraction of two nearly equal numbers is the greatest source of inaccuracy in most calculations, and forms the weakest link in a chain computation where it occurs. Floating-point arithmetic offers little or no protection against this form of accuracy loss. Wherever it may occur, special programming precautions must be taken to avoid the difficulty or at least to make the programmer aware that a dangerous point in the calculation has arisen.

Further to illustrate how this loss of accuracy can jeopardize the results of a calculation, another example will be given, this time of an extreme case where all significant digits are lost.

Example 3.   Compute $a^2 - b^2 - c$, where

$$a = 1.00020000$$
$$b = 1.00010000$$
$$c = 0.00020001$$

The program to perform this calculation might do the steps as follows: square a, then square b, subtract, and then subtract c. This process would work for most sets of values for a, b, and c, but look what would happen in the present case. The floating-point representations are

$$a = .1000200 \times 10^1$$
$$b = .1000100 \times 10^1$$
$$c = .2000100 \times 10^{-3}$$

Succeeding steps in the arithmetic are

$$a^2 = .1000400 \times 10^1$$
$$b^2 = .1000200 \times 10^1$$
$$a^2 - b^2 = .0000200 \times 10^1$$

or, normalized, $.2000000 \times 10^{-3}$.

$$a^2 - b^2 - c = -.0000100 \times 10^{-3}$$

or, normalized, $-.1000000 \times 10^{-7}$. In decimal form, this is $-.00000001$.

If the calculation were done in double precision, so that the fractional part would be 14 digits in length, no significant digits would be lost, and we would have

$$a^2 = .10004000400000 \times 10^1$$
$$b^2 = .10002000100000 \times 10^1$$
$$a^2 - b^2 = .00002000300000 \times 10^1$$

or, normalized, $.20003000000000 \times 10^{-3}$.

$$a^2 - b^2 - c = .00002000000000 \times 10^{-3}$$

or, normalized, $.20000000000000 \times 10^{-7}$. In decimal form this is $.00000002$. The original result was off by a factor of two and had the wrong sign! The answer had no significance whatsoever.

It is interesting to note that the correct result would have been obtained in the single-precision calculation if we had only ordered the calculation differently. We would write $a^2 - b^2 - c$ as $(a + b)(a - b) - c$, and the calculation in single precision would have gone as follows:

$$a + b = .2000300 \times 10^1$$
$$a - b = .0000100 \times 10^1$$

or, normalized, $.1000000 \times 10^{-3}$.

$$(a + b)(a - b) = .2000300 \times 10^{-3}$$
$$(a + b)(a - b) - c = .0000200 \times 10^{-3}$$

or, normalized, $.2000000 \times 10^{-7}$. This is the same answer as that obtained in double precision. Thus subtraction problems can be fatal for one arrangement of steps in a calculation and completely innocous for a slightly different arrangement.

3.53   Error Accumulation in Multiplication

Let u_1 and u_2 be two approximate numbers to be multiplied and let Δu_1 and Δu_2 be upper limits on the absolute value of their absolute errors. Then the true values might be as great as $u_1 + \Delta u_1$ and $u_2 + \Delta u_2$, so that the true value of the product might be as large as

$$(u_1 + \Delta u_1)(u_2 + \Delta u_2) = u_1 u_2 + u_1 \, \Delta u_2 + u_2 \, \Delta u_1 + \Delta u_1 \, \Delta u_2$$

On the other hand, the true values might be as small as $u_1 - \Delta u_1$ and $u_2 - \Delta u_2$, in which case the true value of the product might be as small as

$$(u_1 - \Delta u_1)(u_2 - \Delta u_2) = u_1 u_2 - u_1 \, \Delta u_2 - u_2 \, \Delta u_1 + \Delta u_1 \, \Delta u_2$$

We notice a curious thing about the true product. It can be farther on the high side of the approximate product $u_1 u_2$ than it can be on the low side. Since the term $\Delta u_1 \, \Delta u_2$ has a positive sign in both of the above expressions, it contributes to the error portion of the first expression but detracts from it in the second. It is customary to simplify both of the expressions above by assuming that the errors in the original numbers are small compared to the numbers themselves. That is, $\Delta u_1 / u_1$ and $\Delta u_2 / u_2$ are small compared to one. If that is the case, then the term $\Delta u_1 \, \Delta u_2$ is so small in the above expressions it can be neglected. If this is done, the absolute error of the product becomes $u_1 \, \Delta u_2 + u_2 \, \Delta u_1$. The relative error, according to the formula of Section 3.3, is approximately

$$\frac{u_1 \, \Delta u_2 + u_2 \, \Delta u_1}{u_1 u_2} = \frac{\Delta u_2}{u_2} + \frac{\Delta u_1}{u_1}$$

or is roughly the sum of the relative errors of the numbers being multiplied.

Example 1.   Find the product of 369.7 and .0042131, assuming each of the numbers accurate to the number of figures given.

In a machine having seven-decimal-place precision, the numbers would be stored as

$$.3697000 \times 10^3$$
$$.4213100 \times 10^{-2}$$

The multiplication within the machine would give the 14-place fractional part, .15575830700000, which would be rounded for storage to give

$$.1557583 \times 10^1$$

In multiplication, too, then the machine will impose a roundoff error on the result of the calculation, so that if two numbers having relative errors r_1 and r_2 are multiplied, the relative error of the product is

$$r = r_1 + r_2 + 5 \times 10^{-t} \tag{3-9}$$

3.54   Error Accumulation in Division

Again, let u_1 and u_2 be approximate numbers, and Δu_1 and Δu_2 be the upper limits on the absolute values of their absolute errors. The largest possible true value of the quotient is

$$\frac{u_1 + \Delta u_1}{u_2 - \Delta u_2}$$

and the smallest possible value is

$$\frac{u_1 - \Delta u_1}{u_2 + \Delta u_2}$$

If in the first of these expressions the numerator and denominator are multiplied by $u_2 + \Delta u_2$, that limit can be rewritten as

$$\frac{u_1 u_2 + u_1\,\Delta u_2 + u_2\,\Delta u_1 + \Delta u_1\,\Delta u_2}{u_2{}^2 - (\Delta u_2)^2}$$

In like manner, if the numerator and denominator of the second expression are multiplied by $u_2 - \Delta u_2$, that expression can be written

$$\frac{u_1 u_2 - u_1\,\Delta u_2 - u_2\,\Delta u_1 + \Delta u_1\,\Delta u_2}{u_2{}^2 - (\Delta u_2)^2}$$

If the errors are assumed small, so that the product $\Delta u_1\,\Delta u_2$ and the square $(\Delta u_2)^2$ can be neglected, the upper limit becomes

$$\frac{u_1 u_2 + u_1\,\Delta u_2 + u_2\,\Delta u_1}{u_2{}^2} \quad \text{or} \quad \frac{u_1}{u_2} + \frac{u_1}{u_2}\left(\frac{\Delta u_2}{u_2} + \frac{\Delta u_1}{u_1}\right)$$

and the lower limit

$$\frac{u_1 u_2 - u_1\,\Delta u_2 - u_2\,\Delta u_1}{u_2{}^2} \quad \text{or} \quad \frac{u_1}{u_2} - \frac{u_1}{u_2}\left(\frac{\Delta u_2}{u_2} + \frac{\Delta u_1}{u_1}\right)$$

Thus the absolute error of the quotient is roughly

$$\frac{u_1}{u_2}\left(\frac{\Delta u_2}{u_2} + \frac{\Delta u_1}{u_1}\right)$$

and the relative error, by Section 3.3, is roughly the absolute error divided by u_1/u_2, or

$$\frac{\Delta u_2}{u_2} + \frac{\Delta u_1}{u_1}$$

Thus in division, as in multiplication, the relative error is the sum of the relative errors of the numbers involved. Also, as in multiplication, the machine will round off the quotient to t decimal places, if t is the precision being used. Thus, if two numbers having relative errors r_1 and r_2 are divided one into the other, the relative error of the quotient is

$$r = r_1 + r_2 + 5 \times 10^{-t} \tag{3-10}$$

3.6   ERROR PROPAGATION

In Section 3.5 the behavior of absolute and relative errors for the basic arithmetic operations was discussed. A computer performs its numerical work by doing repeated additions, subtractions, multiplications, and divisions, so the error in a final result can be estimated from the errors in the individual steps, errors we now know how to estimate.

Although the formulas were derived for positive numbers only, it is clear that they are applicable for signed numbers as well. The formulas are summarized in Table 1.

TABLE 1
Relative Error Accumulation for Arithmetic Operations

Numbers u_1 and u_2, Relative Errors r_1 and r_2,
Floating-Point Precision t

Operation	Relative Error
$u_1 + u_2$	$\dfrac{|u_1|}{|u_1 + u_2|}r_1 + \dfrac{|u_2|}{|u_1 + u_2|}r_2 + 5 \times 10^{-t}$
$u_1 - u_2$	$\dfrac{|u_1|}{|u_1 - u_2|}r_1 + \dfrac{|u_2|}{|u_1 - u_2|}r_2 + 5 \times 10^{-t}$
$u_1 u_2$	$r_1 + r_2 + 5 \times 10^{-t}$
u_1/u_2	$r_1 + r_2 + 5 \times 10^{-t}$

The expressions in Table 1 can be used in conjunction with a description of the calculation to be performed in order to obtain error estimates.

3.61   Description of a Calculation

There are many ways to describe a calculation to be performed. The most common way is by means of an arithmetic expression. Expressions such as

$$a + b + c + d$$

or

$$abcd$$

describe arithmetic operations. However, they do not describe the exact order in which the operations are to be performed. This is not necessary in classical arithmetic, because of certain laws of arithmetic. These basic laws are

(*1*)   The commutative law for addition, which says

$$a + b = b + a$$

That is, in adding two numbers, it does not matter which is placed first.

(*2*)   The associative law for addition, which says

$$(a + b) + c = a + (b + c)$$

That is, in adding more than two numbers, it does not matter which pair is added first.

(*3*)   The commutative law for multiplication, which says

$$ab = ba$$

That is, in multiplying two numbers, it does not matter which is placed first.

(*4*)   The associative law for multiplication, which says

$$(ab)c = a(bc)$$

That is, in multiplying more than two numbers, it does not matter which pair is multiplied first.

(*5*)   The distributive law for addition and multiplication, which says

$$a(b + c) = ab + ac$$

That is, in multiplying a sum of terms by a common factor it does not matter whether one adds first, then multiplies, or multiplies each term by the factor first, then adds.

These laws hold rigorously for arithmetic in which all numbers are exact. They do not necessarily hold for approximate numbers, however. Example 3 of Section 3.52 demonstrated a case where the answers were quite different, depending on the order in which the calculations were performed. In computer arithmetic, it is safe to assume that the commutative laws hold.* It is *not* safe to assume that the associative and distributive laws hold. Thus the usual algebraic expressions are not really adequate to describe a computer calculation if we are to be concerned about accuracy, because they do not actually prescribe the order in which the steps are to be taken. One can make them adequate by the inclusion of a sufficient number of parentheses to delineate all the steps. For example, if $a + b + c + d$ were rewritten

$$(a + (b + (c + d)))$$

it would be clear that c and d are to be added first, then b added to that result, then a added to that result. A more graphic description of the order of steps in a calculation is provided by a tree representation. This is discussed next.

3.62   Tree Representations

The concept of a tree, in the mathematical sense, is a development in the mathematical field of graph theory. A tree is a special case of a graph. A graph is a set of points, or nodes, some of which are interconnected by lines. Figure 3-1 gives an example of a connected graph. A path which returns to its

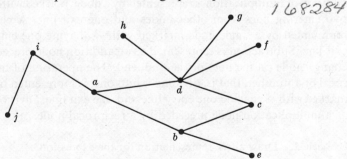

Figure 3-1: Example of a connected graph

starting point is called a circuit. In Figure 3-1, *a-b-c-d-a* is a circuit. If on each line of a graph, we have a specified direction for proceeding, we have a directed graph. Figure 3-2 gives an example of a directed graph. Arrows are used to

---

* Even this is not quite true in some machines. The last digit in a sum can sometimes be different, depending on the order in which the numbers are called into the adder. Most current machines do rigorously obey the commutative laws, however.

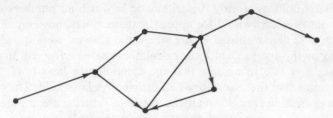

Figure 3-2: Example of a directed graph

indicate the direction of each line. A directed graph which contains no circuits and which has at most one branch exiting each node is called a tree. Figure 3-3 is an example of a tree.

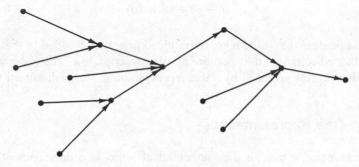

Figure 3-3: Example of a tree

A tree can be used to represent arithmetic calculations as follows. Each addition or multiplication is represented by a node, marked with + or ·, having two entering lines from other nodes and one exit line. A change of sign is represented by a single node, marked with −, having one entry line and one exit line. Subtraction is represented as an addition node preceded by a sign = change node on the line for the minuend. The operation of taking the reciprocal of a number, that is, dividing it into one, is represented by a single node, marked with ÷, having one entry line and one exit line. Division is represented as a multiplication node preceded by a reciprocal node.

Example 1.   Draw a tree representation for the expression

$$(a + b) - c/d$$

Figure 3-4 gives the tree representation; a and b are to be added as one operation, c divided by d as another operation, then the result c/d subtracted from the result $a + b$.

The tree representation of an arithmetic expression can be used in conjunction with the rules from Table 1 to compute error estimates. The process can be made more graphic by noting what will happen to the error at each type of node in the tree, according to Table 1.

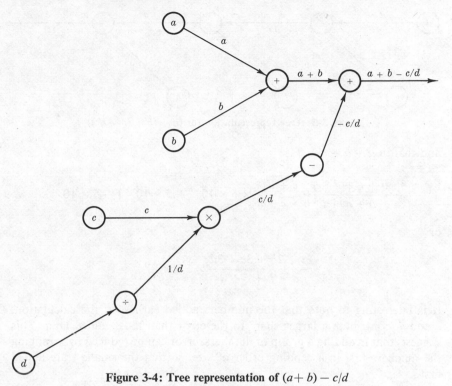

Figure 3-4: Tree representation of $(a+b)-c/d$

(1) At a sign change node or a reciprocal node, nothing will happen to the relative error. It will remain unchanged.

(2) At a multiply node, the relative errors from the two branches are added, and a roundoff error of 5×10^{-t} is added also.

(3) At an add node, the relative errors are weighted according to the ratio of incoming value to sum at the node, then added, and the roundoff error of 5×10^{-t} is added also.

If each line in the tree representation is marked with the numerical value which it represents, it is particularly easy to pick off the relative error behavior at any point. This was done in Figure 3-4.

Example 2. Find an error bound for

$$y = (a + (b + (c + d)))$$

where the initial numbers are assumed to be exact and positive.

A tree representation is shown in Figure 3-5. Since a, b, c, and d have zero relative errors, the relative error in $c + d$ is 5×10^{-t}, that in $b + c + d$ is

$$\frac{c + d}{b + c + d} \times 5 \times 10^{-t} + 5 \times 10^{-t}$$

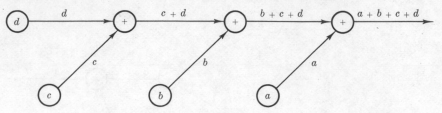

Figure 3-5: Tree representation for $(a + (b + (c + d)))$

and that in $a + b + c + d$ is

$$\frac{b+c+d}{a+b+c+d} \left(\frac{c+d}{b+c+d} \times 5 \times 10^{-t} + 5 \times 10^{-t} \right) + 5 \times 10^{-t}$$

or

$$\frac{a + 2b + 3c + 3d}{a + b + c + d} \times 5 \times 10^{-t}$$

It is interesting to note that the numbers added earliest in the calculation, c and d, contribute a larger share to the error than those added later. This suggests that in adding a group of numbers, error can be reduced by arranging the numbers first in ascending order of size, so that the smallest are added first.

Example 3. Find an error bound for

$$y = (a + b) + (c + d)$$

where the initial numbers are assumed to be exact and positive.

This is the same arithmetic problem as Example 2, with the steps arranged in a different order. A tree representation is shown in Figure 3-6. Since a, b, c, and d have zero relative errors, the relative error in $a + b$ and in $c + d$ is 5×10^{-t}, and that in the final sum is

$$\frac{a+b}{a+b+c+d} \times 5 \times 10^{-t} + \frac{c+d}{a+b+c+d} \times 5 \times 10^{-t} + 5 \times 10^{-t}$$

or

$$\frac{2a + 2b + 2c + 2d}{a + b + c + d} \times 5 \times 10^{-t}$$

or simply 10^{-t+1}. Comparing this result with Example 2, we see that this error is smaller if a is less than $c + d$, and the previous error was smaller

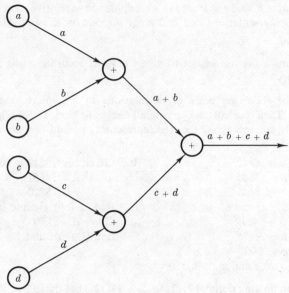

Figure 3-6: Tree representation for $(a + b) + (c + d)$

otherwise. More generally, we see that the error contribution of a number in a sum of many numbers is proportional to the number of add nodes to which it contributes in the tree representation.

EXERCISE 8

1. Using Table 1, compute a bound for the relative error in $a + b$ for each of the following cases, where the numbers are accurate to the number of places given, and seven-place floating decimal arithmetic is to be used.

    a. $a = 9.85, b = .47193$ b. $a = 2.15, b = .3082$
    c. $a = 6.4513, b = -6.4505$ d. $a = 9.316, b = 2.745 \times 10^{-7}$

2. Assuming that a, b, c, and d as stored in a computer have relative errors of 5×10^{-5}, use the tree representations in Figures 3-5 and 3-6 and find error bounds in the quantities $a + (b + (c + d))$ and $(a + b) + (c + d)$.

3. Calculate the sum of the following numbers, using four-place floating decimal arithmetic. First add in the order given and then rearrange for maximum accuracy.

$$.4835 \times 10^3$$
$$.2168 \times 10^1$$
$$.3115 \times 10^0$$
$$.6182 \times 10^0$$
$$.7984 \times 10^0$$
$$.1391 \times 10^1$$

4. Assume that a, b, and c as stored in a computer have relative errors of 5×10^{-t}. Draw tree representations and find error bounds for the following quantities.

    a. $a(b-c)$ b. $ab-ac$

    Which formula do you expect to give the most accurate result in computing this quantity?

5. Perform the following pairs of calculations in four-place floating decimal arithmetic. Then recalculate, carrying all decimal places, to see which answer is more nearly correct. Draw the tree representations and find error bounds for each formula.

    a.   Given $a = .6842$ b.   Given $a = .9812 \times 10^2$
                 $b = .5685$ $b = .4651$
                 $c = .5641$ $c = .8320$
       find $a(b-c)$ and $ab-ac$ find $(a+b)+c$ and $a+(b+c)$
    c.   Given $a = .4712$ d.   Given $a = .1560$
                 $b = .4675$ find $(1+a)^2$ and $1 + (2a + a^2)$
                 $c = .5000$
       find $(a-b)/c$ and $a/c - b/c$

6. Show that in floating decimal $2./2. = 2. \times (1./2.)$ but that $3./3. \neq 3. \times (1./3.)$

3.63   Expected versus Maximum Errors

The formulas in the preceding sections were for the purpose of obtaining bounds on the errors, or estimates of what the errors might be in the worst possible case. This means that they represent a rather pessimistic view, and in most cases the errors will be considerably less than indicated by the bounds. This is particularly true for the roundoff error for addition and multiplication, which in the examples given was estimated to increase by 5×10^{-t} at each step. Actually, if roundoff is performed correctly, the error should be sometimes positive and sometimes negative, and tend to cancel. According to Gauss's law of error from probability theory, the accumulated roundoff error after a number of steps should tend to follow the normal probability distribution. A quantity x is said to follow a normal distribution with mean zero and standard deviation σ if the probability that it takes on a value between given limits $-a$ and a is given by

$$\text{Probability}\ [-a < x < a] = \frac{2}{\sigma\sqrt{2\pi}} \int_0^a e^{-x^2/2\sigma^2}\, dx \qquad \textbf{(3-11)}$$

Table 2 shows for a normal distribution the probabilities that x will lie in various ranges. It is seen that there is a very high probability that the absolute value of x will not exceed 3σ, and an extremely high probability that it will not exceed 6σ.

TABLE 2

Normal Distribution Probabilities

Range of x	Probability x Will Lie in This Range
$-\sigma$ to σ	.6827
-2σ to 2σ	.9545
-3σ to 3σ	.9973
-4σ to 4σ	.99994
-5σ to 5σ	.9999994
-6σ to 6σ	.999999998

The actual application of Gauss's law of errors to the case of roundoff errors can be stated as follows: If N roundoff errors, each lying randomly within a range -5×10^{-t} to 5×10^{-t}, are added, the resulting error is normally distributed with mean equal to zero and standard deviation equal to $\sqrt{N/3} \times 5 \times 10^{-t}$.

By Table 2, this means there is an extremely high probability that the error does not exceed six times this amount, or $\sqrt{12N} \times 5 \times 10^{-t}$. The upper-bound error, from Section 3.5, would be $N \times 5 \times 10^{-t}$. For very large calculations the difference between these two error estimates is quite large. In a computer which performs an addition of two numbers in 1 millionth of a second, calculations involving millions of arithmetic operations are not unusual. If $N = 10^7$, the upper-bound estimate would imply seven significant digits lost by roundoff, whereas the probabilistic estimate would indicate that a loss of more than four digits would be highly unlikely.

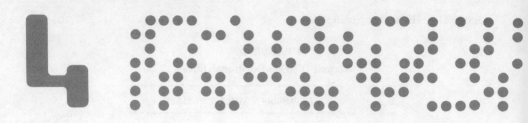

FORTRAN Language

4.1 INTRODUCTION

In Chapter 2 it was indicated that computer programming has progressed to the point at which it is no longer necessary for the programmer to write a set of commands in machine language in order to have a calculation performed. Instead, ready-made compiler programs can be used to convert a set of statements that describe the problem into a set of machine language commands that will actually cause the machine to solve the problem. The task of the programmer, then, is to write the problem in the form of a set of statements which the compiler is capable of converting. He need not know how the compiler does this conversion, but he must know those symbols and statements the compiler can handle, and must adhere religiously to them. One of the earliest compilers for scientific use was the IBM FORTRAN. The original FORTRAN compiler was so successful that FORTRAN compilers now exist for most machines. In this chapter we will describe the FORTRAN language; that is, we will define those symbols and statements the typical FORTRAN compiler is capable of converting to machine language. Unfortunately, even FORTRAN compilers vary somewhat from one machine to another, so the language described below is not really universal. However, the modifications required for particular machines are generally small, so this FORTRAN language will be used in describing solutions to problems in later chapters. In using this material in connection with a particular computer, the reader would be wise to consult the FORTRAN programming manuals for that particular computer so that he can make modifications if any are required.

A FORTRAN program consists of a series of statements of different types. One describes arithmetic operations to be performed, such as $Z = X + Y$. A second type causes the machine to call for input, say from a data card or a magnetic tape, or to provide output, say on a printed page or a magnetic tape. Another specifies the flow of control through the set of statements, that is, the order in which the statements are to be executed. Still another type of statement simply provides information about the procedure without causing any action. The following sections will describe these statements and demonstrate how they are used to form a program.

4.2 ARITHMETIC ASSIGNMENT STATEMENTS

The heart of a FORTRAN program is the set of arithmetic statements which describes the calculations to be performed. These statements involve arithmetic operations on constants and variables stored in the machine.

4.21 Constants

The FORTRAN compiler allows arithmetic to be done in either fixed-point or floating-point form. In the fixed-point arithmetic, the operations are done as if the numbers involved are integers. Any positive or negative integer (not too large to fit in a memory cell of the machine) used in a FORTRAN statement will be stored as an integer and can be used in fixed-point arithmetic. To be recognized as an integer, the constant must be written without a decimal point. For example, the constants

$$5$$
$$0$$
$$-150$$
$$+4361$$

if appearing in FORTRAN statements would be recognized as fixed-point numbers and stored as such.

Floating-point constants are usually referred to as real constants. They may be written in FORTRAN statements in either of two forms. Any number written with a decimal point will be recognized by FORTRAN as a floating-point number and stored as such. For example, the numbers

$$+5.$$
$$0.$$
$$16.793$$
$$-.0001985$$

if appearing in FORTRAN statements would be stored as real constants. If the number is so large or so small that it is inconvenient to write it in the above form, it can be followed by the letter E and a one- or two-digit positive or negative power of 10 by which the number is to be multiplied. For example,

$$5.1E6 \text{ is recognized as } 5.1 \times 10^6$$
$$-.32E-12 \text{ is recognized as } -.32 \times 10^{-12}$$

The allowed sizes for integer and real constants are properties of the actual computer and compiler being used. It was seen in Chapter 2 that a 32-bit word could be used to store a signed integer of at most 31 bits, for a maximum size in decimal form of 2147483647. For real constants, the seven bit positions allotted for the exponent could accommodate magnitudes from 16^{-63} through 16^{63}, or approximately 10^{-75} through 10^{75}. For machines handling different word sizes, the allowed size range for constants will be different. The attempt to use a constant outside the allowable size range will cause an error indication. In a 32-bit word, the 24 bit positions allotted for the fractional part allow for roughly seven significant digits in the number. Ordinarily, if a FORTRAN constant is written with too many digits in the fractional part, the excess digits are ignored.

Example 1. Of the following quantities, which ones are acceptable integer constants, which ones are allowable real constants, and which ones are illegal?

a. 101	b. 101.	c. 1.2E6
d. 98765.2E−2	e. 3.45.E2	f. 1981

The numbers a and f are integer constants; b, c, and d are real constants; e is illegal because of the extra decimal point.

Example 2. Write the following quantities as FORTRAN constants.

$$\text{a.} \quad 3 \times 10^{-4} \qquad \text{b.} \quad 1/8 \qquad \text{c.} \quad \pi$$

The number 3×10^{-4} would be written 3.E−4 (or .3E−3, or .0003, etc.). The quantity 1/8 would be written as .125 (or 1.25E−1, or 125.E−3, etc.). The number π could be written as 3.14159, the number of digits used being determined by the accuracy desired, up to the precision allowed by the machine.

4.22 Variables

Any number to be used in a FORTRAN statement whose value is unknown or whose value may change during the calculation is termed a variable. Just as in an ordinary algebraic calculation we give a name such as x or y or t

to any unknown or any variable, so in FORTRAN we must give names to any variables used. FORTRAN will recognize any combination of six or less letters or digits, beginning with a letter, as a name of a variable, and will save a memory location within the computer for storage of that particular quantity.* The quantities

> A
> B1
> CLOCK
> K
> M589C

are all acceptable names for FORTRAN variables.

As for constants, a FORTRAN variable can be either a fixed-point (or integer) number or a floating-point (or real) number, and the name indicates which it is.

Any variable whose name begins with I, J, K, L, M, or N is treated by FORTRAN as a fixed-point number. A variable whose name begins with any other letter is treated as a floating-point number.

Example 1. Which of the following are allowed FORTRAN integer variable names?

a.	MOON	b.	JAKE	c.	ALPHA
d.	1M34	e.	KK5	f.	LONGITUDE

The names MOON, JAKE, and KK5 are allowed. ALPHA and 1M34 are not, because they do not begin with I, J, K, L, M, or N. LONGITUDE is not, because it contains too many characters.

Example 2. Which of the following are allowable FORTRAN real variable names?

a.	A34	b.	37A	c.	HELIX
d.	ZEBRA	e.	N1	f.	ACCELERATION

The names A34, HELIX, and ZEBRA are allowed. 37A is not because it begins with a number rather than a letter. N1 is not, because N as a first letter is reserved for integer variables. ACCELERATION is not, because it contains too many characters.

4.23   Expressions

In FORTRAN, each of the four basic arithmetic operations is represented by a distinct symbol:

---

* The allowed number of characters in a name is again dependent on the particular machine being used. The limit ranges from as few as five to as many as eight characters.

addition: +
subtraction: −
multiplication: *
division: /

FORTRAN will recognize mathematical expressions made up of mathematically proper combinations of constants, variables, and these symbols. Parentheses can be used to indicate the order in which operations are to be performed but not to replace an operation's symbol. For example, a FORTRAN expression for the algebraic expression $a(b + c)$ would be A*(B+C). A(B+C) would not be a correct expression, since no operation is indicated between A and B. When there are no parentheses to indicate the order of operations, the multiplications and divisions would be performed first, working from left to right, and then the additions and subtractions, also working from left to right. Thus A/B*C would be interpreted to mean $\left(\dfrac{A}{B}\right)C$ and not $\dfrac{A}{BC}$.

Fixed- and floating-point quantities cannot be mixed in the same expression. Thus the expression

$$DOG+CAT$$

will cause the two quantities named DOG and CAT to be added in floating-point fashion. The expression

$$MICE+MOOSE$$

will cause the two quantities called MICE and MOOSE to be added fixed-point fashion. The expression

$$CAT+MICE$$

is illegal because CAT is stored in floating-point form and MICE is in fixed-point form.

When expressions use integer variables and constants, fixed-point arithmetic will be performed. When real variables and constants are used, floating-point arithmetic will be performed. This is particularly important in integer division, because the programmer may get some startling results if he does not pay full attention to the properties of integer division. The expression

$$N/M$$

will give the largest integer which when multiplied by M gives a number less than or equal to N. Thus the expression

$$1/2$$

will have the value zero. On the other hand, the expression

$$1./2.$$

will have the value .5, since these are floating-point constants.

Some compilers do allow mixed-mode expressions, in which both integer and real quantities are involved. Such usage involves a number of special rules, and rather than become involved in them we will avoid mixed-mode expressions in this text and obey the rule that in an expression all quantities are real or all are integer.

Example 1.   Write the following arithmetic expressions as FORTRAN expressions, using the names A for a, B for b, X for x, and N for n.

a. $ax + b$ b. $3x - 16$ c. $2n + 1$

Correct expressions are

a. $A*X + B$ b. $3.*X - 16.$ c. $2*N + 1$

In b, the constants must be written in real form to match the real variable X. In c, the constants must be written in integer form, to match the integer constant N.

Example 2.   The following FORTRAN expressions have been written to represent the arithmetic expressions alongside them. Which ones are correct?

a. $x(a + y)$ X*(A + Y)
b. $b/(c - d)$ B/(C − D)
c. $3x + a$ 3*X + A
d. $a(b - c)$ A(B − C)
e. $1 + \dfrac{a}{bc}$ 1. + A/B*C
f. $1 + \dfrac{ab}{c}$ 1. + A*B/C

Examples a, b, and f are correct. Example c is incorrect, because 3 is integer, while X and A are real. A correct expression would be $3.*X + A$. Example d is incorrect because there is no operation indicated between A and B (the parentheses do not indicate multiplication in FORTRAN). A correct expression would be A*(B−C). Example e is incorrect because in the FORTRAN expression, A would be divided by B and then the result multiplied by C. A correct expression would be $1. + A/(B*C)$.

4.24   Arithmetic Assignment Statements

An arithmetic assignment statement in FORTRAN is written in the form

some FORTRAN variable = some FORTRAN expression

Although it contains an equals sign, the statement is not really an equation. FORTRAN interprets the equals sign to mean: Compute the numerical value of the FORTRAN expression on the right and store the result in the memory location reserved for the variable named on the left. Thus the statement

A = DOG + CAT

means compute the sum of the numbers in the memory locations reserved for the variables named DOG and CAT and store the result in the memory location reserved for the variable named A. The previous contents of that memory location are lost. The statement

M = M + 1

means add one to the integer in the memory location reserved for the variable named M and store the result back in the same location. (It is quite clear that this last statement cannot be interpreted as an equation in the normal sense.) The statement

B = 3.74*A

means multiply by 3.74 the contents of the memory location reserved for the variable named A and store the result in the memory location reserved for the variable named B.

Example 1.   Which of the following arithmetic statements are correct representations of the corresponding mathematical equations?

a. $s = (1/2)gt^2$ S = .5*G*T*T
b. distance = rate × time   D = R*T
c. voltage = current × resistance   E = I*R
d. $y = (3x + 2)/(2z + 1)$ Y = (3.*X+2.)/(2.*Z+1)
e. $w = x(y + (2z + 1))$ W = X*(Y+(2.*Z+1.)))

f.	$x^2 = xy - 3$	$X*X=X*Y-3.$
g.	$x = n - 2$	$X=N-2$
h.	$n = x + 2$	$N=X+2.$

Examples a, b, g, and h are correct. In g, the expression computed is fixed point, and it is placed in a storage location assigned for a real quantity. The compiler will cause a conversion to floating point before storing the result. In h, the expression computed is real, and will be automatically converted to integer before storage as N. Example c is wrong because the quantity I is fixed point and R is floating point. Example d is wrong because the number 1 is integer and the remainder of the expression is real. It could be corrected by placing a decimal point after the 1. Example e is wrong because there is one too many right parentheses at the right-hand end of the statement. Example f is wrong because it is not of the form FORTRAN variable = FORTRAN expression. The quantity to the left of the equals sign must be a single-variable name only.

EXERCISE 9

1. Tell which of the following are allowed integer constants, which are allowed real constants, and which are illegal.

   a. 141.63   b. .002E4   c. 0
   d. .06−E6   e. 110111   f. 93476218354
   g. 93476.218354

2. Evaluate the following FORTRAN expressions.

   a. 3+1/3   b. (3+1)/3
   c. (4*N+2)/N+1 for N=1,3,5,7
   d. N/M+M/N for N=3,M=5 and N=7,M=2
   e. (2*N+3)/(N+4) for N=1,4,7,10

3. Tell which of the following are allowed integer variable names, which are allowed real variable names, and which are illegal.

   a. WHICH   b. X12345   c. N12345
   d. RADIUS   e. INTERSECTION   f. AB$C

4. For which of the following do the FORTRAN expressions correctly represent the mathematical expressions? Rewrite the incorrect FORTRAN expressions.

   a. $a = \pi r^2$ A=3.14159*R(R
   b. $v = r\omega$ V=R*OMEGA
   c. $y = ax^2 + bx + c$ Y=(A*X+B)X+C
   d. $w = 3x + \dfrac{y}{2z}$ W=3.*X+Y/2.*Z
   e. $S = S_0 + vt + (1/2)at^2$ S=S0+VT+.5*A*T*T
   f. $T = 2\pi r^2 + 2\pi rh$ T=6.28318*R*(R+H))
   g. $e = mc^2$ E=M*C*C

5. Write FORTRAN statements to represent each of the following formulas, using any variable names you choose.

a. $E = h\nu$

b. $Z = R + 2\pi f C + \dfrac{1}{2\pi f L}$

c. $R = \dfrac{1}{\dfrac{1}{R_1} + \dfrac{1}{R_2} + \dfrac{1}{R_3}}$

d. $A = (1/2)h(b_1 + b_2)$

e. $y = 1 + x + x^2 + x^3$

f. $y = \dfrac{ax + b}{cx + d}$

4.3 INPUT AND OUTPUT STATEMENTS

Special FORTRAN statements can be used to cause input of data or output of results. The most important of these statements are described below. When they are used, the compiler expects input numbers to be in decimal form and causes them to go through a conversion to binary. Likewise, it causes binary numbers to be converted to decimal before output.

4.31 READ and FORMAT

Input is caused by a statement which begins with the word READ, followed by the names of the variables to be read into the machine. Data input with FORTRAN is normally from punched cards, and each READ statement initiates the reading of a new card. Values for the variables must be punched on the card in the order the names appear in the READ statement. The READ statement then causes these values to be put in memory in the proper locations.

A READ statement by itself does not contain all the information required for reading a number from a punched card. The card contains 80 columns, each of which may contain a decimal digit. The machine must be told which of these columns contain the particular number to be stored, whether this number is fixed or floating point, and where the decimal point is located. This information is placed in a separate statement, called a FORMAT statement. The FORMAT statement consists of the word "FORMAT," followed by parentheses containing the description of each number to be read in by the corresponding READ statement. If the number is fixed point, this description would consist of the letter I followed by a number that indicates the number of card columns reserved for the number. If the number is floating point, it was mentioned earlier that it might appear in either one of two forms: either as an ordinary decimal number with decimal point or as a number followed by the letter E and a power of 10 by which the number is to be multiplied. In the first case, the description in the FORMAT statement is the letter F, then a number representing the number of card columns reserved for the number, then a period, then a number representing the number of digits after decimal point in the number as it appears on the card. In the second case, the letter E is used rather than the letter F. For example, the statement

$$\text{FORMAT(I5,F8.4,E12.2)}$$

would indicate that the first 5 columns of the card contain a fixed-point number, the next 8 contain a floating-point number having four digits after the decimal point, and the next 12 contain a floating-point number having two digits after the decimal point and followed by a power of 10 as described in Section 4.21.

Any FORTRAN statement can be given a statement number that can be used for reference by other statements. The statement number is placed at the beginning of the statement. A FORMAT statement *must* be given a statement number so that READ statements can refer to it. Then each READ statement, immediately after the word "READ," contains the number of the FORMAT statement that describes how the reading is to be done. A few examples will serve to clarify the way in which this combination works.

Example 1.   Write input statements to read from a card values for the variables A, B, and X, where A covers 10 columns of the card and has four digits after the decimal point, B covers 8 columns of the card and has two digits after the decimal point, and X covers 12 columns of the card and has six digits after the decimal point.

A suitable set of statements is

        READ 15,A,B,X
    15 FORMAT(F10.4,F8.2,F12.6)

Note that there are commas after the 15, A, and B in the READ statement and commas separating the three descriptions of numbers in the FORMAT statement. These commas are required, and FORTRAN could not correctly interpret these statements without them.

Example 2.   Write input statements to read from a card values for the variables MAT, A1, A2, A3, A4, where MAT is a fixed-point number occuping 5 columns on the card and A1 through A4 are floating point, each occupying 10 columns and having five digits after the decimal point.

A suitable set of statements is

         READ 123,MAT,A1,A2,A3,A4
    123 FORMAT(I5,F10.5,F10.5,F10.5,F10.5)

In this case we have four identical descriptions F10.5. In such cases these can be grouped and written as 4F10.5. Thus another suitable solution is

        READ 73,MAT,A1,A2,A3,A4
    73 FORMAT(I5,4F10.5)

Example 3. Write input statements that will read values for A1, A2, A3, A4 from a card and then read values for B1, B2, B3, B4 from a second card, where each number occupies 8 columns on the card and has four digits after the decimal point.

A suitable set of statements is

        READ 101,A1,A2,A3,A4
        READ 101,B1,B2,B3,B4
    101 FORMAT(4F8.4)

Note that both READ statements refer to the same FORMAT statement. It is permissible to have many READ statements referring to the same FORMAT statement.

4.32 PRINT

Output is normally accomplished by a PRINT statement, which causes the desired values to be listed on a printer, usually one capable of printing 120 characters in a line. Alternatively, depending upon the particular machine, the PRINT statement may cause the output values to be recorded on magnetic tape for later printing in a separate operation. In either event, the PRINT statement is quite similar to the READ statement, consisting of the word PRINT, followed by the statement number of the FORMAT statement which describes how the printing is to be spaced across the page, followed by the names of the variables whose values are to be printed.

Example 1. Write a set of FORTRAN statements that will cause the machine to print the quantities I and A, where I is a fixed-point number having at most five digits and A is a floating-point number less than 500, and we wish to print the value to four decimal places.

A suitable set of statements is

        PRINT 150,I,A
    150 FORMAT(I5,F8.4)

This set of statements will cause I to be printed in the first 5 character positions and A to be printed in the next 8 (one space will be used to print a decimal point). Since this way of printing would cause the values of I and A to run together on the page, a better FORMAT statement might be

150 FORMAT(I5,F20.4)

With this FORMAT statement, 20 spaces will be reserved for A. Since the number requires only 8 of these, there will be vacant spaces between the value of I and the value of A, making the outputs easier to read. This same result can be obtained by use of another symbol allowed in a FORMAT statement: The symbol X means leave a blank space; 5X means leave five blank spaces, and so on. Thus the statement

150 FORMAT(I5,12X,F8.4)

would also cause space to be left open between the two printed numbers.

4.33   Hollerith Characters

Frequently it is desirable to have the computer print out just a set of characters or words, as page headings or titles, or as indicators that the calculation has reached a certain stage. Such symbols or characters are referred to as "Hollerith" characters, after Hollerith, former director of the Bureau of Census, who introduced the use of binary coding of letters and characters on punched cards. Hollerith characters can be included in a FORMAT statement, and will be printed out just as written in the FORMAT statement whenever a PRINT statement is executed which uses that format. The Hollerith characters must be preceded by a number which is the actual number of characters to be printed, and then the letter H. For example, the statements

PRINT 105
105 FORMAT(15H END OF PROBLEM)

would cause the computer to print "END OF PROBLEM." Note that the number of Hollerith characters is listed as 15. The blank space after H and the two blank spaces between words must be included in the count.

4.4   SIMPLE FORTRAN PROGRAMS

Only two more FORTRAN statements are necessary before we will be able to write simple programs in FORTRAN. The first of these is the STOP statement. The word STOP is used to indicate to the computer that the calculation is over, and the machine should stop and wait for its next task. The second is the END statement. This statement is used to tell the FORTRAN compiler that there are no more statements in the program, so that the compiler can complete its task of constructing an object program.

4.41   Examples

Example 1.   Write a FORTRAN program to read in three numbers A, B, and C, and compute and print their sum and their product.

A suitable program is

```
      READ 101,A,B,C
      X=A+B+C
      Y=A*B*C
      PRINT 101,X,Y
      STOP
  101 FORMAT(3E12.4)
      END
```

This program could be used by punching the statements on cards, one statement per card. The FORTRAN compiler would be placed in memory, and the deck of cards with the above program, or source deck, in the card reader. The compiler would then write a machine language program, or object program, which could either be punched on cards for use at a later time, or left in the machine memory and executed as soon as the compilation was complete. This object program, when executed, would, in effect, cause the machine to execute the FORTRAN statements in the order in which they are listed in the above FORTRAN program.

Since the FORMAT statement is not executed but only provides information, its location in the program does not matter. We have chosen to place it at the end. Note also that the FORMAT statement provides formats for three numbers and the PRINT statement uses only two of these. The format E12.4 requires that the input numbers be in floating-point form and that A be punched in the first 12 columns of the card, B in the next 12 columns, and C in the next 12. Exponential form is used because we have no idea of the sizes of A, B, and C.

If we wanted a fancier printed output, we might replace the above PRINT statement by

      PRINT 102,X,Y

and add another FORMAT statement:

  102 FORMAT(5H SUM=,E12.4,10X,8HPRODUCT=,E12.4)

With these statements, the printed results will be a line containing the words SUM =, followed by the value of the sum, then 10 blank spaces, then the words PRODUCT =, followed by the value of the product.

Example 2. Write a FORTRAN program which will read in the quantities XENGTH, WIDTH, HEIGHT, which are dimensions of a box in inches, and print out the total surface area in square feet and volume in cubic feet.

A suitable program is

```
        READ 101,XENGTH,WIDTH,HEIGHT
        AREA = (XENGTH*WIDTH + WIDTH*HEIGHT + XENGTH
                                            *HEIGHT)/72.
        VOLUME = XENGTH*WIDTH*HEIGHT/1728.
        PRINT 101,AREA,VOLUME
        STOP
    101 FORMAT(3E12.4)
        END
```

(In this program, the word LENGTH could not be used for length of a side because, beginning with " L," it would be treated by the machine as a fixed-point variable.)

EXERCISE 10

1. Write a FORTRAN program that will read in four numbers A, B, C, and D and compute and print their average.

2. Write a FORTRAN program that will read in the radius and height of a cylinder and compute and print its total area and volume.

3. Write a FORTRAN program that will read in two numbers x and y, and compute and print the following quantities:

$$u = 2x + 4y$$
$$v = x^2 + 5xy + 4y^2$$
$$w = 3 + 4x + 5x^2 + 6x^3$$
$$z = (x + 2y)/(6x - 4y)$$

4.5 TRANSFER OF CONTROL

In the FORTRAN programs we have written so far the statements are executed in the same order in which they appear in the program. In Chapter 2 it was indicated that there are many advantages in having machine language commands which can transfer control to other points in the program. There are similar advantages in being able to alter the order in which FORTRAN statements are executed. There are several FORTRAN statements designed for this purpose.

4.51   The "GO TO" Statement

The FORTRAN statement

$$\text{GO TO } n$$

where n is the statement number of some other statement in the program, will cause statement n to be executed next.

Example 1.   Write a program that will read in sets of three numbers A, B, and C, and compute and print their sum and product.

   This is Example 1, Section 4.41, except that the calculation is to be repeated for several sets of numbers A, B, and C. A suitable program is

```
  1 READ  101,A,B,C
    X=A+B+C
    Y=A*B*C
    PRINT  101,X,Y
    GO TO 1
101 FORMAT(3E12.4)
    END
```

This program will read in a card, compute and print X and Y, and then the statement GO TO 1 will cause the calculation to return to statement 1 and read the next card. This process will continue until all cards are used up.

4.52   The Arithmetic "IF" Statement

The "GO TO" statement just discussed is analogous to the unconditional branch in machine language. A FORTRAN statement analogous to the conditional branch is also extremely useful. The arithmetic "IF" statement serves this purpose. It is of the form

$$\text{IF}(e) \; n_1, n_2, n_3$$

where e stands for any expression as described in Section 4.23, and n_1, n_2, and n_3 are statement numbers. If the value of the expression in parentheses is negative, control is transferred to n_1. If it is zero, control is transferred to n_2. If it is positive, control is transferred to n_3.

Example 1. Write a program that will read in pairs of numbers, A and B, and print out the smaller, or, if the numbers are equal, that will print the word "EQUAL."

A suitable program is

```
  1 READ  101,A,B
    IF(A − B)2,3,4
  2 PRINT 101,A
    GO TO 1
  3 PRINT 102
    GO TO 1
  4 PRINT 101,B
    GO TO 1
101 FORMAT(2E12.4)
102 FORMAT(6H EQUAL)
    END
```

Statement 1 reads in a set of values for A and B, and the IF statement computes the difference $A − B$. If this number is negative, control is transferred to statement 2, which causes A to be printed. Then the GO TO 1 statement returns control to statement 1, and another set of values is read in. If $A − B$ is zero, the IF statement transfers control to statement 3, which causes a print according to FORMAT statement 102, which is simply the word "EQUAL." Again there is a GO TO 1 statement to return control to statement 1 to read in another card. If $A − B$ is positive, the IF statement transfers control to statement 4, which causes B to be printed. Again, there is a GO TO 1 statement to return control to statement 1 to read in another card. The "IF" statement serves as a branch point in the program switching control to 2, 3, or 4, depending on the value of $A − B$. The process continues as long as input cards are available.

Another use of the IF statement is to cause the machine to stop properly when the last data card has been executed. In the last few examples we have used the GO TO statement to cause the machine to return to a READ statement to read new input. This was a loop with no end, and we have said that the executive routine, or supervisor, would terminate the program when there was no input available. This is not the most desirable way of stopping, since the supervisor will ordinarily treat the lack of input as an error condition and abort the program with an error message. It is usually wise to terminate a program at a stop statement, so that the supervisor will recognize the condition as normal completion. This can be done with an IF statement as demonstrated in the following example.

Example 2.   Write a program that will read in sets of three numbers A, B, and C, and compute and print their sum and product.

This is Example 1, Section 4.51. Suppose we know that the value of A never exceeds 10^{30}. (For most problems such an assumption would be quite reasonable. At least we will know something about A, and can use that knowledge to construct a test such as the one in the program below.) Then we can use the program below to make the desired computation:

```
  1 READ 101,A,B,C
    IF(A − 1.E30)2,3,3
  2 X = A + B + C
    Y = A*B*C
    PRINT 101,X,Y
    GO TO 1
  3 STOP
101 FORMAT(3E12.4)
    END
```

Here the IF statement is used as an exit from the loop. In order to utilize it, we must place behind the real data cards a false data card with a value for A greater than 10^{30}. Then when the machine reads this data card, the IF statement will cause a transfer to statement 3, which will stop the program.

4.53   The Computed "GO TO" Statement

Another kind of conditional branch statement is available in the FORTRAN language, the computed GO TO statement. This statement has the form

$$GO\ TO\ (n_1, n_2 \ldots, n_m), i$$

where i is a fixed-point variable and $n_1, n_2, \ldots, n_m$ are statement numbers. If the value of i is 1 at the time this statement is reached, control will be transferred to statement n_1. If i is 2, control will be transferred to statement n_2, and so on. This kind of statement is useful when the program may branch in any of several different directions.

Example 1. Write a program that will input a value X, which may be in inches, feet, yards, or miles, and print out its value in yards.

Let us assume that each input card contains a value for X and also a fixed-point constant I, which has the value 1 if X is in inches, 2 if it is in feet, 3 if it is in yards, and 4 if it is in miles. Then a suitable program is

```
  1 READ 101,X,I
    IF(X−1.E30)2,3,3
  2 GO TO(4,5,6,7),I
  4 X=X/36.
    GO TO 6
  5 X=X/3.
    GO TO 6
  7 X=1760.*X
  6 PRINT 101,X
    GO TO 1
  3 STOP
101 FORMAT(E12.4,I1)
    END
```

In this program we have again used an IF statement to stop the problem when all input cards have been used. The computed GO TO statement functions as follows: If X is in inches, $I = 1$, and transfer is to statement 4, which divides X by 36 to convert to yards. Then a transfer to 6 allows the value to be printed. If X is in feet, $I = 2$, and transfer is to statement 5, which divides X by 3 to convert to yards. Again a transfer to 6 allows the value to be printed. If X is in yards, $I = 6$, and transfer is to statement 6 for immediate printing. If X is in miles, transfer is to statement 7, which multiplies X by 1760 to convert to yards, and control proceeds to the next statement, statement 6, where X is printed. Note that the statements in this program are not numbered sequentially. Statement numbers are for identification only and need not appear in order throughout a program.

4.6 SUBSCRIPTED VARIABLES

It is common in mathematics to use subscripts on variable names to economize on notation. For example, if we wish to speak of a set of values of x, we may label them $x_1, x_2, x_3, \ldots, x_n$. Sometimes such a set of quantities is referred to as an array and each individual quantity as an element in the array. FORTRAN provides for the use of subscripts in variable names. The subscript is a positive fixed-point constant or variable placed in parentheses after the variable name, thus:

$$X(1), X(2), \text{etc} \quad \text{or} \quad X(I), X(M), \text{etc.}$$

(Note that zero is not allowed as a subscript.) A subscripted variable name can be used in FORTRAN statements just as any other variable name. For example, a statement to calculate a quantity $y = x_1 + x_2 + x_3$ might be written $Y = X(1) + X(2) + X(3)$. The subscript notation is of no particular value in a statement such as this one, however. Its chief value is that it frequently allows lengthy calculations to be programmed as loops. For example, the quantity $y = x_1 + x_2 + \cdots + x_{100}$ can be computed by the set of statements

```
    Y=0.
    I=1
  5 Y=Y+X(I)
    I=I+1
    IF(101-I)6,6,5
  6 PRINT 10,Y
```

This FORTRAN loop is similar in many respects to the machine language loops described in Chapter 2. The required sum is built up a term at a time in the memory location reserved for Y. The fixed-point number I serves as a counter to keep track of how many terms have been added, as well as a subscript for X to assure that the proper term is added at each step. The statements $Y = 0$ and $I = 1$ initialize the values of Y and I. Statement 5 does the actual calculation, and the next statement advances by one the value of I. The IF statement serves as an exit for the loop, sending the calculation back to statement 5 each time until X(100) has been added into the sum, then sending the calculation to statement 6 for printing the answer.

4.61   The DIMENSION Statement

So far we have written a few statements using subscripted variables, but have written no complete programs using subscript notation. The reason for this is that a FORTRAN compiler is unable to compile a program containing subscripted variables unless it is given some advance information. In compiling the program, it will need to know which variables are subscripted, and how many different values each subscript may have, so that it can save a memory location for each quantity to be stored. This information is provided by a DIMENSION statement. Before any subscripted variable appears in a program, it must be mentioned in a DIMENSION statement. Usually a DIMENSION statement listing all the subscripted variables is placed at the beginning of the program. This statement is of the form

$$\text{DIMENSION } V_1, V_2, V_3, \ldots,$$

where V_1, V_2, V_3, ... are the names of the subscripted variables. Each variable name must be followed, in parentheses, by the maximum value the subscript will have. For example, a program involving a subscripted variable x_i, where i can be any integer from 1 to 100, would require a DIMENSION statement:

$$\text{DIMENSION } X(100)$$

Example 1. Write a dimension statement for a program that will use the subscripted variables X(I), where I may be as large as 50, ZAV(J), where J may be as large as 25, and MOX(K), where K may be as large as 10.

A suitable statement is

$$\text{DIMENSION } X(50),ZAV(25),MOX(10)$$

It should be noted that, when subscripted variables are used, a separate memory location is saved for each value of the subscript. This can use up available memory space very quickly, so subscripted variables should not be used unless they are actually needed. The following example demonstrates an unnecessary use of a subscripted variable.

Example 2. Write a program that will input the quantities x_1, x_2, ..., x_{100} and find and print their sum.

One program which will perform this task is the following:

```
      DIMENSION X(100)
      Y=0.
      I=1
    1 READ 50,X(I)
      Y=Y+X(I)
      I=I+1
      IF(101−I)2,2,1
    2 PRINT 50,Y
      STOP
   50 FORMAT(E12.4)
      END
```

In this program, 100 memory locations are saved for the quantities X(I). Each time statement 1 causes a new value of X(I) to be read in, it will go into a new memory location. Immediately it is added into the value of Y, and then is no longer needed. The following program would do the calculation just as well, and without the wasted memory space:

```
    Y=0.
    I=1
  1 READ 50,X
    Y=Y+X
    I=I+1
    IF(101-I)2,2,1
  2 PRINT 50,Y
    STOP
 50 FORMAT(E12.4)
    END
```

In the following example, the subscripted variable is really needed, since all 100 values are needed in memory at the same time.

Example 3. Write a program that will input 100 quantities $x_1, x_2, \ldots, x_{100}$ and print them out in order of size, from smallest to largest.

The following program will suffice:

```
    DIMENSION  X(100)
    I=1
  1 READ  50,X(I)
    I=I+1
    IF(101-I)2,2,1
  2 A=1.E30
    I=0
  3 I=I+1
    IF(101-I)6,6,4
  4 IF(X(I)-A)5,3,3
  5 J=I
    A=X(I)
    GO TO 3
  6 IF(.9E30-A)8,8,7
  7 PRINT  50,A
    X(J)=1.1E30
    GO TO 2
  8 STOP
 50 FORMAT(E12.4)
    END
```

In this program, first all 100 values x_i are read in. Then a number A is used for comparison. It is first set equal to 10^{30} (we have assumed that all the x_i values are known to be less than this number). Each x_i in turn is compared to A, and, when one smaller than A is found, A is given that value, and the corresponding subscript is stored as the quantity J. After all 100 values x_i have been gone through in this fashion, the value A is equal to the smallest x_i. A is printed, and then the value of that particular x_i is set to 1.1×10^{30}, so that it cannot again be selected as the smallest. A is then reset to 10^{30}, and the process repeated. When all the x_i's have been selected as smallest, printed, and set equal to 1.1×10^{30}, the test in statement 6 will transfer control to 8, where the program will be stopped. In Chapter 5, the subject of machine time is discussed. The astute reader may be able to find a method of rearranging the numbers which is faster than the method in the above program.

4.62   Double and Triple Subscripts

Sometimes it is convenient to have more than one subscript on a variable. For example, the transformation from one rectangular coordinate system to another is conveniently represented by the equations

$$y_1 = a_{11}x_1 + a_{12}x_2 + a_{13}x_3$$
$$y_2 = a_{21}x_1 + a_{22}x_2 + a_{23}x_3$$
$$y_3 = a_{31}x_1 + a_{32}x_2 + a_{33}x_3$$

If the elements in an array have a single subscript, the array is called one-dimensional. If they have two subscripts, the array is two-dimensional, and so on. Thus the nine coefficients a_{11}, a_{12}, etc., for a two-dimensional array. FORTRAN provides for the use of two- and three-dimensional arrays as well as one-dimensional arrays. A doubly subscripted variable is written in the form A(I,J), and a triply subscripted one in the form B(I,J,K). The DIMENSION statement must prescribe the maximum value for each subscript in a subscripted variable. Thus, if a program contains the subscripted variable B(I,J,K), where I may be as large as 10, J as large as 15, and K as large as 20, the DIMENSION statement should appear as

DIMENSION B(10,15,20)

Example 1.   Write a program that will input sets of values x_1, x_2, x_3, and compute and print y_1, y_2, y_3 according to the equation

$$y_1 = a_{11}x_1 + a_{12}x_2 + a_{13}x_3$$
$$y_2 = a_{21}x_1 + a_{22}x_2 + a_{23}x_3$$
$$y_3 = a_{31}x_1 + a_{32}x_2 + a_{33}x_3$$

A suitable program is

```
    DIMENSION  X(3),Y(3),A(3,3)
    READ  101,A(1,1),A(1,2),A(1,3)
    READ  101,A(2,1),A(2,2),A(2,3)
    READ  101,A(3,1),A(3,2),A(3,3)
  1 READ  101,X(1),X(2),X(3)
    IF(X(1)−1.E30)2,3,3
  2 Y(1)=A(1,1)*X(1)+A(1,2)*X(2)+A(1,3)*X(3)
    Y(2)=A(2,1)*X(1)+A(2,2)*X(2)+A(2,3)*X(3)
    Y(3)=A(3,1)*X(1)+A(3,2)*X(2)+A(3,3)*X(3)
    PRINT  101,Y(1),Y(2),Y(3)
    GO TO 1
  3 STOP
101 FORMAT(3E12.4)
    END
```

The above program makes use of subscript notation in identifying the quantities involved but does not take advantage of the capabilities of subscript notation for reducing the amount of writing required in the program. In Section 4.63 a somewhat neater way of programming this calculation will be given.

4.63   The "DO" Statement

It was seen earlier that loops may be constructed in FORTRAN by means of the IF statement. Another extremely useful way of programming loops is provided by the DO statement.

This statement has the form

$$\text{DO } n \, i = m_1, m_2, m_3$$

where n is the statement number of some later statement in the program, i is a fixed-point variable, and m_1, m_2, and m_3 are positive fixed-point constants or variables. This statement is a command to repeat the sequence of FORTRAN statements which follow it, up to and including statement number n, altering the value of i each time. The first time through, i is set equal to m_1. Each successive time through the statements, i is increased by an amount m_3, until it exceeds m_2. At that point control proceeds beyond statement n to the remainder of the program. The quantity m_3 can be omitted, and if it is, FORTRAN assumes the value of m_3 to be 1.

Example 1. Write a program that will input the quantities x_1, x_2, ..., x_{100} and find their sum.

This is Example 2, Section 4.61. Suitable programs were given there. Another suitable program is

```
    Y=0.
    DO 4 I=1,100
    READ 50,X
  4 Y=Y+X
    PRINT 50,Y
    STOP
 50 FORMAT(E12.4)
    END
```

The DO statement sets $I = 1$. Then the program reads a value of X and adds it to Y. After statement 4 is executed, I is increased to 2, and the next value of X is read in and added to Y. When finally $I = 100$, the 100th value of X is read in and added to Y. Then control proceeds to the print statement.

Example 2. Write a program that will input sets of values x_1, x_2, x_3, and compute and print y_1, y_2, y_3 according to the equations

$$y_1 = a_{11}x_1 + a_{12}x_2 + a_{13}x_3$$
$$y_2 = a_{21}x_1 + a_{22}x_2 + a_{23}x_3$$
$$y_3 = a_{31}x_1 + a_{32}x_2 + a_{33}x_3$$

This is Example 1, Section 4.62. A suitable program was given there. Another suitable program is

```
      DIMENSION  X(3),Y(3),A(3,3)
      DO 1 I=1,3
    1 READ  101,A(I,1),A(I,2),A(I,3)
    2 READ  101,X(1),X(2),X(3)
      IF(X(1)-1.E30)3,4,4
    3 DO 5 I=1,3
      Y(I)=0.
      DO 5 J=1,3
    5 Y(I)=Y(I)+A(I,J)*X(J)
      PRINT  101,Y(1),Y(2),Y(3)
      GO TO 2
    4 STOP
  101 FORMAT(3E12.4)
      END
```

4.64   Restriction on DO Loops

Certain rules must be observed in using the DO statement. The more important of these rules are listed below:

(*1*) The last statement in a DO loop must not be GO TO, IF, or another DO statement. In order to assist in obeying this rule, FORTRAN provides a dummy statement which can be used when necessary as the final statement in a DO loop. This statement is simply the word CONTINUE.

(*2*) The statement following a DO statement must not be a DIMENSION or FORMAT statement.

(*3*) DO loops may be nested, but may not overlap. That is, if a DO statement occurs within a DO loop, it must set up a loop which terminates before or on the same statement as the other DO loop.

(*4*) No statement may cause transfer of control from outside a DO loop to the interior of the loop.

(*5*) No statement in a DO loop,

$$DO \quad n \; i = m_1, m_2, m_3$$

may alter the values of i, m_1, m_2, or m_3.

4.65   Implied DO Loops in Read and Print Statements

When it is desired to read or print a series of values, say for a subscripted variable, a form of DO loop can be used to simplify the READ or PRINT statement. The statement

    READ 100,A(1),A(2),A(3),A(4),A(5)

can be abbreviated

    READ 100,(A(I),I = 1,5)

In this statement the parameters to be read are enclosed in parentheses, separated by commas, and followed by the indication of the range of index values to be used. The contents within the parentheses are treated as a DO loop. The expression

    PRINT 100,(A(I),B(I),I = 1,3)

would cause the same action as

    PRINT 100,A(1),B(1),A(2),B(2),A(3),B(3)

The READ statements of Example 2, Section 4.64, could be replaced by the single statement

   1 READ 101,((A(I,J),J=1,3),I=1,3),(X(I),I=1,3)

Care must be used in the arrangement of these implied DO loops. An inadvertent misordering of the indices will cause the inputs to be mixed up. For example, if instead of the above statement we had used

   1 READ 101,((A(I,J),J=1,3),X(I),I=1,3)

we would mix up the A's and X's. The first READ statement expects to find the data in the order A(1,1), A(1,2), A(1,3), A(2,1), A(2,2), A(2,3), A(3,1), A(3,2), A(3,3), X(1), X(2), X(3), whereas the second READ statement expects the order to be A(1,1), A(1,2), A(1,3), X(1), A(2,1), A(2,2), A(2,3), X(2), A(3,1), A(3,2), A(3,3), X(3). The inadvertent shuffling of input or output in this manner can create a source of great puzzlement for the programmer, by causing his calculations to be done with his input numbers assigned to the wrong quantities.

EXERCISE 11

1. Write FORTRAN programs which will read in values of x and compute and print corresponding values of y for each of the following equations.

   a. $y = 3x + 5$ b. $y = x^2 + 6x - 4$ c. $y = 4x^3 + 6x^2 + 3x - 1$

2. Write FORTRAN programs that will read in values of x and compute and print corresponding values of y for each of the following equations. In each case, there is some value or values of x which will make the denominator zero; have your program test for this, and in such cases, instead of trying to compute y, print "DENOMINATOR IS ZERO" and proceed to the next value.

   a. $y = (3x + 2)/(x - 1)$

   b. $y = (x^2 + 2x + 7)/(x^2 - 3x + 4)$

   c. $y = (x^2 - 2x + 3)/(x^2 - x + 1)$

3. Write a program to read in 100 sets of values (x_i, y_i) and print them out in order of increasing value of $x_i^2 + y_i^2$ (i.e., increasing distance from the origin).

4. Write a program to read in sets of values $x_1, x_2, \ldots, x_6$ and compute and print sets of values $y_1, y_2, \ldots, y_6$ according to the equations

$$y_1 = a_{11}x_1 + a_{12}x_2 + \cdots + a_{16}x_6$$
$$y_2 = a_{21}x_1 + a_{22}x_2 + \cdots + a_{26}x_6$$
$$\cdot$$
$$\cdot$$
$$\cdot$$
$$y_6 = a_{61}x_1 + a_{62}x_2 + \cdots + a_{66}x_6$$

4.7 INTRINSIC FUNCTIONS

In the FORTRAN programs discussed so far, the only arithmetic operations involved have been addition, subtraction, multiplication, and division. No mention has been made of square roots, sines, logarithms, and so on, which are encountered so frequently in mathematical problems. The machine language of computers ordinarily contains instructions only for the basic arithmetic operations. By methods which are discussed in Chapter 6, it is possible to calculate square roots, sines, and so on, to any accuracy desired by means of steps involving only the basic arithmetic operations. Thus it is possible to write programs to compute these and other desired mathematical functions. For all the common functions, the required sets of instructions are already included in the FORTRAN compiler, so that they may be used with very little effort on the part of the programmer. Each function is assigned a FORTRAN name, and whenever that name is used in a program, the machine will calculate the value of the function. The argument of the function is placed in parentheses immediately after the function name. For example, the FORTRAN statement

$$Y = EXP(X)$$

will cause $y = e^x$ to be computed. Table 1 lists the common functions which are ordinarily available in all FORTRAN compilers and lists the names we will use for them.

TABLE 1

Mathematical Function	FORTRAN Name
Square root	SQRT
Exponential	EXP
Sine of angle in radians	SIN
Cosine of angle in radians	COS
Arctangent in radians	ATAN
Natural logarithm	ALOG
Absolute value	ABS

The argument of a function does not have to be a single variable. It may may be any floating-point expression. The function value is computed in floating-point form. For example, the equation

$$Y = \sqrt{1 + X^2} + \sqrt{1 - X^2}$$

can be represented by a FORTRAN statement

$$Y = SQRT(1. + X*X) + SQRT(1. - X*X)$$

An additional function or operation that is available in all FORTRAN systems is raising to a power. The FORTRAN expression for the function X^A is

$$X**A$$

This function is usually described in programming manuals along with the arithmetic operations, but strictly speaking it belongs in the same class as the functions just described. If A is not an integer, $y = x^A$ really means $y = e^{A \ln x}$, and it is this latter quantity which is really calculated when the FORTRAN expression X**A is used. This particular type of expression is an exception to the rule that fixed- and floating-point quantities cannot be mixed. When the argument is floating point, the exponent can be either fixed or floating. The expression X**2. would be calculated as $e^{2 \ln x}$, while X**2 without a decimal point would be calculated as X*X.

Example 1. Write a FORTRAN program that will compute and print the value of $y = \sqrt{a + b \sin x}$ for 100 equally spaced values of x from zero to $\pi/2$. The constants a and b are to be read from a card.

A suitable program is

```
      READ  101,A,B
      X = 0.
      DX = .015707963
      DO 2 I = 1,100
      X = X + DX
    1 Y = SQRT(A + B*SIN(X))
    2 PRINT  102,X,Y
      STOP
  101 FORMAT(2E10.4)
  102 FORMAT(F10.5,E10.4)
      END
```

Note that in this program the argument of the square-root function contains another function, the sine. This is permissible in most versions of FORTRAN. If desired, that statement could be replaced by two simpler ones:

```
      U = A + B*SIN(X)
      Y = SQRT(U)
```

EXERCISE 12

1. Write FORTRAN programs to compute and print y for 500 equally spaced values of x for each of the following expressions.

    a. $y = e^x \sin x$ b. $y = e^x + 3e^{4x} + 2e^{5x}$
    c. $y = x \cos x + \tan x$ d. $y = 1/2 \sin x + 4 \sin 2x + 5 \sin 3x$
    e. $y = \ln (1 + \cos x)$ f. $y = 1 + \tan^4 x \, e^2 + \cos^2 x$

2. The transformation from polar to rectangular coordinates is given by

$$x = r \cos \theta$$
$$y = r \sin \theta$$

    Using the above subroutines
    a. Write a program that will print x and y, given r and θ.
    b. Write a program that will print r and θ, given x and y. (*Note:* Be sure that θ is given in the correct quadrant.)

3. By the law of cosines, when two sides and the included angle of a triangle are known, the third side is given by

$$a^2 = b^2 + c^2 - 2bc \cos A$$

    Using the above subroutines, write a program to compute a, given b, c, and A. Under what circumstances do you think this program might give results with large errors?

4. Using the above subroutines, write a program to determine the area of a triangle, given the sides a, b, and c.

4.8   STATEMENT FUNCTIONS

It frequently happens that repeated use must be made of some complicated expression or function which is not one of the standard functions included in the FORTRAN system. To provide for such cases the FORTRAN system allows a function to be defined within the program itself. The programmer gives the function a name, which may be any name allowed for FORTRAN variables. The name of the function is followed by parentheses enclosing the argument or arguments, which are separated by commas if there is more than one. The function is defined in the program by writing a FORTRAN statement $a = b$. The left-hand side is the name given to the function, and the right-hand side is the arithmetic expression which describes the calculation that must be done to obtain the value of the function. For example, if the quantity $\sqrt{1 + x} + \sqrt{1 - x}$ were to be needed frequently during a calculation, one could put the FORTRAN statement

SUSQF(X)=SQRT(1.+X)+SQRT(1.−X)

at the beginning of the program, and then use the symbol SUSQF(X) whenever this quantity is desired thereafter. The above statement merely defines the function. It does not cause computation to take place. It must appear in the program before any executable statement.

Example 1. The quantities

$$y = (x/2)(\sqrt{1+x} + \sqrt{1-x}), \quad z = (y/2)(\sqrt{1+y} + \sqrt{1-y}),$$

and

$$w = (z/2)(\sqrt{1+z} + \sqrt{1-z})$$

are to be computed for a series of values of x between zero and one. Write a program that will compute y, z, and w for each given x.

A suitable program is

```
  1 GIVF(X) = X*(SQRT(1.+X)+SQRT(1.-X))/2.
  2 READ 101,X
    IF(X)4,3,3
  3 Y = GIVF(X)
    Z = GIVF(Y)
    W = GIVF(Z)
    PRINT 101,X,Y,Z,W
    GO TO 2
  4 STOP
101 FORMAT(4E12.4)
    END
```

Statement 1 merely defines the function GIV(X). The X is a dummy variable which stands for the argument of the function. In statement 3 and the two succeeding statements, the function is actually calculated for the arguments X, Y, and Z. A negative number input for X will stop the problem.

4.9 FORTRAN SUBPROGRAMS

Useful as the FORTRAN function statement is, it has some rather significant limitations. First, the function must be something which can be defined by a single FORTRAN statement. Second, it can only compute a single value as output. These limitations are removed in the FORTRAN system by the use of subprograms. There are two types of subprograms, the FUNCTION subprogram and the SUBROUTINE subprogram. Those are discussed in succeeding sections.

4.91   FUNCTION Subprogram

The FUNCTION subprogram removes the first of the limitations listed above by allowing the function definition to consist of many FORTRAN statements. Every time the name of the function is encountered in the main program, a branch is made to this set of statements. After they are executed, control is returned to that point in the main program from which the branch was made.

Two special statements characterize the FUNCTION subprogram. The first statement must consist of the word FUNCTION, followed by the name assigned to the function. The name may be any name allowed for FORTRAN variables. It is followed by parentheses containing the names of all variables used in calculating the function value. This name statement is followed by the FORTRAN statements to perform the required calculation and set the variable having the function name equal to the computed value. At the point at which the function value has been computed, and a return to the main program is desired, the second special FORTRAN statement is used. It is simply the word RETURN. As for an actual FORTRAN program, the FUNCTION subprogram must be terminated by an END statement.

Example 1. Write a FORTRAN program that will compute the value of $w = \sqrt{y + \sin y}$, where y is the larger of the following:

$$y = \cos 1.2x^2 - \cos 1.1x$$

and

$$y = \cos 1.3x^3 - \cos 1.05x$$

for

$$x = 1, 2, \ldots, 100$$

We will do this by first writing a FUNCTION subprogram that will compute y, and then using this subprogram in a main program that will compute w. Let us give the complicated functional relation which determines y, given x, the name COMP. The FUNCTION subprogram can be written

```
      FUNCTION COMP(A)
      Z1 = COS(1.2*A*A) - COS(1.1*A)
      Z2 = COS(1.3*A**3) - COS(1.05*A)
      IF(Z1 - Z2)6,6,8
    6 COMP = Z2
      RETURN
    8 COMP = Z1
      RETURN
      END
```

For an argument of function COMP, we have used the quantity A. A main program using this function would be written as follows:

```
  1 DO 5 I=1,100
    X=I
  2 Y=COMP(X)
    W=SQRT(Y+SIN(Y))
  5 PRINT 101,X,W
    STOP
101 FORMAT(2E12.4)
    END
```

Statement 2 causes function subprogram COMP to be executed. Note that this argument was called A in defining the function but was called X when the function was actually used. The A in the definition is a dummy variable, and the subprogram actually uses whatever variable is specified when the function is called. When function COMP is used, the values Z1 and Z2 are computed, COMP is set equal to the larger of the two. Then a RETURN statement is encountered and control returns to the main program, immediately after statement 2.

Some FORTRAN compilers require that when a FUNCTION subprogram is used, the main program contain a special statement to indicate this situation. The statement is the word EXTERNAL, followed by the name of the subprogram. In the above example, the statement would be

EXTERNAL COMP

4.92   SUBROUTINE Subprogram

The SUBROUTINE subprogram removes both limitations of internal functions indicated in Section 4.9. The SUBROUTINE subprogram, like the FUNCTION subprogram, is characterized by the presence of two special statements. The first statement in the subprogram must consist of the word SUBROUTINE, followed by the name assigned to the subroutine. The name may be any name allowed for FORTRAN variables (see Section 4.2). It is followed by parentheses containing the names of all input and output variables of the subprogram, separated by commas. The second special statement is the same RETURN statement used in FUNCTION subprograms. The SUBROUTINE subprogram must be terminated by an END statement.

In the main program, at any point where it is desired that a subprogram be used, a special statement is inserted. The statement consists of the word "CALL" followed by the name of the subroutine and, in parentheses and separated by commas, the names of the arguments to be used.

Example 1. Write a program that will compute the function $w = r + 2r \sin^2 \theta + r^3 \sqrt{1 - \cos^3 \theta}$, where r and θ are polar coordinates of a point, for given sets of points (X, Y) given in cartesian coordinates.

We can use a subroutine to make the coordinate transformation. A suitable subroutine is

```
      SUBROUTINE TRANS(R,THET,X,Y)
      R=SQRT(X*X+Y*Y)
      THET=ATAN(Y/X)
      IF(X)2,3,3
    2 THET=THET+3.1415927
    3 RETURN
      END
```

Since the arctangent routine gives a principal value, between $-\pi/2$ and $\pi/2$, it was necessary to test to determine the quadrant which θ really belongs in and place it in that quadrant. This is done by testing the abscissa, x, and adding π to θ if x is negative.

The main program which uses the above subroutine might appear as follows:

```
      READ 101,X,Y
      CALL TRANS(R,THET,X,Y)
      W=R+2.*R*(SIN(THET))**2
      W=W+R**3*SQRT(1.-(COS(THET))**3)
      PRINT 101,X,Y,W
      STOP
  101 FORMAT(3E12.4)
      END
```

In this program the subroutine had several arguments. The values of X and Y are needed by the subroutine and must be made available before the subroutine is called. The values of R and THET are determined by the subroutine.

4.93 Communication between Program and Subprograms

Ordinarily if some variable name is used in a subprogram, the compiler assigns a storage location to that variable without regard to whether the name also appears somewhere outside the subprogram. If the same name appears elsewhere in the program, it is assigned some other storage location for that usage, and no connection is recognized between the two.

For example, suppose we were to use FUNCTION COMP(A) of Example 2, Section 4.91, as follows:

        A = .2
        Z1 = 5.
        X = COMP(A)
        PRINT 101,X,Z1
        STOP
    101 FORMAT(2E12.4)
        END

If this program were run, the call on function COMP(A) would cause, within that subprogram, a value for the variable Z1 to be computed as

$$Z1 = \cos(1.2 \times (.2)^2) - \cos(1.1 \times .2)$$
$$\approx .02294$$

Yet when the PRINT statement is reached, the value printed out for Z1 will be 5. Changing Z1 in the subprogram did not change the value in the main program, because the compiler treats the Z1 in function COMP and the Z1 in the main program as entirely different entities. This arrangement is advantageous in most cases, in that while naming variables in a subprogram one does not have to worry about accidental duplication of a name already used in the main program or some other subprogram. On some occasions, however, it is desired to have a name recognized as being the same in subprogram and calling program. A special FORTRAN statement is provided for this purpose, the COMMON statement. The statement consists of the word COMMON, followed by the names of all variables desired to be treated as common values between subprograms and calling programs. It must be placed before any executable statement in both the calling program and subprogram. For example, if the statement

                        COMMON Z1

were placed as the first statement in the above program and as the statement immediately after the FUNCTION statement in function COMP, then upon running the program given above, the value .02294 would be printed out for Z1. If it were desired to have Z2 also in common, the statement

                        COMMON Z1,Z2

would be used. Alternatively, two COMMON statements could be used:

                        COMMON Z1
                        COMMON Z2

When more than one variable is placed in common, the variables must be named in both calling program and subprogram in exactly the same order. Otherwise, a mixup in storage will occur. If a dimensioned variable is placed in common, it is not necessary to use both a dimension statement and a common statement.

For example, if the variable B is to be dimensioned 50 and placed in common, it can be done with the two statements

$$\text{DIMENSION B(50)}$$
$$\text{COMMON B}$$

or simply with the single statement

$$\text{COMMON B(50)}$$

EXERCISE 13

1. Write a FORTRAN statement to define each of the following functions.
   a. $y = e^x \cos 2\pi x$ b. $y = 1 + 2 \cos x + 3 \cos 2x + 4 \cos 3x$
   c. $y = \ln (1 + \sqrt{1 + x^2})$ d. $z = \sqrt{2x^2 + 3y^2 - x + 1}$

2. Write a program that will define the function $x + \ln (1 + 3x + x^2)$ and use it to compute the quantities
$$y = x + \ln (1 + 3x + x^2)$$
$$z = y + \ln (1 + 3y + y^2)$$
$$w = z - \sin x - \ln (1 + 3 \sin x + \sin^2 x)$$

3. Write a FORTRAN subroutine that will convert a point (r, θ, ϕ) in spherical polar coordinates to rectangular coordinates (x, y, z).

4. Write a FORTRAN subroutine that will convert a point (x, y, z) in rectangular coordinates to spherical polar coordinates.

5. Write a FORTRAN subroutine that will convert a pair of values (x, y) to another pair (u, v) given by

   a. $u = \cos x \sin y$ b. $u = e^x \cos y$ c. $u = \ln (y + \sqrt{y^2 + x^2})$
      $v = \sin x - \sin y$ $v = e^x \sin y$ $v = \ln (x + \sqrt{x^2 + y^2})$

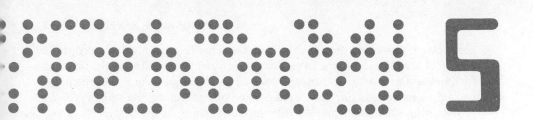

Problem Solving

5.1 INTRODUCTION

Now that we have given some understanding of the functioning of a digital computer and have introduced a language for the programming of a computer, we will turn our attention for most of the remainder of this book to methods for solving particular types of problems. Before doing so, however, it may be well to give some thought to such general questions as:

(*1*) Why does one solve problems on a computer?

(*2*) How does one go about obtaining the computer solution to a problem in an organized and efficient fashion?

One quickly thinks of trivial answers to the first question, such as because one wants the answer. The question merits more profound consideration, however. The cost of computer time is sufficiently great that one must not only be curious to know the answer, but also be willing to pay the price to obtain the answer. By some standards, the answer must have an economic value comparable to its cost or the problem would not be solved. It would lead us too far afield to try to ascribe economic values to computer answers and would probably involve us in a hopeless philosophical quagmire. However, some points can be made that have a bearing on both the choice of material to be presented in this chapter and the approach taken in later chapters.

The first point is that the problems which we solve on a computer using FORTRAN language are problems in mathematics. The answers are found by manipulating numbers. If one were to attempt to make a sweeping classification of the reasons for wanting answers to mathematical problems he could well argue that there are only two basic reasons. The first reason is education, either in the use of computers or in the nature of solutions to particular kinds of mathematical problems. This is certainly an appropriate and valid use for computers. In modern society, a large fraction of the physical facilities and economic resources is devoted to education, a practice which has proved to be sound. For such applications of computers, the steps involved in obtaining the answer may be fully as enlightening as the answer itself.

The second basic reason for wanting answers to mathematical problems is to infer something about the real world. In many fields of science and engineering, the fundamental laws are expressed in mathematical form. Indeed, an area of the physical science is generally considered to be understood when there is a known set of mathematical relationships or equations which can be used to calculate numerical values which agree with measured values in that area. In such technical areas, when one wants to get answers to a physical problem, he simply sets down the mathematical model, that is, the mathematical relationships which govern the physical phenomena of interest, and then treats that mathematical model, on a computer or by hand, to get the answers to his problem. The process sounds quite straightforward, and, in fact, it has been so successful that there are widespread efforts to extend the approach to nearly every area of human knowledge or endeavor: engineering, economics, transportation planning, even behavioral sciences and international relations. The attempt in each area is to define "measurables," quantities for which numerical values can be obtained, and to develop a mathematical model which describes relationships between these measurables in a way which matches observations of the real world. This process has become known by the name of "systems analysis." The problem in systems analysis consists not of simply selecting the proper mathematical model and solving it with the proper input numbers, but of experimenting with the model itself, adding terms to represent different aspects of the problem or deleting terms which on trial have proved inappropriate or unnecessary. Even in physical problems, the mathematical model does not necessarily remain fixed but may be modified during the course of a particular investigation. For example, first calculation of the behavior of an object entering the earth's atmosphere might assume a flat earth and an oversimplified approximation of air density with altitude. Later additions might include round earth, then ellipsoidal earth, then local gravity variation; more realistic air density–altitude relationship, including day–night variation; dependence on sunspot cycle; and so on. In some investigations, answers of adequate accuracy can be obtained by considering only a few major effects. In other cases, however, a combination of a number of secondary effects may have a major influence on results. In these cases, experimentation with the mathematical model is usually required to arrive at a satisfactory formulation.

It would seem that in attempting to answer the first question posed at the beginning of this discussion, we have arrived at two principal uses for computers:

(1) To obtain numerical results for well-defined mathematical problems.

(2) To investigate different possible mathematical models in order to find an appropriate one to describe a real-world system.

The remainder of this chapter will be devoted to answering the second question for each of these computer uses: How does one perform the task efficiently?

5.2 SOLUTION TO MATHEMATICAL PROBLEMS

Classically, the field of numerical analysis has been concerned with finding solution methods for different types of mathematical problems. By a solution method is meant a step by step procedure for obtaining a solution. The step by step procedure, where every step is a well-defined mathematical operation, is termed an algorithm.

5.21 Algorithms

As an example of an algorithm, consider the following.

Example 1. Solve the set of linear equations

$$ax + by = c \qquad\qquad (5\text{-}1)$$

$$dx + ey = f \qquad\qquad (5\text{-}2)$$

where a, b, c, d, e, and f are given constants.

Multiplying the first equation by e and the second by b and subtracting, we have

$$(ae - bd)x = ce - fb \qquad\qquad (5\text{-}3)$$

or

$$x = (ce - fb)/(ae - bd) \qquad\qquad (5\text{-}4)$$

Multiplying the first equation by d and the second by a and subtracting, we have

$$(bd - ae)y = cd - af \qquad\qquad (5\text{-}5)$$

or

$$y = (cd - af)/(bd - ae) \qquad\qquad (5\text{-}6)$$

Expressions (5-4) and (5-6) give the solution in most cases. However, there are certain values of a, b, c, and d for which the steps we took were invalid. In particular, if $ae - bd = 0$, we have divided by zero in going from step 3 to step 4, and from step 5 to step 6, so that the results are invalid. If $ae - bd = 0$, we have a special case in which one of the following conditions apply:

(*1*)   If either c or f is nonzero, the equations are inconsistent, and there is no solution.

(*2*)   If c and f are both zero, then there is no unique solution. Instead there is an infinite family of solutions. These solutions are discussed more fully in Chapter 10.

The above description of the solution constitutes an algorithm for the solution of a pair of simultaneous linear equations in two unknowns. Note that to be complete the algorithm must cover all special cases as well as the general case.

5.22   Selection of an Algorithm

The algorithm used in Section 5.21 for solution of a pair of simultaneous linear equations is by no means the only one possible. For example, if we add the two equations, we obtain the equations

$$(a + d)x + (b + e)y = c + f \tag{5-7}$$

If we subtract the second from the first, we obtain

$$(a - d)x + (b - e)y = c - f \tag{5-8}$$

If we now solve this pair of equations in the fashion used in Section 5.21, we obtain

$$x = \frac{(c + f)(b - e) - (c - f)(b + e)}{(a + d)(b - e) - (b + e)(a - d)} \tag{5-9}$$

$$y = \frac{(c + f)(a - d) - (a + d)(c - f)}{(a - d)(b + e) - (a + d)(b - e)} \tag{5-10}$$

One can calculate the roots by substituting the value of a, b, c, d, e, and f into (5-9) and (5-10), so we have here another algorithm for obtaining the roots. Indeed, if we remove parentheses, combine terms, and cancel in (5-9) and (5-10), these expressions reduce exactly to (5-4) and (5-6). Clearly, this algorithm involves more computation than the first one, so the first one seems to be the preferred choice. In fact, the natural question at this point is: Why even consider other algorithms when we have one which seems quite adequate? The answer is that this algorithm is not adequate in all cases. It will fail for certain sets of input numbers. Consider for example, the set of equations

$$.9998x - y = 1 \tag{5-11}$$

$$x - 1.0002y = 1.0002 \tag{5-12}$$

which by inspection has the solution $x = 0$, $y = -1$. If we were to attempt to use the algorithm of Section 5.21 on a machine which allowed only four decimal places to be carried, we would first compute $ae - bd$, and obtain zero for an answer, and would calculate that the equations are inconsistent. On the other hand, if equations (5-9) and (5-10) are used, the correct answers are obtained. Thus the second algorithm is superior for this particular set of equations. Although the particular example is somewhat contrived, the point to be made is this: For most problems there are several known algorithms, which differ in speed and accuracy, each one of which will fail under some circumstances. There is not a "best" algorithm. In fact, there are not even well-formulated guidelines to say which algorithms should be used in which cases. In later chapters we will describe some of the more common algorithms for solution of various types of problems and try to give some insight into the process of selecting the particular algorithm to be used for a particular problem. To be effective, the problem solver will need to familiarize himself with these algorithms and probably many others. Algorithms are the tools of the problem solver. Knowledge of the contents of the tool chest and understanding of their use are the marks of the craftsman in this field.

5.23   Description of Algorithms

Before the advent of computers, it was customary to describe algorithms, as we have done above, by means of word description and equations. In the computer age, other means of describing algorithms have come into general use. In particular, the computer program itself is a description of the algorithm in terms that can be interpreted by the computer. If the program is in a higher language, such as FORTRAN, it can be used for communication of the algorithm from human to human, as well as from human to machine. For example, to the person acquainted with FORTRAN, the program below describes the solution to Example 1 as adequately as did the written description given above.

```
    READ 100,A,B,C,D,E,F
    DEL=A*E-B*D
    IF(DEL)20,10,20
10  IF(C)30,15,30
15  IF(F)30,16,30
16  PRINT 101
    STOP
20  X=(C*E-F*B)/DEL
    Y=(C*D-A*F)/DEL
    PRINT 102,X,Y
    STOP
```

    30 PRINT 103
        STOP
   100 FORMAT(6E12.4)
   101 FORMAT(19H NO UNIQUE SOLUTION)
   102 FORMAT(3H X=,E12.4,10X,2HY=,E12.4)
   103 FORMAT(27H EQUATIONS ARE INCONSISTENT)
        END

The program first calculates $ae - bd$. If this quantity is zero, it checks c and f. If both are zero, it prints "NO UNIQUE SOLUTION." If one or both are nonzero, it prints "EQUATIONS ARE INCONSISTENT." If $ae - bd$ was nonzero, the program solves for x and y and prints the values.

A correctly written computer program is indeed a precise way to describe an algorithm, for each step assuredly has a clearly defined, unambiguous meaning. However, even a rather simple program, such as the one given above, takes time and study to absorb the meaning and content. Longer programs can become quite tedious in this respect. For human communication somewhat higher visibility and ease of comprehension is provided by another method of presenting an algorithm, the flow chart. The next section describes the construction and use of flow charts.

5.24   Flow Charts

A flow chart is a pictorial representation of an algorithm in which the steps of the algorithm are enclosed in boxes of various shapes and the flow of control is indicated by lines joining the boxes. There is some variation within the literature on the meanings of boxes having different shapes, but this is quite a minor matter, since the boxes are so self-explanatory anyhow. We shall follow the lead of that school which tends to minimize the number of different symbols used. Flow charts of complete programs will begin at a box labeled START and end at a box labeled STOP. Flow charts of segments of programs will begin at a box labeled ENTER and end at a box labeled EXIT. The progression from beginning to end is traced by following the connecting lines through the other boxes in the chart. The other types of boxes are as follows.

(*1*) *The assertion box.*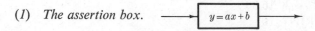

This box asserts that the operations contained therein are to be executed at this time. These operations may involve the calculation of a numerical quantity, the replacement of a quantity by another, the input of certain quantities into the memory, or the printing of certain results. The above assertion box indicates that the value of y is to be computed by taking a times x, plus b, the quantities a, x, and b already stored in the memory.

(2)   *The test box.* $x - 16 < 0$?   yes

no

This box represents the use of a conditional transfer instruction of some type. It has one input line and two output lines. Either output is selected, depending on whether the answer to the question contained in the box is affirmative or negative. It represents a branching point in the program.

(3)   *The remote connector.*   ③

This circle indicates that the logical control is transferred to the point at which another such circle appears with the same number in it. It is used to keep the flow chart uncluttered by connecting lines crisscrossing the diagram to join boxes.

(4)   *The variable remote connector.*   ⓐ

This circle indicates that the point to which control is transferred is set by the program itself. Somewhat prior to the position of this symbol in the flow chart there must be an assertion box which assigns a numerical value to α. Then this connector will indicate a transfer to the entry circle with that particular numerical value in it.

The use of the symbols just described will be demonstrated with a few examples.

Example 1.   Draw a flow chart for the calculation of $y = a^2 + ab + b^3$.

Figure 5-1 shows a suitable flow chart. This flow chart is extremely simple, requiring only assertion boxes. Since each of the boxes is a simple assertion, there is no real advantage in having a separate box for each item. All the statements could be placed in a single box, as in Figure 5–2. When several statements are included in a single box, they should be taken in order from top to bottom.

Example 2.   Draw a flow chart for the calculation of $y = ax + b$ for a set of given values of x.

A suitable flow chart is shown in Figure 5–3.

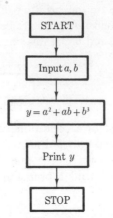

Figure 5-1

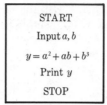

Figure 5-2

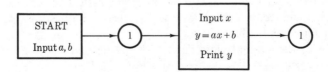

Figure 5-3

Example 3. Draw a flow chart describing the solution of the pair of simultaneous linear equations

$$ax + by = c$$
$$dx + ey = f$$

(See Example 1, Section 5.21.)

A suitable flow chart is shown in Figure 5-4. Comparing this flow chart with the FORTRAN program of Section 5.21, we see that the two do describe the same algorithm, but the flow chart does so a little more clearly and succinctly.

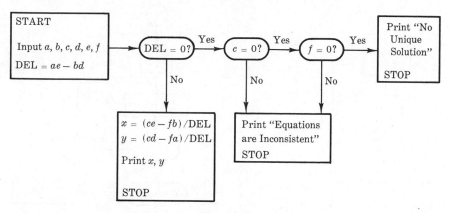

Figure 5-4

Example 4.   Draw a flow chart for a program that will input sets of values x_1, x_2, x_3 and compute and print y_1, y_2, y_3 according to the equations

$$y_1 = a_{11}x_1 + a_{12}x_2 + a_{13}x_3$$
$$y_2 = a_{21}x_1 + a_{22}x_2 + a_{23}x_3$$
$$y_3 = a_{31}x_1 + a_{32}x_2 + a_{33}x_3$$

A FORTRAN program for this problem was given in Example 2, Section 4.63. A suitable flow chart is shown in Figure 5-5. This flow chart contains some features not present in Example 3. One is the use of the arrow in several boxes, as in $j + 1 \rightarrow j$. This expression would be read "$j + 1$ replaces j. Another feature of this flow chart is the presence of a number of loops, indicated by a test box and a return line to an earlier point in the chart.

Comparison of this flow chart and the FORTRAN program of Example 2, Section 4.63, indicates that while the two describe the same basic algorithm, the FORTRAN program in this case manages to do the job more succinctly. This is because the flow chart has described the loops in more detail than does the FORTRAN. Flow charts may be detailed or gross, depending on their purpose. Unless we are programming in machine language, we probably do not really care how the incrementing of i and j and the tests are handled in these loops, as long as the tasks are indeed performed. Figure 5-6 gives a less detailed flow chart for the same calculation. In this flow chart, some shorthand has been used to save space and effort in describing the loops. The loops for input and printing have been described by simple assertions. This practice is frequently followed in drawing flow charts. The loops for calculating the y_i's use a less common, but quite convenient shorthand, the use of a remote connector inside an assertion box. The symbol

$$\rightarrow \textcircled{2}\, i = 1,3$$

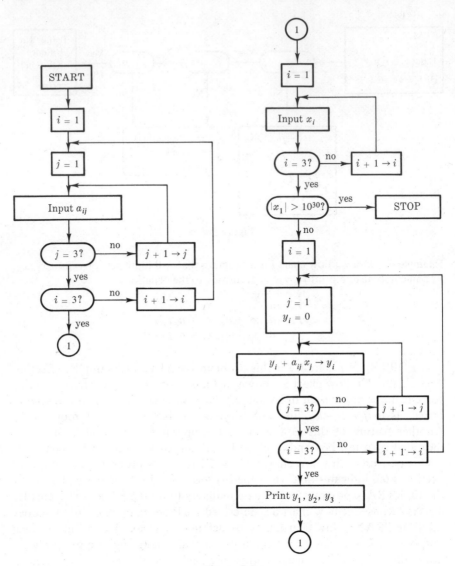

Figure 5-5

means continue through the flow chart using $i = 1$ until remote connector 2 is reached, then return to this symbol and trace the same path with $i = 2$, then with $i = 3$. It is seen that this usage is analogous to the DO loop in a FORTRAN program.

The second symbol,

$$\rightarrow ② j = 1, 3$$

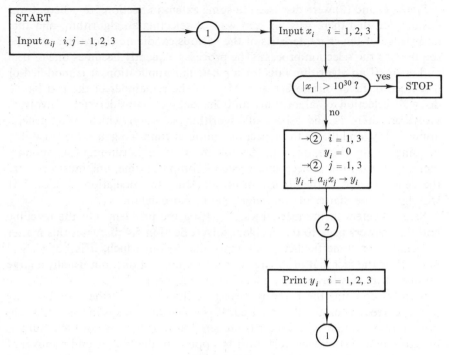

Figure 5 6

indicates a second loop terminating on the same symbol. This inner loop is to be performed three times, for $j - 1$, 2, and 3, for each value of j. We will use this abbreviated representation for loops where convenient in the remainder of the text.

EXERCISE 14

Draw flow charts for the problems of Exercise 11, Chapter 4.

5.3 PROGRAM CONSTRUCTION AND CHECKOUT

The steps in solving a mathematical problem on a computer might be summarized as follows:

(*1*) Select an algorithm.

(*2*) Express the algorithm as a program in FORTRAN or other language acceptable to the machine.

(*3*) Communicate the program to the machine.

(*4*) Perform a checkout calculation to debug the program.

(*5*) Perform the desired calculations.

Steps (*1*) and (*2*) were discussed to some extent in Section 5.2, where it was pointed out that there was no sure way of selecting an algorithm, and that knowledge of the characteristics of the various candidate algorithms is a prerequisite to the selection process. The process frequently becomes one of trial and error. If an algorithm fails for a particular application, it is modified or discarded in favor of another one. Most of the remainder of the text has to do with selection of algorithms and the detailed considerations involved. However, there is one aspect of algorithm selection which is of general applicability, and that is the aspect of computer time. To gauge the feasibility of using an algorithm, one must be able to estimate its running time requirement, and, to avoid unnecessary waste of computer time, one must organize the detailed steps of the algorithm for efficiency in calculation. Section 5.31 will discuss the matter of computer time in more detail.

Step (*3*) refers to the mechanics of getting the problem into the machine and the answers back to the problem solver. Section 5.4 discusses this matter in detail after some further preliminary discussion, which, it is hoped, will make clear the fact that in solving a problem, not just one, but usually a large number of passes through the machine are required.

Step (*4*) contains the perhaps surprising inference that the program may not be correct as originally prepared. It is a fact that this is indeed the situation more often than not. To err is human, but to forgive is not the nature of of a computer. Attentive effort must be applied to the finding and removal of errors. Section 5.32 describes some of the steps which can be taken toward this end.

5.31   Computer Time

If there is a single aspect of the digital computer which is most responsible for its growth and development in the last several years it is the speed with which it can calculate. This speed allows not only standard problems to be solved more quickly and easily than would be the case with hand calculators, but also the solution of complex problems whose solution would not even be attempted or considered without computers. If it takes a man 10 seconds to multiply two eight-digit numbers, the man would not even consider attempting a calculation involving 1 billion multiplications by hand, for it would require working 8 hours a day for nearly 100 years. If a computer can perform a multiplication in 1 microsecond, this same calculation would require less than 20 minutes on the computer, so the problem is quite feasible on the computer. On a slower computer which requires 1 millisecond to perform a multiplication, this same calculation would require 1 million seconds, or a little over 10 days at 24 hours a day, or roughly 1 month utilizing 8-hour days. Thus the time requirement for a particular problem determines whether or not it may be feasible to perform the calculation on a particular machine. Since machine time is usually obtainable at a certain cost per hour (which varies greatly with the size and speed of the machine), the time required for the calculation determines the cost of obtaining the answers, and even for a calculation which is quite feasible the answers may not be worth the cost involved. Hence it is worthwhile to be able to estimate the time required to perform the program.

It would be extremely difficult to make an accurate prediction of the time requirement, and such a prediction is not needed in most cases. A crude estimate, sufficient for most purposes, can be made by simply counting up the number of arithmetic operations to be performed and multiplying by some nominal execution time characteristic of the machine to be used. For the examples in this text, we will assume an average time of 1 microsecond for add, subtract, multiply, and divide operations and ignore all other instructions. This will give a rough gauge as to the relative feasibility of various algorithms on current computers. If one has in mind a particular machine, one can get a similar crude estimate by using the appropriate "average" instruction time for that machine. Internal machine characteristics and operating systems are so varied that any more refined estimate would require specific consideration of the physical installation to be used and might require timing runs on the computer itself for calibration purposes.

Example 1.   Estimate the time required to compute the value of $y = 1 + 2x + 3x^2 + 4x^3 + 5x^4$ for 5000 different values of x.

If one merely computes term by term, there is one multiplication for the second term, two for the third, three for the fourth, and four for the fifth, for a total of 10 multiplications. With the 4 additions, this gives 14 operations. At 1 microsecond per operation, the 5000 values will be computed in .07 second.

Thus far, we have said nothing about the time required for input and output. I/O takes place simultaneously with computation in modern computers, so that I/O time is not a factor unless it becomes excessive. In the above case, none of the peripheral devices listed in Chapter 2 could handle the input of 5000 values of x and output of 5000 values of y in .07 second. The problem is I/O bound, in that the I/O requirements are more time consuming than the computational requirements. In this case, if the supervisory software is arranged to go on to another program any time the processor is not busy on this program (that is, operation in a multiprogramming mode), no time need be wasted. If, however, all the problems available to the machine are I/O bound, the processor will be forced to sit idle some portion of the time while waiting for input or output to take place. With the cautioning word that excessive I/O requirements can cause misestimates of machine time, we will neglect I/O in the running-time estimates in this text.

Looking again at Example 1, we note that the number of arithmetic operations required can be reduced substantially by a regrouping of the terms.

If the relation is written

$$y = 1 + x(2 + x(3 + x(4 + 5x)))$$

it can be evaluated in only eight operations, a saving of almost a factor of two. Whatever else can be said about machine time, it can certainly be said that it should not be wasted on needless calculation. The proper grouping of terms, as demonstrated above, is one way to increase efficiency in use of machine time. Another way is in the programming of loops.

When loops are used in a program, care should be taken to include only required calculations in the loop. The two solutions given for Example 2 below illustrate how the total number of multiplications performed in a program is sometimes drastically affected by the way in which a loop is constructed.

Example 2.   Write a FORTRAN program that will input values of constants a, b, and c, and compute and print $y = a^2 + b^2 + 2cx$ for $x = 1, 2, \ldots, 1000$.

One program is

```
      READ  101,A,B,C
      X=0.
      DX=1.
      DO  2  I=1,1000
      X=X+DX
      Y=A*A+B*B+2.*C*X
    2 PRINT  101,X,Y
      STOP
  101 FORMAT(3E12.4)
      END
```

In this program, the products A*A, B*B, and 2.*C are each formed 1000 times, instead of just once, because they occur inside a DO loop. Nearly 3000 unneeded multiplications are performed by the program. These unneeded multiplications are avoided if the program is written as follows:

```
      READ  101,A,B,C
      SQ=A*A+B*B
      TWOC=2.*C
      X=0.
      DX=1.
      DO  2  I=1,1000
      X=X+DX
      Y=SQ+TWOC*X
    2 PRINT  101,XY
      STOP
  101 FORMAT(3E12.4)
      END
```

5.32   Correctness of the Program

It is obvious that a program containing an erroneous command, or even an erroneous digit in the address portion of some command, will probably produce incorrect results. It might be thought that we should dispose of this problem with the simple statement that one must use great care in programming and must recheck the program very carefully. However, the larger calculations usually involve such a long and complicated sequence of commands, and the human tendency to err in writing such a sequence is so great that no amount of review of the program can be depended upon to ensure that the program actually does what it is supposed to do. This is true even when the program is written in an abbreviated form, as in FOR-TRAN. In nearly every case, it is advisable to check the program by comparing the results of a few cases with the results of hand calculations. This does not mean that it is necessary to duplicate all the machine operations by hand—indeed, this would be impossible from the standpoint of time. Ordinarily, however, the program requires the machine to repeat the same sequence of instructions hundreds or thousands of times, and if they are performed correctly once they can ordinarily be presumed to be performed correctly thereafter. Furthermore, if the program gives correct answers with one set of input numbers, it will generally give correct answers for other sets. (This is not an infallible rule, as we shall see shortly.) Thus a check calculation can usually be prepared which will utilize numbers that make for simple hand calculations but which will serve to check that the machine is performing all the steps of the calculation properly. In preparing such check calculations, it is desirable to choose numbers that make the hand calculation simple, but care must be taken lest the numbers chosen be so simple as to fail to check the entire operation.

No universally applicable set of steps can be given for the construction of check calculations, but the following examples will illustrate some of the considerations involved.

Example 1.   Prepare a check calculation for the problem

$$y = \frac{1 + x}{1 + 2x}$$

This particular illustration is so simple as to be almost trivial, but it will serve to illustrate a few points. The simplest check is to feed $x = 0$ into the machine as the input value for x; the computed result should be $y = 1$. However, the fact that the machine produces $y = 1$ for $x = 0$ is not very much of a check on the correctness of the program. In fact, the machine may be performing a completely wrong operation with the term $+x$ in the numerator and the $+2x$ term in the denominator, but because the test value $x = 0$ eliminates these terms, we do not see the indications of such errors. The program might be doing

$$y = \frac{1}{1} \quad \text{or} \quad y = \frac{1 + x^2}{1 + 4x}$$

or any number of other calculations instead of the desired ones and our check calculation would not indicate anything amiss. Thus it is dangerous to use $x = 0$ as a check value in this case; and in general it is dangerous to use numbers that will cause part of the terms to vanish, unless separate checks are arranged to make sure these terms are computed correctly.

In the present example, one might try $x = 1$ for a check value. This should give the result $y = 2/3$. If this value of y is obtained, we have some confidence in the program. Yet there are some relatively simple mistakes in programming that would produce the correct value of y for this case even with an erroneous program. For example, the expression

$$\frac{1 + x^2}{1 + 2x}$$

would give the correct answer for this check value. Values of x other than zero or 1 are less likely to give a false check on the calculations. For example, we might choose $x = 4$, giving $y = 5/9$. Still the check given by any single value is not absolute, and the use of two or more check values is frequently advisable.

Example 2.   The following program is designed to input values of a quantity u and calculate y from the formulas

$$y = \frac{1 + u}{1 + 2u} \qquad \text{if } u > 0$$

$$y = 1 + 4u + u^2 \qquad \text{if } u \leqslant 0$$

Write input data for a check calculation for this program:

```
1 READ 101,U
  IF(U−1.E30)2,2,3
2 IF(U)4,4,5
4 Y=1.+U*(4.+U)
  GO TO 6
5 Y=(1.+U)/(1.+2.*U)
6 PRINT 101,Y
  GO TO 1
3 STOP
101 FORMAT(E12.4)
  END
```

In this program there are two arithmetic statements, statements 4 and 5. Statement 5 is the same equation discussed in Example 1, in which it was seen that there are a few test values that would be poor choices. A value mentioned as a satisfactory one for checking that calculation was 4. Thus $U = 4$ would be a good test value for the program. At least one other test value should be used also, a negative value of U. This is required for two purposes: first, to check the arithmetic statement 4; and second, to check statement 2. Note that statement 2 decides which equation will be used to compute Y. Even if both statements 4 and 5 are correct, we can obtain a wrong result if statement 2 picks the wrong equation. For example, it happens that, on the standard key punch keyboard, the character U and the numeral 1 are on the same key, U being the lowercase symbol and 1 being the uppercase symbol. Thus it would be quite easy accidentally to print statement 2 as

2 IF(1)4,4,5

If this were done, the program would run but would use the formula in statement 5 for both positive and negative U. The use of both a positive and a negative value of U as check values, say $U = 4$ and $U = -2$, would provide reasonable assurance that the program is correct. These inputs should cause the machine to give the results $Y = 5/9$ and $Y = -3$.

The above examples demonstrate that a good check calculation should check control statements as well as arithmetic statements. In many problems the checking of control statements is a much more imposing task than the checking of arithmetic statements but is one that cannot safely be ignored.

It is frequently advisable to include extra instructions (or FORTRAN statements) within a program for the sole purpose of assisting in the checking of the program for correctness. These instructions most often take the form of extra print instructions which will cause the machine to print out inter-mediate results that can be used for check purposes, or error stop instructions so that the machine will halt at some intermediate point in the program and print a message indicating the stop location. Such aids not only assist in determination of whether the program is right or wrong, but also assist in determining in what section of the program the faulty command or com-mands may be. For lengthy programs this can be a most valuable aid. When the program is finally checked, these extra commands are removed, or in some other fashion ignored, for the "production runs," the final calculations using the real problem input data.

In FORTRAN, many checks are built into the compiler itself. For example, the compiler will ordinarily refuse to compile a program in which there are expressions mixing fixed- and floating-point expressions, or one having improperly nested DO loops, or IF statements referring to non-existent statement numbers, and so on. In fact, a good compiler is so efficient in finding programming errors that it is a temptation to assume that a program which compiles successfully must be correct. This is not a safe assumption, and the experienced programmer has usually learned by sad experience that a program should not be trusted until checked and rechecked in every reasonable way.

EXERCISE 15

1. Using the timing information given in Section 5.22, estimate the machine time requirement for the following problems.

    a. $y = 1 + 3x^2 + 5x^4$ for 5000 unequally spaced values of x.

    b. $y = x_1 + 2x^3 + 7x^5 + 4x^7$ for 10,000 equally spaced values of x.

    c. $y = \dfrac{1 + 6x + 9x^3 + x^6}{2 + 4x^2 + 5x^5}$ for 10,000 equally spaced values of x.

    d. $y = x_1 + 4x_1x_2 + 3x_1{}^2x_2{}^2 + 5x_2{}^3$ for 5000 different *pairs* of values x_1, x_2.

    e. $y = x_1{}^2 + x_2{}^2 + x_3{}^2 + x_4{}^2 + x_5{}^2$ for 1000 different *sets* of values x_1, x_2, x_3, x_4, x_5.

2. Write a FORTRAN program for each of the calculations indicated in problem 1.

3. Write a set of input data that will check both the arithmetic and the book-keeping portions of the calculation for each of the programs in problem 2. Indicate two errors that could exist in your program and not be discovered by your check calculation.

5.4 COMMUNICATING WITH THE COMPUTER

For many years the almost universal way of submitting a problem to the computer was to punch the program statements onto punch cards, which were then read into the machine through the card reader and the results printed out on a line printer. Typically, a backlog of programs waiting to be run would build up during the busy hours of the day, so that the problem solver would wait a few hours or even a day or more for results. Such a turnaround time from program submission to receipt of answers would be tolerable in most cases if the results represented the final answer to the problem, but in most cases they do not. Usually, the problem solver makes some change in the program and immediately resubmits it, for one of four reasons:

(*1*) An error in the program was uncovered, and a corrected version must be run.

(*2*) The answers indicate that another run with different input numbers would be more informative than the run just made.

(*3*) The algorithm proved defective and must be modified or replaced.

(*4*) The mathematical model used was only a trial one, and a trial with a modified model is desired.

Thus before the problem is considered solved, it may pass through the machine tens or hundreds of times, with some changes in the program almost every time. In this circumstance, a long turnaround time starts becoming a serious impediment to solution of the problem, and pressures mount for more rapid ways of accomplishing program change and rerun. These pressures have led to the development of more rapid means of communication between problem solver and computer, ways which allow the program to be treated as a dynamic entity rather than a fixed entity, and to be modified at will within the machine itself rather than physically removed and resubmitted for each change. Further, as described in Chapter 2, supervisor programs have been developed which make more efficient use of the machine time and which handle the queuing of programs at the machine.

5.41   Manipulation of the Source Program

As was disclosed in Chapter 2, it is possible to write a service program which will accept almost any set of symbols in the computer and manipulate them according to any prescribed set of rules. One application of this capability is to the preparation, maintenance, and updating of a program written in FOR-TRAN or other source language.

An interpretive routine can be written which allows these operations to be done from a remote terminal. Several commercial systems based on remote-terminal usage are in existence. Generally speaking, they operate as follows: The user turns on the teletypewriter or other console at the remote site and dials the telephone number of the computer if the connection is by phone line or signals by whatever means is supplied if the connection is a private line. At the computer, a device on one of the I/O channels is activated by this signal and sends an interrupt signal over the channel to the computer. The supervisor program services this interrupt, and causes the interpretive routine to be activated. The interpretive routine generates a message, such as "USER NUMBER?", and sends it back over the channel, where it is transmitted to the remote terminal and printed out or displayed on a cathode-ray scope. Meanwhile, the interpretive routine has sent an interrupt to the supervisor program, indicating that it is through for the moment, and the supervisor puts the processor back to whatever calculation it was performing. The interruption occupied the processor for perhaps a few tens of microseconds. It will probably be several seconds before the user at the other end can respond and require anything else from the computer. Meanwhile, every little while the supervisor will call on the interpretive routine to check and see if a response has come back over the channel. If not, it will simply return control to the supervisor. If a response has come back, it will check the response against the authorized list of user numbers stored in the routine itself. If a match is found, it will send back another message, such as "READY." If no match is found, it will perhaps send back a message such as "UN-AUTHORIZED USER" and sever the connection to the remote terminal. If a "READY" was given, it will also request the supervisor to set aside a section of memory for workspace for the remote terminal. The user can then type entries which the interpretive routine will enter into this workspace. He can type entries which will cause the interpretive routine to bring into the workspace a file or files from auxiliary storage such as a disk file, or to store the contents of the workspace onto the disk, or to list them out at the remote terminal, or to call on the compiler to compile and run an object program from a FORTRAN program stored in the work space, or to perform any number of other duties, depending on the degree of sophistication of the interpretive routine.

The set of commands which the interpretive routine will handle is itself another form of higher language. No standard form for this language has yet emerged. For purposes of demonstration, we will define a set of commands sufficient for the examples for this text. The commands selected are similar to those used in systems such as the General Electric Time Sharing System and to those in the International Business Machines APL Terminal System, but do not correspond exactly to either of these. Specific commands in this language are described in the next two sections.

5.42   A Language for Remote Terminal Operation

Entries are made into the computer by typing the appropriate set of characters on the keyboard at the terminal, followed by a carriage return. Nothing is transmitted to the computer until the carriage return is depressed. When this is done, an interpretive routine within the computer examines the characters transmitted and takes action in accordance with the rules given below.

Desired Computer Action	User Action at Terminal	Notes
Store a line of information in the workspace.	Type a number, 10 digits or less, then the information to be stored, then a carriage return.	This is the method for entering FORTRAN statements into the workspace. The number serves as a sequence number to indicate where in the program the statement belongs (see Example 1, below). If a sequence number is reused, the old entry bearing that number is destroyed and the new entry replaces it.

List the contents of the workspace.	Type the characters LIST, then carriage return.	
Compile and run the program in the workspace	Type the characters RUN, then carriage return.	
Transfer the contents of the workspace to a file in permanent storage for later use.	Type SAVE, followed by a space, followed by one to five letters or numbers, followed by a carriage return.	The one to five letters or numbers form a reference name, which must be used to recall the program. If the reference name has already been used under this user number, the information previously stored is dropped and the new file replaces it. The interpretive routine will return the message "READY" when the task is completed.
Transfer a previously stored file from permanent storage to the workspace.	Type LOAD, followed by the file name, followed by a carriage return.	If the file name given does not correspond to any that have been used under this user number, the interpretive routine will give the message "NO SUCH FILE." If it finds the file it will return the message "READY."
Discard a file previously stored in permanent storage.	Type DROP, followed by a space, followed by the file name, followed by a carriage return.	The interpretive routine returns the message "READY" when action is complete.
List the names of all files currently stored under this user number.	Type LIB, followed by a carriage return.	
Terminate the operation.	Type BYE, followed by a carriage return.	Releases workspace, disconnects line.

Any entry other than those given in the table will cause the interpretive routine to respond with the message "WHAT?". The examples below will serve to clarify the language just defined.

Example 1.   Write the FORTRAN program of Example 2, Section 5.32, as it would be typed in at a remote terminal.

A suitable arrangement would be

```
 1        1 READ 101,U
 2          IF(U−1.E30)2,2,3
 3        2 IF(U)4,4,5
 4        4 Y=1.+U*(4.+U)
 5          GO TO 6
 6        5 Y=(1.+U)/(1.+2.*U)
 7        6 PRINT 101,Y
 8          GO TO 1
 9        3 STOP
10      101 FORMAT(E12.4)
11          END
```

It is understood that a carriage return is given at the end of each line. Also, in listing the program, for easy visibility we have used enough blank spaces in each line to line up the FORTRAN statements. This is not necessary. A single space will suffice. Note that the program appears exactly as it did in Example 2, Section 5.32, with the exception of the sequence numbers appearing in front. The sequence numbers, or line numbers, serve to specify the order in which the statements are to be arranged. If we want to insert another statement between lines 4 and 5, we can simply type a number between 4 and 5, for example, 4.1 or 4.5 or 4.9, and then the statement iself. The statement will automatically be inserted between the proper pair.

Note also that there is no connection between the line numbers and the statement numbers. The two serve entirely different purposes and cannot be combined in any way.

Note further that it would not have been necessary to type the statements in the order listed above. They could have been typed in any order, and as long as the same statements were paired with the same sequence numbers as above, the result would have been the same. A LIST command would still bring the statements out in the order given above.

Example 2.   In typing the program of Example 1, line 5 was inadvertently typed as

$$5 \text{ GO FO XO}$$

How could this be corrected?

The correction can be made by merely typing the line again, correctly:

$$5 \text{ GO TO 6}$$

Example 3.   It is desired to store the program of Example 1 for future use under the name YCAL1. What command would be typed in?

The proper command would be

SAVE YCAL1

Example 4. It is desired to run the program of Example 1 with the following values for U:

$$U = -3., 5.8, \text{ and } -16.71$$

The command to compile and run the program is simply the word RUN. However, before typing in this command, we would have to face the question, how and where will the program get the input numbers? The answer is that the READ statement will cause the computer to look in the workspace immediately after the END statement for the input quantities. Thus we should first type in the line

$$12 \quad -3., 5.8, -16.71, 1.1E30$$

and then type the entry

RUN

The final input number, 1.1E30, is to cause the IF statement, sequence 3, to route us to the STOP.

Note that the format of the input numbers does not match the specification in the FORMAT statement. It is customary in compilers for remote terminal use that the commas override the field length specified in the FORMAT statement and the decimal point overrides that indicated in the FORMAT statement.

The requirement that the data be stored in the workspace immediately behind the program is convenient in some cases but not in others. Modifications to FORTRAN which improve remote terminal capability will be described in Section 5.43. One of these will provide an alternative method of data input.

Example 5. It is desired to terminate the problem and leave the terminal. What action should be taken?

The command BYE should be typed in.

EXERCISE 16

1. You are the user who stored the program of Examples 1 through 5, Section 5.42. You aren't sure, but you think you stored it under the name YCAL5, so you type in "LOAD YCAL5." What message do you get back?

2. You decide to look at the names of the files you have stored. What message do you type in?

3. After determining that the correct file name was YCAL1, you type in "LOAD YCAL1." What message do you get back?

4. Your program is now in active workspace, and you type in "LIST" to list out the program. Will the line with sequence number 12, which was entered in Example 4, appear on this listing? Why?

5. You wish to run the program for the following values of U: 4.38, −8.71, 2.15. What do you type in?

6. You wish to modify the program so that the statement numbered 4 reads Y=3.+U*(5.+U). What do you type in?

7. You wish to add between statements 5 and 6 a statement which reads Y=Y*U. What do you type in?

8. You wish to run this modified program with the same data used in Exercise 5. What do you type in?

9. You wish to terminate the communication. What do you type in?

5.43   Modification of FORTRAN for Terminal Use

It was indicated in Section 5.42 that terminal use could be made somewhat more convenient by a few modifications to FORTRAN. We will introduce three modifications for this purpose.

The first modification will be introduction of an unformatted PRINT statement. It is of the form

PRINT, variable name, variable name, ...

Each variable named will be printed at the terminal: as an integer in I12 format if stored as an integer; as a decimal in F12.x format, where x is whatever value is needed to make seven significant figures be shown if a real number between 10^{-4} and 10^4; and as a floating-point number in E12.7 format if it is a real number outside the range 10^{-4} to 10^4. In addition, the unformatted PRINT will treat everything inside quotation marks as Hollerith characters.

Example 1. If a program has stored the value 1.874651 for the variable X and 300 for the variable N, what printout will be caused by the following statement?

PRINT, "VALUE=",X,"NUMBER=",N

The printout will have the appearance

VALUE=   1.874651 NUMBER=   300

The spacing given by an unformatted PRINT will not be the neatest possible, but the ease of writing the statement makes it very handy.

A second modification to normal FORTRAN will be use of the INPUT statement. The statement

INPUT, variable name, variable name, ...

will cause the machine to print a question mark at the terminal and wait for values for the variables named to be typed from the keyboard. The typing is free field, with the fields separated by commas. The values, integer or real, are typed in accordance with the rules for FORTRAN constants given in Section 4.2. It is usually advisable to precede an INPUT statement with a print statement which will remind the user of what values are required as input. For example, the pair of statements

PRINT, "X,N"
INPUT, X,N

will cause the machine to print

X,N
?

and wait for input from the typewriter. Without the print statement, the input statement alone would have caused the machine to print

?

and this might leave the programmer at a loss as to the input required. Of course, the actual FORTRAN names do not have to be printed as a reminder. Descriptive words are often better. The statement

PRINT, "COST PER PIECE,NUMBER OF PIECES"
INPUT, X,N

will give the message
COST PER PIECE, NUMBER OF PIECES
?

and when the entries are typed in, it will still assign them to the FORTRAN variables names X and N.

Example 2.   Modify the program of Example 1, Section 5.42, now stored under the file name YCAL1 to accept its input from the keyboard and store the modified program under the file name YCAL2.

A suitable set of keyboard entries is

    LOAD YCAL1

And after the machine responds READY,

    1   1 PRINT, "U"
    2   INPUT, U
    9
    SAVE YCAL2

We wiped out the old lines 1 and 2 with the PRINT and INPUT commands.
The IF test at line 2 is not needed since we are controlling each piece of input
individually from the terminal. If the letter S is typed in response to the
INPUT request, instead of a number, this will be accepted as a stop in the
program. Since the STOP instruction at line 9 was no longer needed, we have
removed it by simply typing the number 9, then carriage return, with no
statement.
    If we were to now give the LIST command, we would obtain

    1   1 PRINT, "U"
    2   INPUT, U
    3   2 IF(U)4,4,5
    4   4 Y=1.+U*(4.+U)
    5   GO TO 6
    6   5 Y=(1.+U)/(1.+2.*U)
    7   6 PRINT 101,Y
    8   GO TO 1
    10   101 FORMAT(E12.4)
    11   END

If we now typed RUN, the machine would type

    U
    ?

If we were to type 1., carriage return, the machine would type

        .6666667
    U
    ?

The machine would have taken U = 1, done the IF test, gone to statement 5,
calculated Y, executed the PRINT at statement 6, gone back to statement 1,

and again requested input. If we now type in 2., carriage return, the machine responds

        .5000000
U
?

We can test as many values of U as we wish in this manner and, when we are through, terminate proceedings by typing the letter S when a value for U is requested.

Example 3. Show all commands and computer responses for entering from remote terminal a program which will ask for x and, when given a value, will return the value $\sqrt{1 + x^3}$. Further, show the entries and responses for running this program with $x=1$ and $x=2$, then stopping the program and terminating the use of the computer.

A typical sequence is shown below, with computer responses in boldface type and keyboard entries in ordinary type. The user number is assumed to be 54321.

```
        USER NUMBER? 54321
        READY
1   1   PRINT, "X"
2       INPUT, X
3       Y-SQRT(1. + X**3)
4       PRINT, Y
5       GO TO 1
6       END
RUN
    X
?   1.
            1.414214
    X
?   2.
            3.000000
    X
?   S
READY
BYE
```

A third modification to FORTRAN that will be advantageous for terminal use is one that will allow material on some other file in permanent storage to be used in addition to the material in active workspace. This is accomplished by the statement

    $USE XXXXX

where XXXXX is the name of the file desired. This statement will cause the contents of file XXXXX to be inserted at the point where the $USE statement appears when the program is compiled and run.

Example 4. Five hundred numbers have been stored in a file named NUM. We wish to run the program of Example 1, Section 5.42, now stored under the name YCAL1, with all these numbers as inputs. What commands must be typed in at the keyboard?

The command LOAD YCAL1 will bring the program to the active workspace. When the reply READY is received, we would type

```
12 $USE NUM
13 1.1E30
RUN
```

The statement 12 $USE NUM would cause the 500 numbers in file NUM to be inserted after the end of the program, where they would be picked up as data by the READ statement. They would be followed by a final item of data, the value 1.1E30 inserted at line 13, which will cause the IF statement at line 2 to terminate the program.

Example 5. Under a file named UTIL, we have stored some utility subprograms, including one named FUNCTION CDIS(XLAT1, XLON1, XLAT2, XLON2) which, given the latitude and longitude of two points on the surface of the earth, computes the great circle distance between them. Write a program for remote terminal use which will allow one to feed in successive locations and obtain the accumulated distance along the path.

A suitable program is

```
 1   1 PRINT, "INPUT INITIAL LAT AND LONG"
 2     INPUT, XLAT1, XLON1
 3     DIS=0
 4   2 PRINT, "NEXT LAT AND LONG"
 5     INPUT, XLAT2, XLON2
 6     IF(XLAT2-500.)3,1,1
 7   3 DIS=DIS+CDIS(XLAT1,XLON1,XLAT2,XLON2)
 8     XLAT1=XLAT2
 9     XLON1=XLON2
10     PRINT, "DIS= ",DIS
11     GO TO 2
12     END
13     $USE UTIL
```

The $USE statement is employed to make function CDIS available to this program. A slight touch of elegance has been added at line 6. An IF test has been included that, if the user types in a latitude of 500 or greater, will cause a jump to 1 and a restart of the problem. This is quicker and easier than stopping the program and forcing the machine to recompile by using another RUN command.

5.44   Languages for Direct Interaction

As we have described the interpretive routine for handling data terminal use, it handles only a very few commands dealing with the preparation and submission of FORTRAN programs and does not actually handle any direct computation. It would be possible to expand the routine to the point where it can recognize and interpret arithmetic statements directly. For example, if we typed in

$$X = 4$$

the routine would immediately set up a memory location for storing X and store in the value 4. If we then typed

$$Y = 1 + X * X$$

the routine would compute and store $Y = 17$. If we then typed

PRINT, Y

the routine would cause the value of Y to be printed. An interpretive routine which works in this way gives a very high degree of responsiveness in performing calculation, particularly those of a somewhat elementary nature. It turns out that FORTRAN is not the most convenient language for this direct interaction. Other languages, such as Iverson's APL language* or the Rand Corporation's JOSS language, have been devised for this type of operation. For our purposes of examining algorithms for problem solving, the FORTRAN language as modified for terminal use will suffice without the use of languages geared for more direct man-machine interaction.

EXERCISE 17

1. Write FORTRAN programs for the following problems for remote terminal use, giving sequence numbers to the statements and utilizing unformatted

* K. E. Iverson, *A Programming Language*, John Wiley & Sons, Inc., New York, 1962.

PRINT and INPUT statements instead of the conventional PRINT and READ statements.

    a.   Example 1, Section 4.41.
    b.   Example 2, Section 4.41.
    c.   Example 1, Section 4.8.

2.   Under a file named CPINT you have stored a subroutine TINT(P,R,T,A) which will compute A, the total amount paid back on a loan of P dollars at interest rate R, if the loan is paid back in T equal installments and the interest is compounded at each installment date. Write a remote terminal program which will call for inputs of P, R, and T; call subroutine TINT to obtain A; print A; and again call for inputs of P, R, and T until stopped.

3.   List all the commands and machine responses, up to the answer itself that you would give and receive in running a case for P = 10000, R = .05, and T = 240.

4.   Write a remote-terminal program which will call for an input number and, after you type it in, will call for another input number, and will continue this until you type in a number greater than 1.E10, at which point it will print out the sum of all the numbers up to the last one.

5.   Do problem 4 for the average of the input numbers.

6.   Do problem 4 for the square root of the sum of the square of the input numbers.

7.   Do problem 4 for the square root of the sum of the squares of the sines of the input numbers.

8.   The population of the 400 largest cities have been stored in order from largest down, one word per city in a file named POP.

    a.   Write a remote-terminal program which will use POP and find the total population of the first N cities, where N is requested as an input number.
    b.   Write a remote-terminal program which will use POP and find the number of cities with population greater than P, where P is requested as an input number.

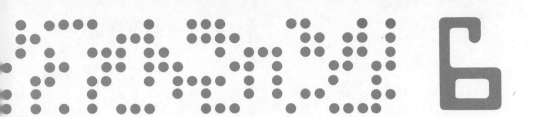

Evaluation of Functions

6.1   INTRODUCTION

In Chapter 4 it was pointed out that while computers actually perform addition, subtraction, multiplication, and division, they can be made to provide values of sines, cosines, logarithms, exponentials, or even more sophisticated functions. These functions can be made available either by storing a table of values within the memory of the machine and writing a program which will cause the machine to choose the correct value from this table, or by writing a program which will compute the required value directly from the series expansion or some other approximate relation. The computer performs arithmetic so rapidly that direct calculation is usually preferred to table look-up; however, both methods are used. This chapter will deal with the problem of computing of values of sines, cosines, and other functions directly by series or polynomial approximations.

6.2   TAYLOR'S FORMULA

In the calculus it was learned that any function of a single, real variable x, which obeyed certain rather general rules, could be expanded into a power series by means of Taylor's formula. Two standard forms of this formula are

$$f(x) = f(a) + (x - a)f'(a) + \frac{(x - a)^2}{2!}f''(a) + \cdots + \frac{(x - a)^{(n-1)}}{(n - 1)!}f^{(n-1)}(a)$$

$$+ \frac{(x - a)^n}{n!}f^{(n)}[a + \theta(x - a)], \qquad 0 < \theta < 1 \qquad \textbf{(6-1)}$$

171

$$f(x + h) = f(x) + hf'(x) + \frac{h^2}{2!}f''(x) + \cdots + \frac{h^{n-1}}{(n-1)!}f^{(n-1)}(x)$$

$$+ \frac{h^n}{n!}f^{(n)}(x + \theta h), \qquad 0 < \theta < 1 \qquad \textbf{(6-2)}$$

On putting $a = 0$ in (6-1) we obtain Maclaurin's formula:

$$f(x) = f(0) + xf'(0) + \frac{x^2}{2!}f''(0) + \cdots + \frac{x^{n-1}}{(n-1)!}f^{(n-1)}(0)$$

$$+ \frac{x^n}{n!}f^{(n)}(\theta x), \qquad 0 < \theta < 1 \qquad \textbf{(6-3)}$$

The last term in these expressions may not look quite familiar to the survivor of a course in elementary calculus. Frequently the significance of that term, or even its existence, is not stressed at that level of mathematical training. It is called the remainder term. If one wishes to generate an infinite series of $f(x)$, one keeps making n larger and larger so that the remainder term moves farther and farther out in the series. If the remainder term tends to zero as n tends to infinity, then the series converges and gives an exact representation of the function $f(x)$. If we wish to approximate $f(x)$ for a particular value of x, we compute several of the terms of the series and sum these terms to get the approximation. The question then arises: How large is the error introduced by taking only a finite number of terms for the sum? That question is answered in each of the expressions (6-1), (6-2), and (6-3). Directing our attention to expression (6-1), we see that the first n terms, which involve the zeroth derivative through the $(n - 1)$st derivative, are just the first n terms of a standard Taylor's series, and the remaining term, which has a somewhat different appearance, is the term we have called the remainder term. We see that expression (6-1) states that $f(x)$ is not just approximately equal to the sum of these terms, but that $f(x)$ is *precisely* equal to the sum of these terms. This is why we refer to expression (6-1) as "Taylor's formula" rather than "Taylor's series." It is an exact relation between two quantities, both of which are expressed in quite finite form, without any infinite series being involved. If we wish to approximate $f(x)$ by the first n terms of the right-hand side of expression (6-1), the error we make in so doing is given exactly by the remainder term. There is one catch. The quantity θ in this remainder term is some number we do not know and usually cannot even calculate. All we do know about it is that it is some number between zero and one. Hence we cannot usually say what the exact value of the error is. This is not a great concern, however, if we can somehow estimate the maximum possible size of the remainder term. Then, by proper choice of n, the number of terms, we may be able to make the maximum possible error sufficiently small that it can be tolerated.

6.21   Examples of the Use of Taylor's Formula

As a review in the use of Taylor's formula we will develop a few functions according to expressions (6-1), (6-2), and (6-3).

Example 1.   Write the Maclaurin expansion for e^x.

Here

$$
\begin{aligned}
f(x) &= e^x & f(0) &= 1 \\
f'(x) &= e^x & f'(0) &= 1 \\
f''(x) &= e^x & f''(0) &= 1 \\
f'''(x) &= e^x & f'''(0) &= 1 \\
f^{iv}(x) &= e^x & f^{iv}(0) &= 1 \\
&\vdots & & \\
f^{(n-1)}(x) &= e^x & f^{(n-1)}(0) &= 1 \\
f^{(n)}(x) &= e^x & f^{(n)}(\theta x) &= e^{\theta x}
\end{aligned}
$$

Substituting these quantities in expression (6-3), we obtain

$$
e^x = 1 + x + \frac{x^2}{2!} + \frac{x^3}{3!} + \cdots + \frac{x^{n-1}}{(n-1)!} + \frac{x^n}{n!}\,e^{\theta x}, \qquad 0 < \theta < 1
$$

Example 2.   Write the Taylor expansion for $\ln x$ about the point $x = 1$.

Here

$$
\begin{aligned}
f(x) &= \ln x & f(1) &= 0 \\
f'(x) &= 1/x & f'(1) &= 1 \\
f''(x) &= -1/x^2 & f''(1) &= -1 \\
f'''(x) &= +2/x^3 & f'''(1) &= 2 \\
f^{iv}(x) &= -2 \times \frac{3}{x^4} & f^{iv}(1) &= -2 \times 3 \\
&\vdots & &\vdots \\
f^{(n-1)}(x) &= (-1)^n (n-2)!/x^{n-1} & f^{(n-1)}(1) &= (-1)^n (n-2)! \\
f^{(n)}(x) &= (-1)^{n+1}(n-1)!/x^n & f^{(n)}[1 &+ \theta(x-1)] \\
& & &= (-1)^{n+1}(n-1)!/[1 + \theta(x-1)]^n
\end{aligned}
$$

Substituting these values in expression (6-1), remembering that $a = 1$ in this case, we obtain

$$
\ln x = (x-1) - \frac{1}{2!}(x-1)^2 + \frac{2}{3!}(x-1)^3 - \frac{2 \times 3}{4!}(x-1)^4 + \cdots
$$

$$
+ \frac{(-1)^n (n-2)!(x-1)^{n-1}}{(n-1)!} + \frac{(-1)^{n+1}(n-1)!(x-1)^n}{n![1 + \theta(x-1)]^n}, \qquad 0 < \theta < 1
$$

or, simplifying the expression,

$$\ln x = (x - 1) - \frac{(x-1)^2}{2} + \frac{(x-1)^3}{3} - \frac{(x-1)^4}{4} + \cdots$$

$$+ \frac{(-1)^n(x-1)^{n-1}}{n-1} + \frac{(-1)^{n+1}(x-1)^n}{n(1-\theta+\theta x)^n}, \qquad 0 < \theta < 1$$

Example 3.   Write the Maclaurin expansion for $\sin x$.

Here

$$
\begin{array}{ll}
f(x) = \sin x & f(0) = 0 \\
f'(x) = \cos x & f'(0) = 1 \\
f''(x) = -\sin x & f''(0) = 0 \\
f'''(x) = -\cos x & f'''(0) = -1 \\
f^{iv}(x) = \sin x & f^{iv}(0) = 0 \\
\quad\vdots & \\
f^n(x) = ? &
\end{array}
$$

Here the pattern of the derivatives is quite clear—the sequence $\sin x$, $\cos x$, $-\sin x$, $-\cos x$, is repeated over and over. However, since the value of $\sin 0$ is zero, there is no particular point in stopping the series when the next term is a sine term, since no calculational effort is saved by doing so. Let us agree for now, then, that n will be an odd number, $n = 2p + 1$, where p is an integer. Then we can write

$$f^{(n)}(x) = (-1)^{(n-1)/2} \cos x = (-1)^p \cos x$$

Now, utilizing expression (6-3), we can write

$$\sin x = x - \frac{x^3}{3!} + \frac{x^5}{5!} + \cdots + \frac{(-1)^p x^{2p+1} \cos \theta x}{(2p+1)!}, \qquad 0 < \theta < 1$$

6.22   A Program for the Exponential Function

In order to determine the error involved in approximating a function by its Taylor expansion, it is necessary to establish the maximum value of the remainder term. For the exponential function of Section 6.21, this remainder was $x^n e^{\theta x}/n!$, where θ is some number between 0 and 1. If we are to establish an upper limit on the value of this quantity without further knowledge of the value of θ, we must choose that value of θ which will maximize the expression. For this expression the value $\theta = 1$ is the proper choice. If we use n terms of the Taylor expansion to approximate e^x, the error will certainly not be greater than $x^n e^x / n!$. If we are interested in writing a computer routine which will compute e^x to eight significant figures, then, by Theorem 2 of Chapter 3, we should insist on a relative error of less than $1/(2 \times 10^8)$. Thus we wish to have

$$\frac{\dfrac{x^n}{n!} e^x}{e^x} < \frac{1}{2 \times 10^8}$$

or

$$\frac{x^n}{n!} < \frac{1}{2 \times 10^8}$$

It is clear that the value of n required depends on how big a value of x we desire to be able to handle. For example, if $x = 1$ is the largest value required, we must have

$$\frac{1}{n!} < \frac{1}{2 \times 10^8}$$

or $n! > 2 \times 10^8$. This is satisfied by $n = 12$. If we desired to have a series expansion suitable for values of x up to $x = 10$, we would need

$$\frac{10^n}{n!} < \frac{1}{2 \times 10^8}$$

which will require $n = 41$, or 41 terms. This would be a much longer calculation. On the other hand, it is not necessary to use this many terms if we simply apply the laws of exponents. For example, $e^{5.632} = e^5 \times e^{.632}$. The factor e^5 can be computed by multiplying e by itself five times. The factor $e^{.632}$ can be computed to sufficient accuracy with 12 terms of a Maclaurin expansion. To the accuracy required,

$$e^x = 1 + x + \frac{x^2}{2!} + \frac{x^3}{3!} + \frac{x^4}{4!} + \frac{x^5}{5!} + \frac{x^6}{6!} + \frac{x^7}{7!} + \frac{x^8}{8!} + \frac{x^9}{9!} + \frac{x^{10}}{10!} + \frac{x^{11}}{11!}$$

or, in a form requiring a smaller number of multiplications to compute,

$$e^x = 1 + x\left(1 + x\left(\frac{1}{2!} + x\left(\frac{1}{3!} + x\left(\frac{1}{4!} + x\left(\frac{1}{5!} + x\left(\frac{1}{6!} + x\left(\frac{1}{7!} + x\left(\frac{1}{8!} + x\left(\frac{1}{9!}\right.\right.\right.\right.\right.\right.\right.\right.\right.$$
$$\left.\left.\left.\left.\left.\left.\left.\left.\left. + x\left(\frac{1}{10!} + \frac{x}{11!}\right)\right)\right)\right)\right)\right)\right)\right)\right)$$

In FORTRAN, the constants for this expression could be determined by the statements

```
DIMENSION A(11)
A(1)=1.
DO 2 J=2,11
FJ=J
2 A(J)=A(J-1)/FJ
```

With these coefficients, e^x could be calculated by the loop

```
Y=X*A(11)+A(10)
DO 4 I=1,9
J=10-I
4 Y=Y*X+A(J)
Y=Y*X+1.
```

or by the single lengthy statement

```
Y=(((((((((X*A(11)+A(10))*X+A(9))*X+A(8))*X+A(7))*X
    +A(6))*X+A(5))*X+A(4))*X+A(3))*X+A(2))*X
    +A(1))*X+1.
```

6.23   A Program for the Sine Function

The remainder term in the Maclaurin expansion of the sine function, from Section 6.21, was $(-1)^p x^{2p+1} \cos \theta x/(2p+1)!$. If we desire the relative error to be less than $1/(2 \times 10^8)$, we note that $\cos \theta x < 1$, so that we require

$$\frac{1}{\sin x} \frac{x^{2p+1}}{(2p+1)!} < \frac{1}{2 \times 10^8}$$

Again the value of p, the number of terms required, depends on how large a value of x must be accommodated. Because of the periodicity of the sine function, it is certainly sufficient to consider only $x < 2\pi$. Also, since $\sin(\pi + x) = -\sin x$, values of x between π and 2π can be replaced by values between 0 and π. Further, since $\sin(\pi - x) = \sin x$, angles between $\pi/2$ and π can be replaced by angles between 0 and $\pi/2$. For values of x between 0 and $\pi/2$, the $\sin x$ in the denominator is no problem, since $1 \leqslant x/(\sin x) \leqslant \pi/2$ for x in this range. The largest value of the left-hand side of the inequality occurs when $x = \pi/2$. For this value of x, the inequality is satisfied by

$2p + 1 = 15$, or $p = 7$. Thus the value of the sine function will be given to eight significant figures by the expression

$$\sin x = x - \frac{x^3}{3!} + \frac{x^5}{5!} - \frac{x^7}{7!} + \frac{x^9}{9!} - \frac{x^{11}}{11!} + \frac{x^{13}}{13!}$$

or, in a form which can be evaluated with a minimum of multiplication,

$$\sin x = x\left(1 - x^2\left(\frac{1}{3!} - x^2\left(\frac{1}{5!} - x^2\left(\frac{1}{7!} - x^2\left(\frac{1}{9!} - x^2\left(\frac{1}{11!} - \frac{x^2}{13!}\right)\right)\right)\right)\right)\right)$$

A pair of FORTRAN statements which will perform this calculation is

```
U = X*X
Y = −((((((U*A(13)−A(11))*U−A(9))*U−A(7))*U−A(5))*U
    −A(3))*U−A(1))*X
```

provided the constants A(J) have already been calculated and stored, as in Section 6.22.

6.24 Alternating Series

Quite frequently the Taylor series takes the form of a series of terms which alternate in sign. When this is the case, the error estimate is quite easily made by taking advantage of the following theorem:

Theorem 1. *Let $\sum_{k=1}^{\infty}(-1)^{k-1}a_k$ be an alternating series with the properties that an $a_{n+1} < a_n$ for each n, and limit $_{n\to\infty}\, a_n = 0$. Then the series converges. Further, if $S = \sum_{k=1}^{\infty}(-1)^{k-1}a_k$ and $S_n = \sum_{k=1}^{n}(-1)^{k-1}a_k$, then S_n differs from S by an amount less than a_{n+1}.*

Paraphrased in less precise terminology, the theorem states that the error in truncating a convergent alternating series of decreasing terms is less than the first term neglected.

Example 1. The series for $\tan^{-1} x$ is

$$\tan^{-1} x = x - \frac{x^3}{3} + \frac{x^5}{5} - \cdots + (-1)^{n-1}\frac{x^{2n-1}}{2n-1} + \cdots$$

Find the number of terms sufficient to determine $\tan^{-1} 1$ to eight significant digits.

For $|x| \leqslant 1$, the series satisfies the conditions of Theorem 1, so the error is less than the first term neglected. Thus we require

$$\frac{(1)^{2n-1}}{2n-1} \leqslant .5 \times 10^{-8}$$

which is satisfied by $n = 1 \times 10^8 + 1$.

6.25 Improvement of Convergence of Alternating Series

For series such as the arctangent series given in Section 6.24, the convergence can be improved in a rather straightforward manner. If we multiply the series by x^2, we have

$$x^2 \tan^{-1} x = 0 + x^3 - \frac{x^5}{3} + \frac{x^7}{5} - \cdots + (-1)^{n-2} \frac{x^{2n-1}}{2n-3}$$

$$+ (-1)^{n-1} \frac{x^{2n+1}}{2n-1} + \cdots$$

Both this and the preceding series for $\tan^{-1} x$ are absolutely convergent for $|x| < 1$, so we may add this series to the other, obtaining

$$(1 + x^2) \tan^{-1} x = x + (2/3)x^3 - (2/15)x^5 + (2/35)x^7 - \cdots$$

$$- (-1)^n \frac{2x^{2n-1}}{(2n-1)(2n-3)} + \cdots$$

After the first two terms, this is an alternating series and by Theorem 1 of Section 6.24 the error incurred by truncating it is less than the first term neglected. It converges faster than the original one for $\tan^{-1} x$ because the denominator increases like $2n^2$ instead of $2n$. If we wish to determine $\tan^{-1} 1$ to eight significant figures, we would require that

$$\frac{2(1)^{2n-1}}{(2n-1)(2n-3)} \leqslant .5 \times 10^{-8}$$

which is satisfied by $n = 10^4 + 1$, still an exorbitant number of terms for calculational purposes.

The process can be continued to reduce still further the number of terms required. We have

$$x^4 \tan^{-1} x = 0 + 0 + x^5 - \frac{x^7}{3} + \cdots + (-1)^{n-3} \frac{x^{2n-1}}{2n-5} + \cdots$$

$$x^6 \tan^{-1} x = 0 + 0 + 0 + x^7 + \cdots + (-1)^{n-4} \frac{x^{2n-1}}{2n-7} + \cdots$$

We can now ask: What combination of the series for $\tan^{-1} x$, $x^2 \tan^{-1} x$, $x^4 \tan^{-1} x$, and $x^6 \tan^{-1} x$ will produce the smallest coefficient for x^{2n-1}? If we form the combination

$$\tan^{-1} x + Ax^2 \tan^{-1} x + Bx^4 \tan^{-1} x + Cx^6 \tan^{-1} x$$

the coefficient of x^{2n-1} is

$$(-1)^{n-1}\left[\frac{1}{2n-1} - \frac{A}{2n-3} + \frac{B}{2n-5} - \frac{C}{2n-7}\right]$$

$$= (-1)^{n-1}\left[\frac{(2n-3)(2n-5)(2n-7) - A(2n-1)(2n-5)(2n-7) + B(2n-1)(2n-3)(2n-7) - C(2n-1)(2n-3)(2n-5)}{(2n-1)(2n-3)(2n-5)(2n-7)}\right]$$

The denominator is a polynomial of degree four in n. The numerator is of third degree, but A, B, and C can be chosen to make the coefficients of n^3, n^2, and n be zero, and then the coefficient of x^{2n-1} will behave like $1/n^4$ and so should decrease rapidly as n increases. The equations for determining A, B, and C are

$$n^3 \text{ coefficient, } 8 - 8A + 8B - 8C = 0$$
$$n^2 \text{ coefficient, } -30 + 26A - 22B + 18C = 0$$
$$n \text{ coefficient, } 142 - 94A + 62B - 46C = 0$$

These three equations are satisfied by

$$A = 3, \qquad B = 3, \qquad C = 1$$

For these values, the expression for the coefficient of x^{2n-1} becomes

$$(-1)^n \frac{48}{(2n-1)(2n-3)(2n-5)(2n-7)}$$

and the series is

$$(1 + x^2)^3 \tan^{-1} x = x + (8/3)x^3 + (11/5)x^5 + (16/35)x^7 + (16/315)x^9 - \cdots$$

$$- \frac{(-1)^n 48 x^{2n-1}}{(2n-1)(2n-3)(2n-5)(2n-7)} + \cdots$$

Note that the expression given for the nth term works only for n greater than three terms, that is, when n is large enough that each of the four series used in the sum actually makes a contribution. The number of terms required for eight-digit accuracy at $x = 1$ is given by

$$\frac{48(1)^{2n-1}}{(2n-1)(2n-3)(2n-5)(2n-7)} \leqslant .5 \times 10^{-8}$$

which is satisfied by $n = 150$. This is still an unmanageably large number of terms, but clearly the process is leading toward reasonable numbers of terms. It would appear that, if the denominator of the x^{2n-1} term were to behave like n^8, then a value of about $n = 10$ should be required and the calculation will have become quite manageable. This result can be accomplished by multiplying the $\tan^{-1} x$ series by x^8, x^{10}, x^{12}, and x^{14}, and then proceeding as above to combine the series. Without the actual steps in the process, the result is

$$(1 + x^2)^7 \tan^{-1} x = x + (20/3)x^3 + (283/15)x^5 + (1024/35)x^7 + (1199/45)x^9$$

$$+ (1388/99)x^{11} + (1013/429)x^{13} + \cdots$$

$$+ \frac{(-1)^n \times 45 \times 2^{10}}{(2n-1)(2n-3)(2n-5)(2n-7) \cdots (2n-15)}$$

$$\times x^{2n-1} - \cdots$$

and terms through $n = 14$ would be required to give eight-place accuracy.

EXERCISE 18

1. Write the Maclauren expansion with remainder for each of the following functions. Determine the number of terms required for eight-place accuracy.

    a. $\cos x$ b. $\log(1 + x)$ c. $\sqrt{1 + x}$

    d. $\tan^{-1} x$ e. e^{-x^2} f. $\displaystyle\int_0^x e^{-t^2/2} \, dt$

2. Following the method of Section 6.25, construct a series for $(A + Bx + \cdots) \times \ln(1 + x)$ whose terms decrease (a) like $1/n^4$, (b) like $1/n^8$.

3. For each of the following, write a subroutine which will return the function value to eight-place accuracy for x in the range indicated, assuming that your computer is performing its arithmetic to sufficient accuracy. Check your program by using the identities indicated.

a. $\sin x, -\dfrac{\pi}{2} \leqslant x \leqslant \dfrac{\pi}{2}: \sin^2(.5) - \sin^2\left(\dfrac{\pi}{2} - .5\right) - 1 = 0$

b. $\ln(1+x), -1/2 \leqslant x \leqslant 0: \log(.64) - 2\log(.8) = 0.$

c. $e^x, -1 \leqslant x \leqslant 1: (e^{.3})(e^{-.3}) - 1 = 0.$

4.   For each of the following, assume you have a subroutine which will return the correct function value for x in the range indicated. Write a subroutine which will accept any value of x, make an appropriate conversion to require x only in the range for which the given subroutine works, and find and return the proper function value.

   a.   GSIN(x) returns $\sin x$ for $-\dfrac{\pi}{2} \leqslant x \leqslant \dfrac{\pi}{2}$

   b.   GLOG(x) returns $\ln x$ for $1/2 < x \leqslant 1$

   c.   GATAN(x) returns $\tan^{-1} x$ for $-1 \leqslant x \leqslant 1$

   d.   GEXP(x) returns e^x for $-1 \leqslant x \leqslant 1$.

6.3   CHEBYSHEV SERIES

The examples of Section 6.2 should have served to make clear one fact about approximations using Taylor's formula. Formula (6-1) tends to give a good approximation to $f(x)$ by using only a few terms as long as x is near to a. As x departs from a, the factor $(x - a)^n$ in the remainder term grows quite rapidly, and the approximation tends to become poor. Rather than have this polynomial approximation which requires few terms for some values of x and large numbers of terms for other values of x to attain the required accuracy, it would be preferable to have an approximating polynomial which is uniformly good over a range of values of x. The Chebyshev series can be used to obtain such an approximation. The Chebyshev polynomials $T(x)$ are defined by

$$T_n(x) = \cos(n\cos^{-1} x) \tag{6-4}$$

For example,

$$T_0(x) = \cos 0 = 1 \tag{6-5}$$

$$T_1(x) = \cos(\cos^{-1} x) = x \tag{6-6}$$

$$T_2(x) = \cos(2\cos^{-1} x) = 2\cos^2(\cos^{-1} x) - 1 = 2x^2 - 1 \tag{6-7}$$

and so forth. By means of trigonometric identities it is possible to establish a general recursion relation for the Chebyshev polynomials. From the relation

$$\cos a \cos b = 1/2 \cos(a + b) + 1/2 \cos(a - b) \tag{6-8}$$

we can say that

$$2 \cos n\theta \cos \theta = \cos(n + 1)\theta + \cos(n - 1)\theta \qquad \textbf{(6-9)}$$

Now, letting $x = \cos \theta$, we have

$$2x \cos(n \cos^{-1} x) = \cos((n + 1)\cos^{-1} x) + \cos((n - 1)\cos^{-1} x) \quad \textbf{(6-10)}$$

or

$$2x T_n(x) = T_{n+1}(x) + T_{n-1}(x) \qquad \textbf{(6-11)}$$

or

$$T_{n+1}(x) = -T_{n-1}(x) + 2x T_n(x) \qquad \textbf{(6-12)}$$

Given any two successive Chebyshev polynomials, the relation (6-12) can be used to find the next one. For example, from (6-6), (6-7), and (6-12) we find that

$$T_3(x) = -T_1(x) + 2x T_2(x) = -x + 2x(2x^2 - 1) = 4x^3 - 3x \quad \textbf{(6-13)}$$

Note that from the recursion relation (6-12), $T_n(x)$ will indeed be a polynomial in x of degree n.

6.31   Orthogonality of Chebyshev Polynomials

Any two Chebyshev polynomials $T_n(x)$ and $T_m(x)$, where $n \neq m$, have the property that

$$\int_{-1}^{1} \frac{T_m(x)T_n(x)}{\sqrt{1 - x^2}} \, dx = 0 \qquad \textbf{(6-14)}$$

To show that this is so, let $x = \cos \theta$. Then the integral can be written

$$\int_{0}^{\pi} \frac{\cos m\theta \cos n\theta}{\sin \theta} (-\sin \theta \, d\theta) = \int_{0}^{\pi} \cos m\theta \cos n\theta \, d\theta = 0$$

Because of this property the Chebyshev polynomials are said to be orthogonal over the interval $(-1, 1)$ with weighting $1/\sqrt{1 - x^2}$.

For the case $m = n$, we have

$$\int_{-1}^{1} \frac{(T_m(x))^2}{\sqrt{1-x^2}} \, dx = \begin{cases} \pi & \text{for } m = 0 \\ \dfrac{\pi}{2} & \text{otherwise} \end{cases} \tag{6-15}$$

6.32   Expansion of a Function in a Chebyshev Series

Because of the properties of Chebyshev polynomials stated in the relationships (6-14) and (6-15), it is possible to express any properly behaved function $f(x)$ as Chebyshev series,

$$f(x) = (1/2)c_0 + \sum_{n=1}^{\infty} c_n T_n(x) \tag{6-16}$$

where the coefficients c_n are given by

$$c_n = \frac{2}{\pi} \int_{-1}^{1} \frac{f(x)T_n(x) \, dx}{\sqrt{1-x^2}} \tag{6-17}$$

This can be seen as follows. Suppose that for a given set of coefficients $c_0, c_1, \ldots, c_n, \ldots$, and some values of x, the series

$$(1/2)c_0 + \sum_{j=1}^{\infty} c_j T_j(x)$$

is absolutely convergent, and denote its sum by $f(x)$. Form the integral

$$\int_{-1}^{1} \frac{f(x)T_n(x)}{\sqrt{1-x^2}} \, dx = \int_{-1}^{1} \frac{1}{2} \frac{c_0 T_n(x)}{\sqrt{1-x^2}} \, dx + \int_{-1}^{1} \frac{\sum_{j=1}^{\infty} c_j T_j(x)T_n(x)}{\sqrt{1-x^2}} \, dx$$

$$= (1/2)c_0 \int_{-1}^{1} \frac{T_n(x)}{\sqrt{1-x^2}} \, dx$$

$$+ \sum_{j=1}^{\infty} c_j \int_{-1}^{1} \frac{T_j(x)T_n(x)}{\sqrt{1-x^2}} \, dx \tag{6-18}$$

Now, by (6-14) and (6-15), all the integrals in the above expression are zero except for the one in which $j = n$. For that one the value is $\pi/2$ (or π, if $n = 0$).

Hence

$$\int_{-1}^{1} \frac{f(x)T_n(x)}{\sqrt{1-x^2}} \, dx = c_n(\pi/2),$$

from which (6-17) follows.

Relation (6-17) says that the coefficients c_n depend on the values of $f(x)$ only for x between -1 and 1, so we can expect the series to represent $f(x)$ only for x in this range. Indeed, from (6-4), the $T_n(x)$ are only defined for values of x between -1 and 1, since the cosine of a real angle cannot exceed 1.

If for computational purposes one uses a finite number of terms of expansion (6-16) to represent a function $f(x)$, again, as in Taylor's formula, one has a polynomial in x to represent $f(x)$, since each of the $T_n(x)$ is a polynomial in x. The coefficients of this polynomial will be different from those in the Taylor's formula expansion, and thus the approximation is a different one, having different accuracy properties.

In practice, it is difficult to generate a Chebyshev series for a given function $f(x)$ by using (6-17) to compute the coefficients. For any but the simplest cases, the integrals are impossible to evaluate analytically. They can be integrated numerically, using one of the techniques from Chapter 7, but this requires that the values of the function $f(x)$ be known from some other source for values of x within the range of integration. The case where $f(x)$ is given as a table of values is discussed in Chapter 12.

6.33   Rearrangement of a Polynomial as a Sum of Chebyshev Polynomials

One way of generating terms of a Chebyshev series, at least approximately, is to first use Taylor's formula to obtain a polynomial approximation and then use this polynomial to represent $f(x)$ in (6-17) to compute the Chebyshev coefficients. Since the $T_n(x)$ are themselves polynomials in x, the same result should be obtained by merely rearranging terms in the Taylor polynomial. Suppose we have a Taylor expansion

$$f(x) = b_0 + b_1 x + b_2 x^2 \cdots b_n x^n + R_n(x) \qquad \text{(6-19)}$$

where n has been chosen sufficiently large that $R_n(x)$ is negligible for our purpose. We wish to convert this into the first $n + 1$ terms of a Chebyshev series

$$f(x) = (1/2)c_0 + c_1 T_1(x) + c_2 T_2(x) + \cdots + c_n T_n(x) \qquad \text{(6-20)}$$

To do so, let us first represent the Chebyshev polynomial $T_n(x)$ in the form

$$T_n(x) = \sum_{i=0}^{n} h_n(i)x^i \tag{6-21}$$

That is, $h_n(i)$ stands for the coefficient of x^i in $T_n(x)$. The coefficient of x^n is $h_n(n)$. Hence we will have matched up the x^n powers between (6.19) and (6-20) if we choose c_n so that

$$c_n h_n(n) = b_n \quad \text{or} \quad c_n = b_n/h_n(n) \tag{6-22}$$

Then we can say that

$$b_n x^n = c_n T_n(x) - c_n \sum_{i=0}^{n-1} h_n(i)x^i \tag{6-23}$$

If we substitute this into (6-19), we have

$$f(x) = d_0 + d_1 x + d_2 x^2 + \cdots + d_{n-1} x^{n-1} + c_n T_n(x) + R_n(x) \tag{6-24}$$

where

$$d_i = b_i - c_n h_n(i) \tag{6-25}$$

for $i = 0$ to $n - 1$. We now have the x^n term converted to a Chebyshev polynomial. The process can then be repeated to convert the x^{n-1} term, then the x^{n-2}, and so on. However, we may have already obtained a very useful result without continuing further. From (6-4) it can be seen that

$$|T_n(x)| \leqslant 1 \quad \text{for all } x \tag{6-26}$$

so that if c_n is small enough to be within the acceptable error limits, we may discard the $T_n(x)$ term and we now have a lower-degree polynomial to represent $f(x)$. This process is termed "telescoping" a polynomial.

Example 1.   Write a Taylor series for $\sin(\pi x/2)$ accurate to eight significant figures. Telescope the polynomial into a one of lower degree, retaining seven significant figures.

From Section 6.23, a series for $\sin x$ accurate to eight significant digits is

$$\sin x = x - \frac{x^3}{3!} + \frac{x^5}{5!} - \frac{x^7}{7!} + \frac{x^9}{9!} - \frac{x^{11}}{11!} + \frac{x^{13}}{13!}$$

Hence

$$\sin \frac{\pi}{2} x = \frac{\pi}{2} x - \left(\frac{\pi}{2}\right)^3 \frac{x^3}{3!} + \left(\frac{\pi}{2}\right)^5 \frac{x^5}{5!} - \left(\frac{\pi}{2}\right)^7 \frac{x^7}{7!} + \left(\frac{\pi}{2}\right)^9 \frac{x^9}{9!}$$

$$- \left(\frac{\pi}{2}\right)^{11} \frac{x^{11}}{11!} + \left(\frac{\pi}{2}\right)^{13} \frac{x^{13}}{13!}$$

$$
\begin{aligned}
= \ & 1.5707963x - .64596410x^3 + .079692626x^5 \\
& - .0046817541x^7 + .00016044118x^9 \\
& - .0000035988432x^{11} + .0000000569217x^{13}
\end{aligned}
\tag{6-27}
$$

This expansion corresponds to the relation (6-19) with $f(x) = \sin(\pi x/2)$. This approximation would be guaranteed to give $\sin(\pi x/2)$ with an error no worse than 5×10^{-9} for $-1 \leqslant x \leqslant 1$ if the coefficients were given exactly. The error introduced by using approximate numbers for the coefficients will be discussed in Section 6.5. The terms in (6-27) are the beginning of an alternating series with decreasing terms, so the error is no larger than the first term neglected. If the term involving x^{13} were neglected, the error could be as large as .0000000569217. The error would be slightly larger than 5×10^{-8}, so if the x^{13} term were dropped, only six significant digits could be assumed correct.

Now let us telescope the series. To do so we need first the expression for $T_{13}(x)$. By repeated application of (6-12) to (6-13) we have

$$
\begin{aligned}
T_4(x) &= 8x^4 - 8x^2 + 1 \\
T_5(x) &= 16x^5 - 20x^3 + 5x \\
T_6(x) &= 32x^6 - 48x^4 + 18x^2 - 1 \\
T_7(x) &= 64x^7 - 112x^5 + 56x^3 - 7x \\
T_8(x) &= 128x^8 - 256x^6 + 160x^4 - 32x^2 + 1 \\
T_9(x) &= 256x^9 - 576x^7 + 432x^5 - 120x^3 + 9x \\
T_{10}(x) &= 512x^{10} - 1280x^8 + 1120x^6 - 400x^4 + 50x^2 - 1 \\
T_{11}(x) &= 1024x^{11} - 2816x^9 + 2816x^7 - 1232x^5 + 220x^3 - 11x \\
T_{12}(x) &= 2048x^{12} - 6144x^{10} + 6912x^8 - 3584x^6 + 840x^4 - 72x^2 + 1 \\
T_{13}(x) &= 4096x^{13} - 13312x^{11} + 16640x^9 - 9984x^7 + 2912x^5 - 364x^3 + 13x
\end{aligned}
\tag{6-28}
$$

We are now in a position to proceed as in equation (6-22). We have

$$
\begin{aligned}
c_{13} = b_{13}/h_{13}(13) &= .0000000000138969 \\
&= 1.38969 \times 10^{-11}
\end{aligned}
$$

Applying (6-25), we have

$$d_0 = d_2 = d_4 = d_6 = d_8 = d_{10} = d_{12} = 0$$
$$d_1 = 1.5707963 - 13c_{13} = 1.5707963$$
$$d_3 = -.64596410 + 364c_{13} = -.64596410$$
$$d_5 = .079692626 - 2912c_{13} = .079692585$$
$$d_7 = -.0046817541 + 9984c_{13} = -.0046816154$$
$$d_9 = .00016044118 - 16640c_{13} = .0001602099$$
$$d_{11} = -.0000035988432 + 13312c_{13} = -.0000034138477$$

so that $\sin(\pi x/2)$ can be written

$$\sin(\pi x/2) = 1.5707963x - .64596410x^3$$
$$+ .079692585x^5 - .0046816154x^7$$
$$+ .0001602099x^9 - .0000034138477x^{11}$$
$$+ 1.38969 \times 10^{-11} T_{13}(x) \qquad \textbf{(6-29)}$$

The maximum additional error which would be introduced by neglecting the term involving T_{13} is 1.38969×10^{-11}. This is about 1 four-thousandth as large as the error that would have been incurred by dropping the x^{13} term in (6-27).

Let us check the effect of eliminating the x^{11} term in (6-29). Renaming these coefficients as $b_1, b_3, \ldots, b_{11}$, we have

$$c_{11} = b_{11}/h_{11}(11) = -.00000000333384$$

Applying (6-25), we have

$$d_0 = d_2 = d_4 = d_6 = d_8 = d_{10} = 0$$
$$d_1 = 1.5707963 - 11c_{11} = 1.5707963$$
$$d_3 = -.64596410 + 220c_{11} = -.64596337$$
$$d_5 = .079692585 - 1232c_{11} = .079688478$$
$$d_7 = -.0046816154 + 2816c_{11} = -.0046722273$$
$$d_9 = .0001602099 - 2816c_{11} = .0001508218$$

and $\sin(\pi x/2)$ can be written

$$\sin(\pi x/2) = 1.5707963x - .64596337x^3$$
$$+ .079688478x^5 - .0046722273x^7$$
$$+ .0001508218x^9 - 3.33384 \times 10^{-9} T_{11}(x) \qquad \textbf{(6-30)}$$

The error introduced by neglecting the $T_{11}(x)$ term is less than 3.4×10^{-9}, so that seven significant digits are retained if the $T_{11}(x)$ term is dropped.

Let us check the effect of telescoping the expression (6-30) once more. Denoting the coefficients in (6-30) by $b_1, b_3, \ldots, b_9$, we have

$$c_9 = b_9/h_9(9) = .000000589148$$

The error involved in neglecting this term is greater than 5×10^{-8}, too large to leave seven correct significant digits, so that we cannot telescope the series further and retain the desired accuracy.

The process of telescoping a polynomial can be accomplished by suitable FORTRAN programs. To do so, the coefficients in the $T_n(x)$, equations (6-28), must either be stored or computed as needed. In order to make the programs as general as possible, we will describe a pair of subroutines in which the coefficients are computed from recursion relations. In finding the $T_n(x)$ for a particular value of n, the recursion relations require that the coefficients be computed for all lower values of n. Hence, rather than telescope a term at a time, as would be done for hand calculation, it is just as convenient to convert the entire polynomial to a sum of Chebyshev polynomials. Then after dropping those terms small enough to be neglected, the remaining sum of Chebyshev polynomials can be rewritten as a single polynomial.

To perform the first task, let us consider the use of the integral (6-17) to find the coefficients a_n for $f(x) = 1,\ x,\ x^2,\ x^3$, etc. Once we have these, the Chebyshev expansion for any polynomial $f(x)$ can be found merely by combining terms properly. Let $a_j(i)$ stand for the coefficient of $T_j(x)$ in the Chebyshev polynomial representation of x^i. Then, by (6-17), we can write

$$a_j(i) = \frac{2}{\pi} \int_{-1}^{1} \frac{x^i \cos(j \cos^{-1} x)}{\sqrt{1 - x^2}}\, dx \qquad (6\text{-}31)$$

or, letting $x = \cos \theta$,

$$a_j(i) = \frac{2}{\pi} \int_0^\pi (\cos \theta)^i \cos j\theta\, d\theta \qquad (6\text{-}32)$$

If we use the identity

$$\cos j\theta = \cos(j-1)\theta \cos \theta - \sin(j-1)\theta \sin \theta$$

we have

$$a_j(i) = \frac{2}{\pi} \int_0^\pi (\cos \theta)^{i+1} \cos(j-1)\theta\, d\theta - \frac{2}{\pi} \int_0^\pi (\cos \theta)^i \sin \theta \sin(j-1)\theta\, d\theta$$

$$(6\text{-}33)$$

The second expression can be integrated by parts. Let

$$u = \sin(j-1)\theta \qquad\qquad dv = (\cos\theta)^i \sin\theta \, d\theta$$

$$du = (j-1)\cos(j-1)\theta \, d\theta \qquad v = -\frac{(\cos\theta)^{i+1}}{i+1}$$

Then

$$\int_0^\pi (\cos\theta)^i \sin\theta \sin(j-1)\theta \, d\theta = -\left.\frac{\sin(j-1)\theta(\cos\theta)^{i+1}}{i+1}\right|_0^\pi \qquad (6\text{-}34)$$

$$+\frac{j-1}{i+1}\int_0^\pi (\cos\theta)^{i+1}\cos(j-1)\theta \, d\theta$$

The first term is zero at both limits of integration. Thus if we combine (6-34) with (6-33), we obtain

$$a_j(i) = \frac{2}{\pi}\int_0^\pi (\cos\theta)^{i+1}\cos(j-1)\theta \, d\theta - \frac{2}{\pi}\frac{j-1}{i+1}\int_0^\pi (\cos\theta)^{i+1}\cos(j-1)\theta \, d\theta$$

$$= \left[1 - \frac{j-1}{i+1}\right]\frac{2}{\pi}\int_0^\pi (\cos\theta)^{i+1}\cos(j-1)\theta \, d\theta$$

From (6-32), we see that the expression following the brackets can be called $a_{j-1}(i+1)$. Hence the above relation reduces to

$$a_j(i) = \left(1 - \frac{j-1}{i+1}\right)a_{j-1}(i+1) \qquad (6\text{-}35)$$

and we now have a recursion relation for finding $a_j(i)$. If the values of $a_0(k)$ are known for each k, (6-35) can be used to generate values for any other $a_n(m)$. By (6-32), $a_0(k)$ is given by

$$a_0(k) = \frac{2}{\pi}\int_0^\pi (\cos\theta)^k \, d\theta \qquad (6\text{-}36)$$

Using integration by parts, one can show that

$$\int_0^\pi (\cos\theta)^k \, d\theta = \frac{k-1}{k}\int_0^\pi (\cos\theta)^{k-2} \, d\theta$$

Thus

$$a_0(k) = \frac{k-1}{k} a_0(k-2) \tag{6-37}$$

Now, by direct integration of (6-32),

$$a_0(0) = 2, \qquad a_0(1) = 0 \tag{6-38}$$

so that any Chebyshev coefficient for x^i can be generated by application of the relations

$$a_0(0) = 2, \qquad a_0(1) = 0$$

$$a_0(k) = \frac{k-1}{k} a_0(k-2)$$

$$a_j(i) = \left(1 - \frac{j-1}{i+1}\right) a_{j-1}(i+1) \tag{6-39}$$

A subroutine to produce the Chebyshev coefficients c_i of (6-20) given the Taylor coefficients b_i of (6-19) [or given any other set of coefficients b_i of any polynomial approximation of $f(x)$] can be produced by a straightforward application of these recursion relations. Each power of x can be replaced by its corresponding expansion,

$$x^i = (1/2)a_0(i) + \sum_{j=1}^{i} a_j(i)T_j(x) \tag{6-40}$$

If this substitution is made into equation (6-19), and the coefficients of $T_j(x)$ from each power of x are combined, we see that the total coefficient of $T_j(x)$ is given by

$$c_j = \sum_{i=0}^{n} a_j(i)b_i \tag{6-41}$$

We see that to compute all the c_j's up to c_n, we need all the $a_j(i)$'s up to $a_n(n)$, each $a_j(i)$ to be multiplied by the corresponding b_i. Figure 6-1 demonstrates how the computation can be organized. First the quantities $a_0(0)$ through $a_0(2n)$ in the left-hand column are computed using the relation for $a_0(k)$ from (6-39). Then $a_0(0)$ through $a_0(n)$ are combined with b_0 through b_n to give c_0. Next the quantities $a_1(0)$ through $a_1(2n-1)$ in the second column are computed using (6-39), and $a_1(0)$ through $a_1(n)$ are combined

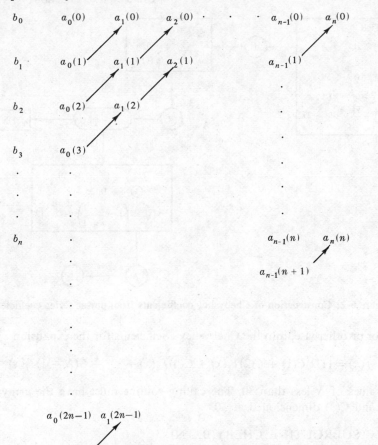

Figure 6-1: Pattern for construction of Chebyshev coefficients of x^n

with b_0 through b_n to give c_1. The process is continued until the right-hand column is computed and used to find c_n. Figure 6-2 gives a flow chart for the process. In the flow chart, the subscripts are left off the $a_j(i)$'s. The reason is that one column of Figure 6-1 is no longer needed once the next one is computed, and to save it is a waste of storage space. The flow chart implies that the new a's can be stored in the space formerly occupied by the old ones.

There is a problem in constructing a FORTRAN subroutine to implement this flow chart, in that zero subscripts are not allowed in many versions of FORTRAN. To get around this difficulty, we can displace all the subscripts by one. The FORTRAN subroutine given below will convert the coefficients of the polynomial expansion,

$$f(x) = B(1) + B(2)x + B(3)x^2 + \cdots + B(N + 1)x^N$$

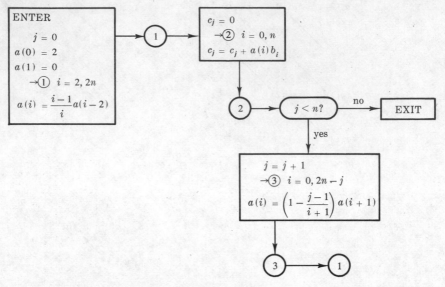

Figure 6-2: Construction of Chebyshev coefficients from power series coefficients

Taylor or otherwise, into the Chebyshev coefficients for the expansion

$$f(x) = (1/2)C(1) + C(2)T_1(x) + C(3)T_2(x) + \cdots + C(N+1)T_n(x)$$

for values of N less than 30. The calling routine must have the arrays for $B(J)$ and $C(J)$ dimensioned at 30.

```
        SUBROUTINE CHEBY(B,C,N)
        DIMENSION  B(30),C(30),A(60)
        J=1
        NN=2*N+1
        NP=N+1
        A(1)=2.
        A(2)=0.
        DO 1 I=3,NN
        FI=I
        A(I)=A(I-2)*(FI-2.)/(FI-1.)
     1  CONTINUE
        C(J)=0.
        DO 2 I=1,NP
        K=NP+1-I
     2  C(J)=C(J)+A(K)*B(K)
        IF(J-N)10,10,20
    10  J=J+1
        FJ=J
```

```
      K = NN + 1 - J
      DO 3 I = 1,K
      FI = I
    3 A(I) = A(I + 1)*(1. - (FJ - 2.)/FI)
      GO TO 1
   20 RETURN
      END
```

Note that the DO loop ending in statement 2 has been arranged so that in finding C(J), the quantity A(NP)*B(NP) is added first, then A(NP−1) *B(NP−1), and so on, and finally A(1)*B(1). This is done in deference to the situation pointed out in Section 3.6—that roundoff errors in sums are minimized if the smaller numbers are added first. If 20 terms are being used, this may improve the accuracy of the result by one significant digit. Actually, it is advisable to do a conversion such as this one in double precision, since accumulated roundoff error in the coefficients is highly undesirable. This can be accomplished by another kind of FORTRAN statement not introduced previously, a type-declaration statement. To declare the A(I), B(I), and C(I) values in double precision, one would insert the statement

DOUBLE A,B,C,

before any executable statement in the subroutine. A similar statement would also be required in the calling program. Type-declaration statements can be used to give other desired attributes to FORTRAN variables. For example, if one wants a variable named ITEM to be treated as a floating-point variable rather than as fixed point, as its name implies, one would use the type-declaration statement

REAL ITEM

On the other hand, if one wished the variable XRAY to be treated as fixed point rather than floating point, he could do so by means of the type-declaration statement

INTEGER XRAY

6.34 Rearrangement of a Sum of Chebyshev Polynomials as a Polynomial in X

Once one has the Chebyshev coefficients for a function and has set to zero the coefficients of any terms to be neglected, it is usually desirable to regroup and collect like powers of x, that is, take the reverse step, from expression (6-20) to (6-19). The recursion formula (6-12) can be used to convert the c_i's to be b_i's. Let $h_j(i)$ stand for the coefficient of x^i in $T_j(x)$. Then from formula (6-12) we can say that equating coefficients of x^i on both sides of the formula and taking $j = n - 1$,

$$h_j(i) = -h_{j-2}(i) + 2h_{j-1}(i-1) \qquad (6\text{-}42)$$

In applying this to the constant term, $i = 0$, we must make the understanding $h_j(-1) = 0$ for all j.

The coefficient of x^i in (6-19) is the sum of all the coefficients of x^i in (6-20). Thus

$$b_i = \sum_{j=0}^{k} c_j h_j(i) \qquad (6\text{-}43)$$

From (6-5) and (6-6), we have

$$T_0(x) = 1 \qquad \text{so } h_0(0) = 1 \qquad (6\text{-}44)$$

$$T_1(x) = x \qquad \text{so } h_1(0) = 0,\ h_1(1) = 1 \qquad (6\text{-}45)$$

We note also that the highest power of x in $T_j(x)$ is the jth power, so that

$$h_j(i) = 0 \qquad \text{for } i > j \qquad (6\text{-}46)$$

This fact, combined with (6-42), gives

$$h_j(j) = 2h_{j-1}(j-1) \qquad (6\text{-}47)$$

Also, we note from (6-28) that $T_j(x)$ contains only even powers of x if j is even and only odd powers of x if j is odd. This fact can be expressed by the statement

$$h_j(i) = 0 \qquad \text{for } i + j \text{ odd} \qquad (6\text{-}48)$$

These facts, along with (6-42), allow the efficient determination of the $h_j(i)$'s as needed. An efficient way to proceed is first to compute $h_2(0)$, $h_4(0)$, ..., $h_n(0)$, and then compute b_0 from (6-43). Next (6-42) can be used to compute $h_1(1)$, $h_3(1)$, $h_5(1)$, etc., and then (6-43) used to determine b_1, etc. Figure 6-3 illustrates the calculation of the $h_j(i)$. The arrows indicate which coefficients are used in computing which others. For example, $h_4(2)$ uses $h_2(2)$ and $h_3(1)$. It is seen that one row in this figure depends only on the preceding one, so it is not necessary to save all rows. Storage in the computer can be conserved by saving only the needed quantities.

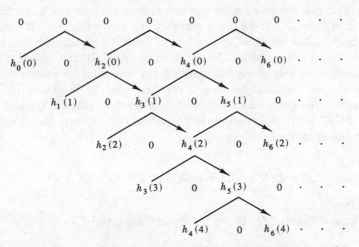

Figure 6-3: Computation of coefficients in Chebyshev polynomials

Figure 6-4 is a flow chart for the calculation. Careful study will probably be required to confirm that the flow chart does indeed represent the calculation. At first glimpse, it appears that it must be omitting something. The quantity h_j has only the single subscript, and there is no indication as to whether at any time the value is $h_j(1)$ or $h_j(2)$, etc. Actually the value of i

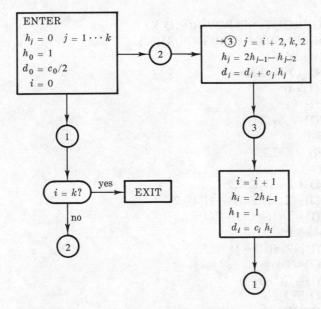

Figure 6-4: Rearrangement of Chebyshev polynomial in powers of x

in $h_j(i)$ is set just before entering the box at remote connector 1. In the main equation, $h_{j-2} = 2h_{j-1} - h_j$, the quantity h_{j-2} is really $h_{j-2}(i)$, and h_j is $h_j(i)$. The h_{j-1} has not been updated from the last pass through this box, so it is still $h_{j-1}(i-1)$, and so the equation is exactly that of (6-42). In this calculation, as in very many calculations, the computation part of the work is quite simple and straightforward compared to the bookkeeping job of ensuring that the quantities are available in the right place at the right time.

The FORTRAN subroutine given below will return the coefficients $D(I)$ in the expansion

$$f(x) = D(1) + D(2)x + D(3)x^2 + \cdots + D(K+1)x^k$$

given the coefficients $C(I)$ in the Chebyshev expansion

$$f(x) = (1/2)C(1) + C(2)T_1(x) + C(3)T_2(x) + \cdots + C(K+1)T_k(x)$$

The dimension statement allows for a maximum of 30 terms ($K < 30$) and must be matched by a dimension statement in the calling programs. The subscripts have been shifted by one from the flow chart to compensate for the lack of a zero subscript in FORTRAN.

```
      SUBROUTINE POWR(C,D,K)
      DIMENSION  C(30),D(30),H(30)
      KK=K+1
      DO 10 J=2,KK
   10 H(J)=0
      H(1)=1.
      D(1)= .5*C(1)
      I=1
    1 IF(I-K)2,2,20
    2 I2=I+2
      DO 3 J=I2,KK,2
      H(J)=2.*H(J-1)--H(J-2)
    3 D(I)=D(I)+C(J)*H(J)
      I=I+1
      H(I)=2.*H(I-1)
      H(2)=1.
      D(I)=C(I)*H(I)
      GO TO 1
   20 RETURN
      END
```

EXERCISE 19

1. Write the following as polynomials in x.

    a. $2T_0(x) + 3T_1(x) + T_2(x)$
    b. $T_0(x) - (1/2)T_2(x) + (1/4)T_4(x)$
    c. $T_1(x) - (1/4)T_3(x) + (1/16)T_5(x)$

2. Write the following as a sum of Chebyshev polynomials.

    a. $1 - x^2 + x^4$ b. $x - \dfrac{x^3}{4} + \dfrac{x_5}{8}$ c. $2 + x - x^2 + x^3$

3. Show that the coefficient c_n in the Chebyshev series

$$f(x) = (1/2)c_0 + \sum_{n=1}^{\infty} c_n T_n(x)$$

is given by

$$c_n = \frac{2}{\pi} \int_0^{\pi} f(\cos\theta)\cos n\theta \, d\theta$$

4. Write the first five terms and the nth term of the Chebyshev series for $\cos^{-1} x$. How many terms would be required for seven-digit accuracy for all values of x?

5. By the binomial theorem, write the first four terms of the expansion of $(1 - x)^{1/2}$. Telescope to three terms.

6. Telescope the polynomial

$$1 + .1x + .01x^2 + .001x^3 + .0001x^4 + .00001x^5$$

to terms through x^4.

7. Write a program which will call for input of an integer K and call SUBROUTINE CHEBY of Section 6.33 to find the coefficients of $T_i(x)$ in the Chebyshev expansion of x^k. Check SUBROUTINE CHEBY by hand for $K = 1, 2, 3$.

8. Write a program which will call for input of an integer K and coefficients $A(1), A(2), \ldots, A(K+1)$, and call SUBROUTINE CHEBY to compute the coefficients of a Chebyshev expansion. Then call SUBROUTINE POWR to rearrange the Chebyshev expansion as a polynomial, and print the resulting coefficients. How would these coefficients relate to $A(1), A(2), \ldots, A(K+1)$?

9. Run the program of problem 8 for $K = 6$ and $A(I) = 10^{-I}$. If your final answers are not the same as the starting $A(I)$'s, explain the discrepancy. If your FORTRAN compiler includes double-precision capability, rerun in double precision and again compare results.

10. Modify the program of problem 2 to set all coefficients of the Chebyshev expansion which are less than 5×10^{-8} equal to zero before calling SUB-ROUTINE POWR. Now input $K = 13$ and the coefficients of $\sin(\pi/2x)$ from Example 1, Section 6.33. Compare the answers with those obtained in Example 1, Section 6.33. Can you trust your routine to do automatic telescoping of power series?

6.4 RATIONAL APPROXIMATIONS

In using the Taylor series for $\tan^{-1} x$ in Section 6.25, it was found that faster convergence was obtained if $\tan^{-1} x$ was first multiplied by a polynomial in x. This actually had the effect of representing $\tan^{-1} x$ as a rational function, that is, as a quotient of two polynomials in x. In general, rational functions can be found which give better accuracy than polynomials for the same number of terms, and which give much better accuracy than any of the approximations discussed in the preceding sections. The so-called rational Chebyshev approximations are of the form

$$f(x) \approx R_{mk}(x) = \frac{\sum_{j=0}^{m} a_j x^j}{\sum_{j=0}^{k} b_j x^j}$$

where the a_j's and b_j's are chosen to minimize the maximum error in estimating $f(x)$ over some range of values for x. The process for determining the coefficients is too involved to reproduce here.*

The rational Chebyshev approximations are usually used for the intrinsic functions such as SIN, COS, EXP, etc., included in FORTRAN compilers, but are sufficiently difficult to generate that they are not of particular value to the analyst who desires to generate an approximating function that will receive only limited use.

6.5 ERROR ACCUMULATION IN EVALUATING POLYNOMIALS

In all the approximations discussed in the preceding sections, the actual computation involved in obtaining a function value is the evaluation of a polynomial (or possibly the quotient of two polynomials) of the form

$$P(x) = a_1 + a_2 x + \cdots + a_{n+1} x^n \tag{6-49}$$

* See, for example, Anthony Ralston and H. S. Wilf, eds., *Mathematical Methods for Digital Computers*, Vol. II, John Wiley & Sons, Inc., New York, 1967.

The degree of this polynomial was determined by consideration of the error involved in truncating a series and neglecting higher-order terms. In these error considerations, it was assumed that this truncation error was the only source of error; that is, the a_i's and x were exact numbers and the arithmetic was performed exactly. In fact, this is not the case. In a computer calculation, the a_i's and x will be approximate numbers and the arithmetic will be performed approximately. As was seen in Chapter 3, the errors from these sources can be of importance and must be considered. Indeed, it was shown that on a computer the value computed for an expression such as (6-49) could be different depending upon the order in which the terms are combined. The normal way of performing the calculation is to group the terms as

$$P(x) = a_1 + x(a_2 + x(a_3 + x(a_4 + \cdots + x(a_{n-1} + x(a_n + a_{n+1}x)) \cdots)))$$

$$(6\text{-}50)$$

This grouping has two advantages. First, it requires a near-minimum number of multiplications and additions, so it is fast. Second, it tends to add the smallest numbers first, since the higher powers of x in the Taylor or Chebyshev series tend to have small coefficients. It was seen in Chapter 3 that this tends to work for improved accuracy.

The error propagation in evaluating (6-50) can be inferred from the rules of Section 3.6. The required calculation can be represented by the steps

$$S_{n+1} = a_{n+1} \qquad\qquad (6\text{-}51)$$

$$S_i = a_i + xS_{i+1} \qquad \text{for } i = n, n-1, n-2, \ldots, 2, 1 \qquad (6\text{-}52)$$

Let ΔS_i be the absolute error in S_i, r the roundoff error in the a_i's and the machine arithmetic, and Δx the absolute error in x. Then by the rules of Section 3.6, the relative error in the product xS_{i+1} is

$$\frac{\Delta x}{|x|} + \frac{\Delta S_{i+1}}{|S_{i+1}|} + r$$

and the relative error in S_i is

$$\frac{\Delta S_i}{|S_i|} = \frac{|a_i|}{|S_i|} r + \frac{|xS_{i+1}|}{|S_i|} \left(\frac{\Delta x}{|x|} + \frac{\Delta S_{i+1}}{|S_{i+1}|} + r \right) + r$$

or

$$\Delta S_i = |a_i|r + \Delta x|S_{i+1}| + |x|\,\Delta S_{i+1} + r|x||S_{i+1}| + r|S_i| \qquad (6\text{-}53)$$

Let us first apply this relationship for $i = 1$, obtaining a relation for ΔS_1, which is ΔP, the error in the final polynomial

$$\Delta P = \Delta S_1 = r(|a_1| + |S_1| + |x||S_2|) + \Delta x |S_2| + |x|\, \Delta S_2$$

Now let us apply it again, with $i = 2$, to the $|x|\, \Delta S_2$ term in this expression, obtaining

$$\Delta P = r(|a_1| + |S_1| + |x||S_2|) + \Delta x |S_2|$$
$$+ r|x|(|a_2| + |S_2| + |x||S_3|) + |x|\, \Delta x |S_3| + |x|^2 \Delta S_3$$

and again with $i = 3$, $i = 4$, etc., until we finally obtain

$$\Delta P = r(|a_1| + |S_1| + |x||S_2|) + \Delta x |S_2|$$
$$+ r|x|(|a_2| + |S_2| + |x||S_3|) + |x|\, \Delta x |S_3|$$
$$+ r|x|^2(|a_3| + |S_3| + |x||S_4|) + |x|^2 \Delta x |S_4|$$
$$+ \cdots$$
$$+ r|x|^{n-1}(|a_n| + |S_n| + |x||S_{n+1}|) + |x|^{n-1} \Delta x |S_{n+1}| + |x|^n \Delta S_{n+1}$$

Now

$$|S_1| \leqslant |a_1| + |a_2||x| + |a_3||x^2| + \cdots + |a_{n+1}||x^n|$$

and

$$\Delta S_{n+1} = r|a_{n+1}|$$

Using these relations and regrouping terms in the above relations, we have

$$\Delta P \leqslant r(2|S_1| + 2|x||S_2| + \cdots + 2|x|^{n-1}|S_n| + |x^n||S_{n+1}|)$$
$$+ \Delta x(|S_2| + |x||S_3| + \cdots + |x|^{n-1}|S_{n+1}|)$$

If we arbitrarily add in a term $|x^n||S_{n+1}|$ and a term $\dfrac{\Delta x}{|x|}|S_1|$ to the right-hand side,

$$\Delta P \leqslant \left(2r + \frac{\Delta x}{|x|}\right)(|S_1| + |x||S_2| + |x^2||S_3| + \cdots + |x^n||S_{n+1}|)$$

Now

$$|S_1| \leqslant |a_1| + |a_2||x| + |a_3||x^2| + \cdots + |a_{n+1}||x^n|$$

and

$$|S_2| \leqslant |a_2| + |a_3||x| + \cdots + |a_{n+1}||x^{n-1}|$$

etc., and if S_1, S_2, and so on, are replaced by these approximations, the above becomes

$$\Delta P \leqslant \left(2r + \frac{\Delta x}{|x|}\right)(|a_1| + 2|a_2 x| + 3|a_3 x^2| + \cdots + (n+1)|a_{n+1}x^n|)$$

$$\frac{\Delta P}{|P|} \leqslant \left(2r + \frac{\Delta x}{|x|}\right)\frac{|a_1| + 2|a_2 x| + 3|a_3 x^2| + \cdots + (n+1)|a_{n+1}x^n|}{|a_1 + a_2 x + a_3 x^2 + \cdots + a_{n+1}x^n|}.$$

$$(6\text{-}54)$$

It is seen that the relative error introduced by roundoff is dependent on the sizes and the signs of the a_i, and the number of terms. Consider the case where x is subject to the same sort of roundoff errors as the a_i, so that $\Delta x/|x| = r$ (in actual application x may be subject to additional errors from other sources, which is why it was treated separately in developing the formula). Also, let us limit our consideration for the moment to the case where all a_i's are positive. For this case, the above formula can be written

$$\frac{\Delta P}{P} \leqslant 3r \frac{(d/dx)(xP(x))}{P(x)} \qquad (6\text{-}55)$$

or

$$\frac{\Delta P}{P} \leqslant 3r\left(1 + \frac{xP'(x)}{P(x)}\right) \qquad (6\text{-}56)$$

If, for example, the polynomial $P(x)$ is approximating the exponential function, then $P(x) \approx e^x$, $P'(x) \approx e^x$, and we have

$$\frac{\Delta P}{P} \approx 3r(1 + x)$$

If we restrict ourselves to values of x less than one, the relation error in P can be as much as six times that of the coefficients used, so that the value obtained for e^x will have about one place less accuracy than the coefficients used in the calculation.

If the a_i's are not all positive, relation (6-56) is not guaranteed to give an upper bound on the relative error. It will give a smaller value than relation (6-54). Sometimes it can be informative to apply (6-56) to cases where the a_i's are not all positive. If it indicates a large error, then (6-54) would indicate an even larger error, and we know that a problem situation exists. If it indicates a small error, then we cannot be convinced that there is no problem, however. Consider the case where $P(x)$ is approximating $\sin x$. Then application of (6-56) would give $P(x) \approx \sin x$, $P'(x) \approx \cos x$, and

$$\frac{\Delta P}{P} \leqslant 3r(1 + x \cot x)$$

The maximum value of $x \cot x$ is one, so that we obtain

$$\frac{\Delta P}{P} \leqslant 6r$$

indicating no serious accuracy problem in computing $\sin x$ from the polynomial expression. As we pointed out, this is not an actual upper bound, so it does not provide positive assurance that the error is small. In this particular case, an actual upper bound for the error in computing $\sin x$ can be obtained by noting that

$$\sinh x = x + \frac{x^3}{3!} + \frac{x^5}{5!} + \cdots$$

so that the expansion for the hyperbolic sine is the same as that for $\sin x$, except that all signs are positive. Hence we can say with rigor that for the approximation of $\sin x$ by a polynomial $P(x)$, relation (6-54) can be written

$$\frac{\Delta P}{P} \leqslant 3r \frac{(d/dx)(x \sinh x)}{\sin x}$$

$$= 3r \frac{\sinh x + x \cosh x}{\sin x}$$

The value of this expression increases as x increases. If we restrict our attention to values of x less than $\pi/2$, then the largest value, at $x = \pi/2$, is 3.94, or about 4. Hence

$$\frac{\Delta P}{P} \leqslant 12r$$

Hence the relative error in the sine can be roughly a factor of 10 higher than that of the coefficients, or the values of the sine can have one less correct significant figure than the input coefficients used.

As mentioned earlier, several overestimates were made in deriving (6-54) as an upper bound for the error. A closer bound for the error can be found by using (6-53) directly. The error estimate can be included in a FORTRAN program right with the computation of the value itself. Let $A(1)$, $A(2)$, ..., $A(N + 1)$ be the coefficients for the polynomial

$$P = A(1) + A(2)x + A(3)x^2 + \cdots + A(N + 1)x^n$$

and let R be the relative error in the coefficient and DX the absolute error in X. Then the statements

```
      P=A(N+1)
      AX=ABS(X)
      AP=ABS(P)
      DP=R*AP
      DO 10 J=1,N
      I=N+1-J
      APOLD=AP
      P=A(I)+P*X
      AP=ABS(P)
   10 DP=R*(ABS(A(I))+AP+AX*APOLD)+DX*APOLD+AX*DP
```

will compute P, the value of the polynomial, and DP, the maximum error in this value.

EXERCISE 20

1. The Taylor series for $\ln(1 + x)$ is

$$\ln(1 + x) = x - \frac{x^2}{2} + \frac{x^3}{3} - \cdots$$

Using relation (6-54), estimate the error in using 25 terms of this series to estimate $\ln(1 + x)$ for $\theta < x < 1/2$, if seven-place arithmetic is used.

2. Write a program which will
    a. Input the quantities $A(1) = A(3) = A(5) = A(7) = A(9) = A(11) = 0$, $A(2) = 1$, $A(4) = -1/3!$, $A(6) = 1/5!$, $A(8) = -1/7!$, $A(10) = 1/9!$, $A(12) = -1/11!$
    b. Calculate P and DP from the FORTRAN statements given in Section 6.5.
    c. Call the library sine routine to compute $\sin x$.
    d. Compute and print $\sin(x)$, $P(x)$, $(\sin x - P(x))$, and $DP(x)$ for $x = .1$, $.2, .3, \ldots, 1.5$.
        What do you conclude about the error estimate DP compared to actual errors $\sin x - P(x)$ for this case?

3. Perform problem 2, using inputs of

$$A(1) = 1, \; A(I) = \frac{1}{(I-1)!} \qquad \text{for } I = 2, 3, \ldots, 11.$$

Compare with the library routine EXP(x) for $x = .1, .2, \ldots, 1.0$.

6.6 ERROR PROPAGATION THROUGH FUNCTIONS

Thus far in this chapter we have concentrated on two sources of error in the evaluation of functions. The first was the truncation error in discarding the high-order terms in the approximating polynomial and the second was the roundoff error associated with the use of approximate values for coefficients and approximate arithmetic. Let us now assume that adequate measures have been taken to make these errors acceptably small, and concern ourselves with a different source of error, the intrinsic error in the inputs to a calculation. That is, assume that given x, we can find $f(x)$ accurately, and we wish to know the error in $f(x)$ when x is in error. To make the considerations more general, consider the case where we have several input quantities $u_1, u_2, u_3, \ldots, u_n$, which are to be used to calculate some quantity N. We can indicate this relationship by writing

$$N = f(u_1, u_2, u_3, \ldots, u_n)$$

Now if small changes are made in u_1, u_2, etc., by amounts Δu_1, Δu_2, etc., we can calculate a quantity called the differential of N by the relation

$$dN = \frac{\partial f}{\partial u_1} \Delta u_1 + \frac{\partial f}{\partial u_2} \Delta u_2 + \frac{\partial f}{\partial u_3} \Delta u_3 + \cdots + \frac{\partial f}{\partial u_n} \Delta u_n \qquad \textbf{(6-57)}$$

This quantity dN is approximately equal to ΔN, the error in N when u_1 is replaced by $u_1 + \Delta u_1$, u_2 by $u_2 + \Delta u_2$, etc. To demonstrate the meaning of this formula, we will use it to rederive the error rules for addition, subtraction, multiplication, and division given in Chapter 3.

For addition, we have

$$N = u_1 + u_2$$

so that

$$dN = \Delta u_1 + \Delta u_2$$

the situation expressed in Section 3.51.

For subtraction, we have

$$N = u_1 - u_2$$

so that

$$dN = \Delta u_1 - \Delta u_2$$

We must remember that Δu_1 and Δu_2 can be either positive or negative, so if we are interested in the maximum error, it is $|\Delta u_1| + |\Delta u_2|$.

For multiplication,

$$N = u_1 u_2$$

or, if we take logarithms, we can write

$$\ln N = \ln u_1 + \ln u_2$$

If now we take the total differential, we have

$$dN/N = \Delta u_1/u_1 + \Delta u_2/u_2$$

This rule corresponds to the statement in Section 3.53 concerning relative errors.

For division, if

$$N = u_1/u_2$$

then

$$\ln N = \ln u_1 - \ln u_2$$

and

$$dN/N = \Delta u_1/u_1 - \Delta u_2/u_2$$

As in subtraction, to obtain an estimate of the maximum possible error, we must allow for the case where Δu_1 and Δu_2 are of opposite sign, so we must consider

$$|\Delta u_1/u_1| + |\Delta u_2/u_2|$$

as the correct expression for estimating the error.

For more complicated expressions, equation (6-57) can be applied directly to give an expression for the error, remembering that in each case signs of the individual errors Δu_1, Δu_2, etc., should be chosen in such a way as to give the maximum result.

6.61   Error Accumulation for the Exponential Function

As a demonstration of the problem of error accumulation, let us apply relation (6-57) to the function

$$y = e^x$$

Applying relation (6-57) with $f(x) = e^x$, we have

$$dy = e^x \Delta x$$

where dy is the absolute error in y. The relative error is

$$dy/y = \Delta x$$

Hence the *relative* error in the computed value of e^x is equal to the *absolute* error in x itself. The disturbing feature of this result can be seen from the following example:

Example 1. Suppose $x = 100$, to three correct significant figures. What is the relative error in $y = e^x$?

The limit of the absolute error in x is

$$\Delta x = .5$$

Hence the relative error in y is .5, or 50%. The value of y has *no* significant figures!

The above example demonstrates that, even though our subroutines may be designed to compute to many correct significant digits, the problem of error accumulation is still with us when we use these subroutines.

6.62   Error Estimate by Formula

The example of Section 6.61 was indicative of the problem associated with the evaluation of any function of one or more independent quantities or approximate numbers. No calculation of this sort can be considered complete until some sort of assessment of the error has been made. For functions which are not too complex, the relation of Section 6.6 can be used for this purpose. Further examples will be given to illustrate its use.

Example 1.   The function $y = a \sin b$ is to be calculated, where $a = 30.0$ and $b = .45$, the numbers being correct to the number of significant digits shown. Find the absolute and relative errors in y.

By the formula of Section 6.6,

$$dy = \frac{\partial y}{\partial a} \Delta a + \frac{\partial y}{\partial b} \Delta b$$

$$= \sin b\, \Delta a + a \cos b\, \Delta b$$

$$= (.435)(.05) + (30.0)(.900)(.005)$$

$$= .022 + .14 = .16$$

or the absolute error is .16.

Since $y = (30.0)(.435) = 13.05$, the relative error is about .16/13, or roughly 1%.

Example 2.   The function $y = a \sin b$ is to be calculated, where $a = 30.0$ and $b = \pi/6$, the number a being correct to three significant digits and the number b being exact. Find the absolute and relative errors in y.

As before, we may write

$$dy = \frac{\partial y}{\partial a} \Delta a + \frac{\partial y}{\partial b} \Delta b$$

but since b is exact, $\Delta b = 0$, so the term $(\partial y/\partial b)\, \Delta b$ will drop out. This points up the fact that, whenever the function under consideration involves *exact* numbers, they can be treated as constants throughout, and the expression need not be differentiated with respect to them. All quantities which may be in error, whether constants or variables, should be treated as variables in applying the error formula of Section 6.6. (As indicated earlier, even the constants are subject to machine roundoff error. In the present case, and in many cases, the truncation error involved in roundoff is so small compared to other sources of error that it can safely be ignored.)

For the present problem, then,

$$dy = \frac{dy}{da} \Delta a$$

$$= \sin b\, \Delta a$$

$$= (.5)(.05)$$

$$\approx .025$$

The absolute error is .025 and the relative error is .025/15, or about 0.2%.

Example 3. The function $y = 2.0 \sin x + 3 \ln x$ is to be evaluated for $x = 1.26$. The constant 2.0 and the value of x are correct only to the number of significant digits shown. The constant 3 is exact. Find the absolute and relative errors in y.

Since the number 2.0 may be in error, it is best to replace it by a symbol before applying the error formula. Thus

$$y = a \sin x + 3 \ln x$$
$$dy = \Delta a \sin x + (a \cos x + 3/x) \, \Delta x$$
$$= (.05)(.952) + [(2.0)(.306) + 3/1.26](.005)$$
$$= .048 + [.612 + 2.38](.005)$$
$$= .048 + .015 = .063$$

The absolute error is $.063$. Since

$$y = (2.0)(.952) + 3(.231)$$
$$= 1.90 + .69 = 2.59$$

the relative error is $.063/2.59$, or about 2%.

Example 4. Perform the calculation of Example 3 for $x = .65$.

Substituting in the formula of the previous exercise, we have

$$dy = (.05)(.605) + [(2.0)(.796) + 3/.65](.005)$$
$$= .030 + [1.59 + 4.62](.005)$$
$$= .030 + .031 = .062$$

Again, the absolute error is about $.062$.
   However, since

$$y = (2.0)(.605) + (3)(-.431)$$
$$= 1.21 - 1.29 = -.08$$

the relative error is about $.06/.08 = .75$!

Although all the numbers used in this case were accurate to 2% or better, the final result had a 75% error! Closer inspection shows that this error came from the operation remarked as dangerous in Chapter 3, the subtraction of two nearly equal quantities. For $x = .65$, $\ln x$ is negative and the quantities $2 \sin x$ and $3 \ln x$ are very nearly equal in absolute value. The subtraction involved in finding y above resulted in loss of the two leading significant figures.

Example 5.   The function $y = ke^{-\mu x}/x^2$ is to be evaluated for 100 values of x, ranging from 100 to 5000, and subject to an experimental error of one unit. The constants are $\mu = 3.0 \times 10^{-3}$ and $k = 1.3 \times 10^7$, each accurate to the number of significant digits indicated. Find the absolute and relative errors in y for a low, medium, and high value of x (use $x = 100$, 700, and 5000).

For functions such as this, where only multiplications, divisions, and powers are involved, it is convenient to take logarithms and then differentiate, thus obtaining relative error directly. Thus

$$\ln y = \ln k - \mu x - 2 \ln x$$
$$dy/y = \Delta k/k - \mu\,\Delta x - x\,\Delta\mu - 2\,\Delta x/x$$

For $x = 100$,

relative error

$$= \frac{dy}{y} = \frac{.05 \times 10^7}{1.3 \times 10^7} - (3.0 \times 10^{-3})(-1) - (100)(-.05 \times 10^{-3}) - \frac{2(-1)}{100}$$

(signs of Δk, Δu, and Δx were chosen to maximize the error)

$$= .038 + .003 + .005 + .02$$
$$= .066 \quad \text{or} \quad 7\%$$

Since

$$y = 1.3 \times 10^7 e^{-(3.0 \times 10^{-3})(100)}/(100)^2$$

$$= 9.6 \times 10^2 \quad \text{or} \quad 960$$

the absolute error is

$$(.07)(9.6 \times 10^2) = .7 \times 10^2 \quad \text{or} \quad 70$$

For $x = 700$,

relative error

$$= \frac{dy}{y} = \frac{.05 \times 10^7}{1.3 \times 10^7} - (3.0 \times 10^{-3})(-1) - (700)(-.05 \times 10^{-3}) - \frac{2(-1)}{700}$$

$$= .038 + .003 + .035 + .003 = .079 \quad \text{or} \quad 8\%$$

$$y = 1.3 \times 10^7 e^{-(3.0 \times 10^{-3})(700)}/(700)^2 = 3.3$$

so the absolute error is

$$(.08)(3.3) \approx .3$$

For $x = 5000$,

relative error

$$= \frac{dy}{y} = \frac{.05 \times 10^7}{1.3 \times 10^7} - (3.0 \times 10^{-3})(-1) - (5000)(- .05 \times 10^{-3}) - \frac{2(-1)}{5000}$$

$$= .038 + .003 + .25 + .0004$$

$$= .29 \quad \text{or} \quad 29\%$$

$$y = 1.3 \times 10^7 e^{-(3.0 \times 10^{-3})(5000)}/(5000)^2$$

$$= 1.6 \times 10^{-7}$$

so the absolute error is about

$$(.29)(1.6 \times 10^{-7}) = .5 \times 10^{-7}$$

Comparing the values of the errors for the three different values of x gives us some feel for the errors throughout the range of values of x. The error is between 5 and 10% for the smaller values of x, and increases to about 30% at the extremely large values of x. We cannot be sure that the percentage error remains in the ranges indicated for all values of x, since we have studied only three particular values. If we wish a surer picture of the behavior of the error throughout the entire range of x, we can look at the expression for the relative error with numerical values substituted for all quantities except x:

$$\frac{dy}{y} = \frac{.05 \times 10^7}{1.3 \times 10^7} - (3.0 \times 10^{-3})(-1) - x(- .05 \times 10^{-3}) - 2(-1)/x$$

$$= .038 + .003 + .00005x + 2/x$$
$$= .041 + .00005x + 2/x$$

We can now study this expression as a function of x, making a plot of it if desired, and thus obtain a more complete picture of the relative error throughout the range of values of x. Figure 6-5 indicates this behavior. The relative error is .066 for $x = 100$, decreases to .061 at $x = 200$, then increases continuously to .29 at $x = 5000$.

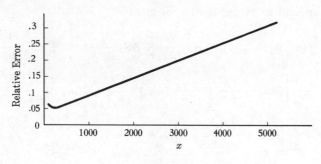

Figure 6-5

6.63   Error Estimate by Computer Trial

The application of the error formula (6-57) of Section 6.6 is straightforward as long as the functions involved are simple enough to be differentiated easily. For extremely complex functions, however, the process may be impracticable because of the difficulties involved in finding the derivatives or in evaluating the derivatives once found. In such cases the process of estimating the errors is often ignored completely. This is indeed unfortunate, since these are just the cases in which error accumulation is most likely to have some unexpected effect on the accuracy of the final answer. Instead of neglecting the problem, one should attempt to estimate the error by other methods. One method, quite adaptable for computer use, is to perform the calculation several times, each time varying one or more of the quantities which may be in error, and observing the effect on the final answer. Used properly, this method can give a more valid index of the error than does the error formula of Section 6.6. Again, as in that section, assume that the quantities u_1, u_2, u_3, ..., u_n are to be combined to form some resulting number N, where

$$N = f(u_1, u_2, \ldots, u_n) \tag{6-58}$$

Suppose now that small changes $\Delta u_1, \Delta u_2, \ldots, \Delta u_n$ are made in the quantities $u_1, u_2, \ldots, u_n$. Then N will be changed to a new value $N + \Delta N$, given by

$$N + \Delta N = f(u_1 + \Delta u_1, u_2 + \Delta u_2, \ldots, u_n + \Delta u_n) \tag{6-59}$$

The Taylor expansion for a function of one variable given in Section 6.2 has its analogue for functions of several variables, the chief difference being that the ordinary derivatives are replaced by partial derivatives. This expansion applied to expression (6-58) gives

$$N + \Delta N = f(u_1, u_2, \ldots, u_n) + \Delta u_1 \frac{\partial f}{\partial u_1} + \Delta u_2 \frac{\partial f}{\partial u_2} + \cdots + \Delta u_n \frac{\partial f}{\partial u_n}$$

$$+ \frac{1}{2} \left[(\Delta u_1)^2 \frac{\partial^2 f}{\partial u_1{}^2} + \cdots + (\Delta u_n)^2 \frac{\partial^2 f}{\partial u_n{}^2} + 2\Delta u_1 \, \Delta u_2 \frac{\partial^2 f}{\partial u_1 \, \partial u_2} + \cdots \right]$$

$$+ \frac{1}{3!} \left[(\Delta u_1)^3 \frac{\partial^3 f}{\partial u_1{}^3} + \cdots \right] + \cdots \qquad \textbf{(6-60)}$$

If the errors Δu_1, Δu_2, $\ldots$, Δu_n are so small that we can neglect their squares, products, and higher powers, we can write (6-60) as

$$N + \Delta N \approx f(u_1, u_2, \ldots, u_n) + \Delta u_1 \frac{\partial f}{\partial u_1} + \Delta u_2 \frac{\partial f}{\partial u_2} + \cdots + \Delta u_n \frac{\partial f}{\partial u_n} \quad \textbf{(6-61)}$$

or, subtracting (6-58) from (6-61),

$$\Delta N \approx \Delta u_1 \frac{\partial f}{\partial u_1} + \Delta u_2 \frac{\partial f}{\partial u_2} + \cdots + \Delta u_n \frac{\partial f}{\partial u_n} \qquad \textbf{(6-62)}$$

This is just the error formula (6-57) of Section 6.6. Thus that error formula is merely an approximation to the value of ΔN defined by relation (6-59). Direct application of relation (6-59) should give a better estimate of the error, since it does not neglect squares or products of errors. The task, once the computer program for evaluating the function f is prepared, is quite straightforward in concept. We merely run the calculation twice, once with input values u_1, u_2, $\ldots$, u_n, and once with input values $u_1 + \Delta u_1$, $u_2 + \Delta u_2$, $\ldots$, $u_n + \Delta u_n$. The difference in the results is then the absolute error. There is one difficulty, however. To obtain the maximum error we must choose the signs of the errors Δu_1, Δu_2, $\ldots$, Δu_n so as to combine in the worst possible way. It is not usually possible to do this by inspection. Consequently, it is necessary first to change each one separately and observe how much N is increased or decreased. In a sense, this procedure is somewhat analogous to applying relation (6-62). Since by the definition of a partial derivative

$$\frac{\partial f}{\partial u_1} = \lim_{u_1 \to 0} \frac{f(u_1 + \Delta u_1, u_2, u_3, \ldots, u_n) - f(u_1, u_2, \ldots, u_n)}{\Delta u_1}$$

then, for well-behaved functions,

$$\frac{\partial f}{\partial u_1} \Delta u_1 \approx f(u_1 + \Delta u_1, u_2, u_3, \ldots, u_n) - f(u_1, u_2, \ldots, u_n)$$

Hence the difference between the value of N when $u_1 + \Delta u_1$, u_2, u_3, $\ldots$, u_n are used in the calculation and that when u_1, u_2, $\ldots$, u_n are used is roughly

$(\partial f/\partial u_1)\,\Delta u_1$. If we do this for each variable and then add the absolute values of the resulting errors in N from all the calculations, we have an error estimate of the same type as is given by relation (6-62). To determine if the higher-order terms contained in relation (6-60) but ignored in relation (6-62) are important, it is usually wise to make a final calculation changing all the variables simultaneously. In each of the calculations in which only one variable has been changed, we observe whether N is increased or decreased, and then make a final calculation in which all variables are changed in directions chosen to produce the same direction change in N. The following set of steps outline the procedure:

(1)   Calculate $N = f(u_1, u_2, \ldots, u_n)$.

(2)   Calculate $N_i = f(u_1, u_2, \ldots, u_i + \Delta u_i, \ldots, u_n)$ for $i = 1$ to n.

(3)   Calculate $N + \Delta N = f(u_1 + a_1\,\Delta u_1,\ u_2 + a_2\,\Delta u_2,\ \ldots,\ u_n + a_n\,\Delta u_n)$, where $a_i = +1$ if $N_i > N$ and $a_i = -1$ if $N_i < N$.

A word of caution should be given in connection with the use of the above procedure. The computation of the values of N and each N_i, and $N + \Delta N$, will be subject to the normal errors associated with the use of approximate numbers. If too small a value is chosen for the Δu_i, the change in N caused by this deliberate alteration may be disguised by the change induced by different roundoff errors. The value of the Δu_i must be chosen large enough that its effect is not lost in the "noise" of roundoff errors.

Example 1.   Use the method just described to estimate the error for Example 3, Section 6.62.

In order to make the procedure clearer, the problem will be rewritten in the notation used in the description above. We wish to find

$$N = f(u_1, u_2)$$

where

$$f(u_1, u_2) = u_1 \sin u_2 + 3 \ln u_2$$

and

$$u_1 = 2.0 \qquad u_2 = 1.26$$
$$\Delta u_1 = .05 \qquad \Delta u_2 = .005$$

Following the steps above, we calculate:

(1) $N = f(u_1, u_2) = 2.0 \sin 1.26 + 3 \ln 1.26 = 2.59$.

(2) $N_1 = f(u_1 + \Delta u_1, u_2) = 2.05 \sin 1.26 + 3 \ln 1.26 = 2.64$,
   $N_2 = f(u_1, u_2 + \Delta u_2) = 2.0 \sin 1.265 + 3 \ln 1.265 = 2.61$.

(3)   Since $N_1 > N$, $a_1 = +1$. Since $N_2 > N$, $a_2 = +1$. Hence $N + \Delta N = f(u_1 + a_1\,\Delta u_1,\ u_2 + a_2\,\Delta u_2) = 2.05 \sin 1.265 + 3 \ln 1.265 = 2.66$, so that $\Delta N = 2.66 - 2.59 = .07$.

Example 2. The expression $y = \ln(a + \sqrt{b + e^{c\,\tan^{-1}x}})$ is to be calculated for values of x from 0 to 10 in order to make a graph. The values of the constants are $a = 2.0 \pm .1$, $b = 3.5 \pm .2$, $c = 1.0 \pm .1$. Draw a flow chart for the calculation, which includes an error estimate for each value of x.

Since in this problem we are allowed to choose the values of x, we may assume them to be precise, so that no error in x need be considered. The quantities a, b, and c are subject to errors $\Delta a = .1$, $\Delta b = .2$, and $\Delta c = .1$. Figure 6-6 shows the flow chart only for a single value of x. Additions to the chart to cause the calculation to be performed for a sequence of values of x are left to the reader.

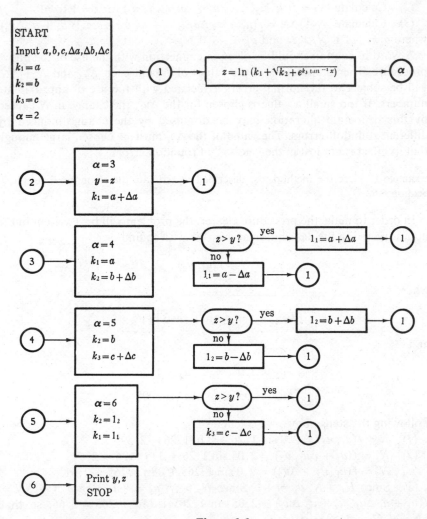

Figure 6-6

In the discussion of flow charts it was pointed out that the charts could be made detailed or crude as the occasion required. In the example just given, the calculation of the very complex function y was relegated to a single box. Most of the chart is devoted to outlining the selection of the values of k_1, k_2, and k_3 to be used in the calculation. The variable connector symbol was used to good advantage in this chart to indicate the reuse of the basic formula for y several times with different values for k_1, k_2, and k_3.

The FORTRAN program given below follows the flow chart rather closely but does include using a sequence of values of x:

```
      READ 101,A,B,C,DA,DB,DC,DX
      X=0
      L=10./DX
      DO 14 I=1,L
      FK1=A
      FK2=B
      FK3=C
      J=1
    1 Z=LOGF(FK1+SQRTF(FK2+EXPF(FK3*ATANF(X))))
      GO TO (2,3,6,9,13),J
    2 J=2
      Y=Z
      FK1=A+DA
      GO TO 1
    3 J=3
      FK1=A
      FK2=B+DB
      IF(Z-Y)4,4,5
    4 FL1=A-DA
      GO TO 1
    5 FL1=A+DA
      GO TO 1
    6 J=4
      FK2=B
      FK3=C+DC
      IF(Z-Y)7,7,8
    7 FL2=B-DB
      GO TO 1
    8 FL2=B+DB
      GO TO 1
    9 J=5
      IF(Z-Y)10,10,11
   10 FL3=C-DC
      GO TO 12
```

```
11  FL3=C+DC
12  FK1=FL1
    FK2=FL2
    FK3=FL3
    GO TO 1
13  PRINT 101,Y,Z
14  X=X+DX
    STOP
101 FORMAT(7E10.4)
    END
```

This estimate by computer trial can be done much more simply in an inter-active fashion from a remote terminal. In the language of Section 5.42, a suitable remote-terminal program is

```
1   1 PRINT, "A,B,C,X"
2     INPUT, A,B,C,X
3     Z=ALOG(A+SQRT(B+EXP(C*ATAN(X))))
4     PRINT, "Z=",Z
5     GO TO 1
6     END
```

Specific values of A, B, C, and X can be run as desired, using the computer as a supercalifragilistic desk calculator.

6.64   Limitations on Validity of the Error Estimate

It might seem that the procedure outlined above and illustrated in the example would provide an absolutely certain means of obtaining an upper estimate on the error. Surprisingly enough, such is *not* the case. It is possible, although unusual, for the error to be much larger than indicated by the error estimate. If the function f happens to behave in a sufficiently erratic fashion, it may be that the quantities

$$N = f(u_1, u_2, \ldots, u_n)$$

and

$$N + \Delta N = f(u_1 + \Delta u_1, u_2 + \Delta u_2, \ldots, u_n + \Delta u_n)$$

may be nearly equal, but that for some set of values of the u's intermediate between $u_1, u_2, \ldots, u_n$ and $u_1 + \Delta u_1, u_2 + \Delta u_2, \ldots, u_n + \Delta u_n$, the function f has a very different value. More advanced studies show that the estimate of

ΔN given above can be depended upon, that is, N will change only slowly as the u's change, if the following conditions are satisfied:

(*1*) All the partial derivatives $\partial f/\partial u_i$ exist and are continuous at the point $(u_1, u_2, \ldots, u_n)$.

(*2*) The errors $\Delta u_1, \Delta u_2, \ldots, \Delta u_n$ are sufficiently small. (We shall not try to define what is meant by "sufficiently." This is properly a subject for an advanced calculus course.)

An example will illustrate what may happen when these conditions are not satisfied.

Example 1. Compute $y = (1/16) \ln(\tan \sqrt{1 + x^2})^2$ for $x = 1.211$, and estimate the error, where the constants in the expression are exact, and x is accurate to the number of digits shown.

Let us attempt to estimate the error by computing y for the value x and for the value $x + \Delta x$, where $\Delta x = .0005$. For $x = 1.211$, $y = 1.02$. For $x + \Delta x = 1.2115$, $y + \Delta y = 1.14$. Hence we would be led to believe that the maximum error is $\Delta y = .12$. However, if we believe this, we are badly misled. For example, if the true value of x were 1.2113633, the value of y would be 1.39, a value differing from our original value by $.37$. Hence the maximum error is clearly more than $.12$. As a matter of fact, there is a value of x between 1.211 and 1.2115 for which y is infinite, so the error in y may be infinite! We can see this as follows: Since $\tan \pi/2 = \infty$, y is infinite when $\sqrt{1 + x^2} = \pi/2$. This is true when $x = \sqrt{-1 + (\pi/2)^2}$. The exact value of this number is between the two approximate numbers 1.2113633 and 1.2113634. The peculiar nature of this function in the region of interest is apparent from its graph, Figure 6-7. It has a vertical asymptote at the value $x = \sqrt{-1 + (\pi/2)^2}$, and the two values of x, 1.211 and 1.2115, happen to give nearly equal values of y on opposite sides of this asymptote. In this example the difficulty arises from the fact that the derivative dy/dx becomes

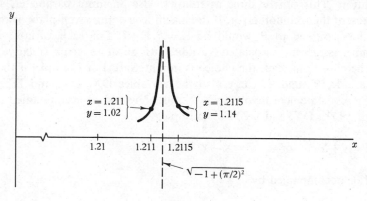

Figure 6-7

infinite within the interval x to $x + \Delta x$. This gives some idea as to what is meant by errors "sufficiently small" in condition (2). Δx must be small enough that the derivative dy/dx does not become inordinately large in the interval x to $x + \Delta x$.

The above example was included to demonstrate that calculation of an error estimate is no sure defense against the accidental acceptance of an answer grossly in error. The only sure protection is a detailed knowledge of the behavior of the function involved. This is no excuse, however, for failure to attempt to evaluate the effects of errors in the constants and variables involved in a calculation, both theoretically and experimentally by additional computer runs varying the values of any uncertain quantities. Stated another way, *part of the performance of any calculation is the testing of the sensitivity of the results to variations in the parameters involved in the problem.* In some cases, as in the preceding example, this testing of the sensitivity may be misleading, but such cases are fortunately rare. (They do however, tend to follow Gumperson's law, which, stated roughly, is: Those events which have a low probability of occurence tend to occur at the least opportune time. This law has been cited as the reason for the ringing of the telephone when one is in the bathtub, or failure of the car to start when one is about to drive to an important engagement. Gumperson reportedly met his death by being struck by an automobile. He was walking down the left side of the road in order to face traffic but was struck down from behind by a car driven by a visiting foreigner who was accustomed to driving on the left-hand side.)

6.65   Detailed Error Bounding in the Program

In any program it is possible to include error-estimation equations based on Table I of Section 3.6, and carry an error estimate right along with the calculation. This can be done as follows. Use an input parameter, say R, to represent the roundoff error. If the machine is doing seven-place arithmetic the value assigned to R would be $R = 5.E-7$. For each variable in the program, assign an associated variable, its absolute error (relative error could be chosen instead, the choice is immaterial). For example, if the variables are X, Y, and Z, carry also the variables DX, DY, and DZ. Each FORTRAN statement involving X, Y, and Z would be accompanied by one involving DX, DY, and DZ. For example,

$$Z = X + Y \quad \text{or} \quad Z = X - Y$$

would be accompanied by

$$DZ = DX + DY + R*ABS(Z)$$

and

$$Z = X*Y \quad \text{or} \quad Z = X/Y$$

would be accompanied by

$$DZ = ABS(Z)*(DX/ABS(X) + DY/ABS(Y)) + R$$

Use of an implicit function, such as

$$Z = SIN(X)$$

would be accompanied by a statement based on equation (6-57), or in this case,

$$DZ = ABS(COS(X))*DX$$

Statements involving combinations of these quantities would be broken down into the individual operations and treated as above.

Example 1. Write a program with error bounding for the evaluation of the expression

$$y = ax + b \cos x + c$$

A FORTRAN expression for the calculation of y is

$$Y = A*X + B*COS(X) + C$$

The compiler will create a program which will evaluate this from left to right, first finding ax, then $b \cos x$ and adding, then adding c. The error bounding should be done in the same fashion, and can be done by a series of FORTRAN statements rather than a single one. A suitable program is

```
READ 101,A,B,C,X
READ 101,DA,DB,DC,DX,R
U = A*X
DU = ABS(U)*(DA/ABS(A) + DX/ABS(X) + R)
V = COS(X)
DV = ABS(SIN(X))*DX
W = B*V
DW = ABS(W)*(DB/ABS(B) + DV*ABS(V) + R)
V = U + W
DV = DU + DW + R*ABS(V)
```

```
        Y=V+C
        DY=DV+DC+R*ABS(Y)
        PRINT 101,Y,DY
101     FORMAT(5E10.4)
        STOP
        END
```

Clearly, this program is longer, slower, and more painful to prepare than a simple one which computes y by a single FORTRAN statement and prints the answer. The time and trouble involved would be warranted only where one had reason to suspect serious accuracy problems.

6.7   LOSS OF SIGNIFICANT DIGITS IN SUBTRACTION

In Chapter 3 it was pointed out that the primary cause of loss of accuracy in calculations was the introduction of leading zeros in subtraction of two nearly equal numbers. In that chapter it was stated that, whenever such an event might occur, special programming precautions must be taken to avoid the difficulty or at least to make the programmer aware that a dangerous point in the calculation has arisen.

An error-bounding program as described in Section 6.6 will ordinarily detect the problem, although, as indicated there, such a technique is expensive in machine time and effort and is not guaranteed to flag the accuracy problems. In some cases it is possible to anticipate where accuracy loss during subtraction may occur, and make provision at those points to provide protection without encumbering the entire program with additional FORTRAN statements for error bounding. Some techniques for accomplishing this will be discussed.

6.71   Programmed Warning of Accuracy Loss

The first problem in protecting against accuracy loss in subtraction is to recognize when such an error may occur in a program. Any subtraction command (or addition command, since the machine adds algebraically) may be guilty if the numbers being handled happen to be of the right size. A program may work beautifully for certain sets of input and yet produce worthless answers for other sets because of loss of leading digits in subtractions. Sometimes it is possible to recognize during programming that such a danger exists, and in other cases it may be virtually impossible to recognize a danger spot. When a potential danger spot in the program can be recognized, programming to provide warning of accuracy loss may be advisable.

Suppose, for example, that a part of our program contains the statement

$$Y = A - B$$

Suppose further that we know that this part of the program will work satisfactorily for most sets of input numbers, but we suspect that in some cases the values of A and B at this point may be nearly equal. We fear that if as many as four leading zeros are produced by the subtraction, our final answer will not be trustworthy. We would like the machine to warn us if four figures are lost at this point in the calculation. It is an easy matter to write a section for the program which will accomplish this. We note first that, if four digits are lost, then the difference obtained as the result of the subtraction is roughly 10^{-4} times the minuend or subtrahend. The following statements will test for this occurrence and print out a warning if it does happen:

```
  2 Y=A-B
    IF(ABS(Y)-.0001*ABS(A))9,3,3
  9 PRINT 101
101 FORMAT(26H ACCURACY LOSS,STATEMENT 2)
  3 (continuation of program)
```

After the calculation of $A - B$, a test is inserted which will determine if the difference is less than 10^{-4} times A. If it is, the print command is executed. If there is no excessive accuracy loss, the program continues without the printout. A section of flow chart which describes this operation might appear as in Figure 6-8.

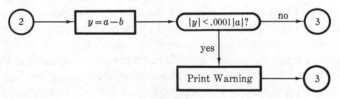

Figure 6-8

Example 1. The program given below will compute the third side a of a triangle, given sides b and c and the included angle A, the angle being denoted by the FORTRAN variable AA. The formula used is the law of cosines. Rewrite this program to give warning when subtraction results in the loss of two or more digits.

```
    READ 101,B,C,AA
    A=SQRT(B*B+C*C-2.*B*C*COS(AA))
    PRINT 101,A
101 FORMAT(3E12.4)
    STOP
    END
```

A program which will give warning is

```
      READ 101,B,C,AA
      U=B*B+C*C
      V=2.*B*C*COS(AA)
      IF(ABS(U-V)-.01*ABS(U))2,2,3
    2 PRINT 102
      STOP
    3 A=SQRT(U-V)
      PRINT 101,A
      STOP
  101 FORMAT(3E12.4)
  102 FORMAT(29H ACCURACY LOSS IN COMPUTING A)
      END
```

In this example we arbitrarily settled on the loss of two leading digits as the danger point, the point at which we desire warning. It is fair to ask how such a requirement might come about. In a practical problem the loss of digits we could tolerate would be determined by accuracy of our knowledge of the input numbers b, c, and A and the required accuracy of the result. A careful error analysis using the general error formula of Section 6.6 would be quite difficult, but a loose line of reasoning following the accuracy theorems of Chapter 3 is sufficient to indicate how the error in the final result will depend on that of the input numbers. For example, suppose b, c, and A are each known to 1 %. Then b^2, c^2, and bc have a relative error of about 2 %. The absolute error in cos A is sin A ΔA, so the relative error in cos A is tan A ΔA. A little study of the expression $b^2 + c^2 - 2bc \cos A$ discloses that the quantities $(b^2 + c^2)$ and $(2bc \cos A)$ are nearly equal only when A is near zero and b is nearly equal to c. When A is near zero, tan A is small, so the relative error in cos A is small. Hence the term $2bc \cos A$ has a relative error of about 2 %. From these values of relative error, we see that the terms $(b^2 + c^2)$ and $(2bc \cos A)$ each have about two significant figures. If one is lost in the subtraction, a^2 has one significant figure, or is accurate to about 10 % (which means a is accurate to about 5 %). If two significant digits are lost, a^2 may have no significant digits.

An error analysis of the type just given, while not at all precise, is usually sufficient to give guidance as to the acceptability of loss of leading significant digits in subtraction.

6.72   Programming to Avoid Accuracy Loss in Subtraction

In Section 6.71 it was demonstrated that it is sometimes possible to program a machine to give automatic warning in the event of serious accuracy loss in subtraction. It would be desirable to have the machine take automatic corrective action instead of merely issuing a warning. This can be done in many cases. Let us first consider the case in which we have only one uncertain input number, x. Suppose we are evaluating the expression

$$y = f_1(x) - f_2(x)$$

where f_1 and f_2 are functions calculable to a high degree of accuracy by standard computer subroutines. Then by formula (6-57) of Section 6.6 the absolute error in y is

$$dy \leqslant |f_1'(x)\, \Delta x| + |f_2'(x)\, \Delta x|$$

and the relative error is

$$\frac{dy}{y} \leqslant \frac{|f_1'(x)| + |f_2'(x)|}{|f_1(x) - f_2(x)|} \Delta x$$

The relative error will be large due to loss of leading significant digits in subtraction when $f_1(x)$ and $f_2(x)$ are nearly equal. This will ordinarily occur near some value of x for which $f_1(x)$ and $f_2(x)$ are exactly equal. For example, suppose that for $x = a$

$$f_1(a) = f_2(a)$$

Then for values of x near $x = a$, say, for example, $x = a + h$ where h is small,

$$f_1(a + h) \approx f_2(a + h)$$

and for small values of h we have subtraction problems. Now by Taylor's formula

$$f_1(a + h) = f_1(a) + h f_1'(a) + \text{terms involving higher powers of } h$$

and

$$f_2(a + h) = f_2(a) + h f_2'(a) + \text{terms involving higher powers of } h$$

Since h is small, we do not ordinarily need to carry these expansions past the first power in h to achieve sufficient accuracy. Thus, for h small, that is, for x near a, we have

$$y = f_1(x) - f_2(x) \approx h[f_1'(a) - f_2'(a)]$$

If the first-order terms were equal, $f_1'(a) = f_2'(a)$, it would be necessary to take the terms involving second powers of h in order to have a useful approximation for y. In cases in which a painstaking error analysis is warranted, the complete Taylor formula with remainder as given in Section 6.2 should be used.

Example 1.   The function $y = 1 - e^{x-1}$ is to be calculated for values of x very near 1. Write an approximation which will have good accuracy for x sufficiently near 1.

Let us consider

$$y = f_1(x) - f_2(x)$$

where

$$f_1(x) = 1$$
$$f_2(x) = e^{x-1}$$

Let

$$x - 1 = h$$

Then

$$h[f_1'(1) - f_2'(1)] = h[0 - 1] = -h$$

so that the approximation is

$$y = 1 - x$$

Example 2.   If an observer takes horizontal sighting over a smooth sea-level surface, how high is the line of sight at a distance of x miles from the observer? (The distance x is to be measured along the curved surface.) Write a program which will perform this calculation.

It can be seen from Figure 6-9 that the correct formula is

$$H = a \sec x/a - a$$

where a is the radius of the earth. In order to use the standard functions of

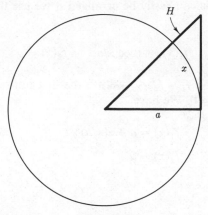

Figure 6-9

Chapter 4, we could write this as

$$H = \frac{a}{\cos x/a} - a$$

This formula appears quite straightforward, yet in using it we are apt to have accuracy difficulties. For example, using 4000 miles for the radius of the earth, consider the case where $x = 4$ miles. Then to seven correct significant digits,

$$\cos x/a = .9999995$$

The values of the various quantities involved, then, as they would be carried inside the computer, are

Quantity	Value Stored
a	$.4000000 \times 10^4$
$\dfrac{a}{\cos x/a}$	$.4000002 \times 10^4$
$\dfrac{a}{\cos x/a} - a$	$.2?????? \times 10^{-2}$

The six leading digits are lost in the subtraction, leaving at most one correct significant digit.

Note that in this case the accuracy problem results from the fact that the computer has only seven significant digits available, and not necessarily from inaccurate input data. It would not be at all unreasonable to ask for a program that would produce better accuracy for values of x on the order of a few miles, and this can easily be arranged if we use the method described above, taking

$$f_1(x) = a \sec x/a, \qquad f_2(x) = a$$

When $x = 0$, we have $f_1 = f_2$, so we use the first nonzero term of a Taylor expansion about $x = 0$. We have

$$f_1(x) = a + a(x/a)^2/2$$
$$f_2(x) = a$$

Therefore,

$$H_1 = f_1(x) - f_2(x) = x^2/2a$$

A use of the remainder term to calculate the error would show that we obtain a more accurate value for H with this formula than with the original formula when $x < .01a$, or when $x < 40$ miles. Figure 6-10, then, indicates a good way of setting up the problem to work for all reasonable values of x.

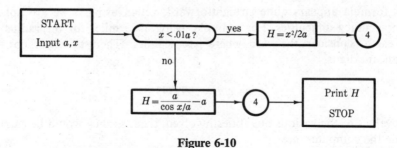

Figure 6-10

A FORTRAN program for this calculation is

```
      A=4000.
    1 READ  101,X
      IF(ABS(X)− .01*ABS(A))2,3,3
    2 FH=X*X/(2.*A)
      GO TO 4
    3 FH=A/COS(X/A)−A
    4 PRINT  102,X,FH
  101 FORMAT(E12.4)
  102 FORMAT(2E12.4)
      STOP
      END
```

An analogous procedure can be followed when several variables are involved. For Example 1, Section 6.61, we can derive an approximate relation for use when subtraction error is a problem as follows:

We need the approximation when A is nearly zero and b and c are nearly equal. Let us write

$$A = 0 + \Delta A$$
$$b = d + \Delta b$$
$$c = d + \Delta c$$

and assume henceforth that ΔA, Δb, and Δc are small. Then

$$a^2 = (d + \Delta b)^2 + (d + \Delta c)^2 - 2(d + \Delta b)(d + \Delta c) \cos \Delta A$$

Expanding, and discarding all terms having powers higher than the *second* in small quantities (all first-order terms drop out in this case, so we must keep the second-order terms):

$$a^2 \approx d^2 + 2d\,\Delta b + \Delta b^2 + d^2 + 2d\,\Delta c + \Delta c^2$$
$$-2(d^2 + d\,\Delta b + d\,\Delta c + \Delta b\,\Delta c)\left(1 - \frac{\Delta A^2}{2}\right)$$
$$\approx \Delta b^2 + \Delta c^2 - 2\Delta b\,\Delta c + d^2\,\Delta A^2$$
$$= (\Delta b - \Delta c)^2 + d^2\,\Delta A^2$$

or, adding and subtracting d within the parentheses, and using $A = \Delta A$,

$$a^2 = (b - c)^2 + d^2 A^2$$

This relation says that, when b and c are nearly equal and A is small, we can determine a by considering it to be the hypotenuse of a right triangle, one of whose legs is $b - c$ and the other of whose legs is dA (or, to the same order of accuracy, bA or cA). Figure 6-11 illustrates this approximation.

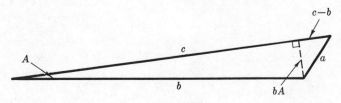

Figure 6-11

EXERCISE 21

1. Find the absolute and relative error in y for the following functions. The constants are accurate to the number of digits shown.

    a. $y = 1.00 + \ln x$ $\qquad\qquad\qquad$ $x = 2.71 \pm .01$
    b. $y = 2.00 \cos x + 3.00\, x^5$ $\qquad$ $x = 4.00 \pm .005$ (the exponent is exact)
    c. $y = x^{-1.2}$ $\qquad\qquad\qquad\quad$ $x = 1, 10, 100, 1000$ (exact values)
    d. $y = e^x \sin x$ $\qquad\qquad\qquad$ $x = 6.3 \pm .05$
    e. $y = 1.00 \cos x + .60 \cos 2x$
        $\qquad + .30 \cos 3x$ $\qquad\qquad\quad$ $x = 1.0 \pm .05$
        $\qquad\qquad\qquad\qquad\qquad\qquad\quad 1.5 \pm .05$
        $\qquad\qquad\qquad\qquad\qquad\qquad\quad 2.4 \pm .05$

2. In a circle of radius a, a chord is drawn which subtends a central angle θ. Write the expression for the distance from the center of the chord to the edge of the circle, x. Draw a flow chart for a program that will calculate to four significant digits for *all* values of θ less than $\pi/2$.

3. Draw a flow chart for a calculation that will find y accurate to four significant figures for *all* values of x between 0 and 1. Write a FORTRAN program to calculate and print y for 1000 equally spaced values of x.

    a. $y = x - \sin x$ $\qquad\qquad\qquad$ b. $y = \tan x - \sin x$
    c. $y = e^x - \cos x$ $\qquad\qquad\qquad$ d. $y = \cos x - 2 \ln(1 + x)$

    e. $y = \dfrac{\tan x - 2 \ln(1 + x)}{\sin x - \ln(1 + x)}$ $\qquad$ f. $y = \dfrac{\cos \pi x/2 - 1}{x - 1}$

4. Write a program for Example 2, Section 6.63, with detailed error bounding.

5. Write a program for determining a side of a triangle from the cosine law

    $$a^2 = b^2 + c^2 - 2bc \cos \theta$$

    with detailed error bounding. Compute a and Δa from this program for $b = 1.$, $\Delta b = .1$, $c = 1$, $\Delta c = .1$, $\theta = .1$, $\Delta\theta = .01$.

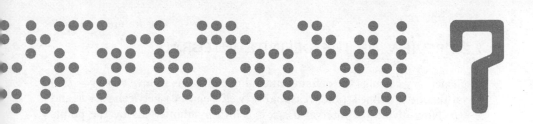

Quadrature

7.1 INTRODUCTION

The preceding chapters have been devoted largely to describing the digital computer and the types of operations it can perform. The remaining chapters are devoted to topics ordinarily treated in a numerical analysis course.

Strangely enough, it seems proper to make quadrature, or integration, the first such topic to be covered. There are two reasons for doing this. First, quadrature as ordinarily done on the computer is a very direct extension of the material of Chapters 5 and 6. Second, quadrature is one of the fields of applied mathematics most markedly affected by the advent of the computer.

In elementary calculus the methods for differentiation and integration of various functions are taught. Generally speaking, differentiation turns out to be the more easily performed of the two operations. Physicists and engineers, then, sometimes find it strange that mathematicians usually consider integration to be the "nicer" process. In particular, the mathematician is inclined to regard a problem as solved once he presents the answer in terms of a quadrature, that is, a definite integral of a known function, between known limits. After all, such an integral merely represents a number. To the physicist or engineer, however, the numerical value of this number may be a matter of considerable concern. Before the advent of the computer, the task of evaluating any but the most simple definite integrals was imposing, to say the least, and was insurmountable in many cases. The digital computer has produced a marked change in this situation. Numerical evaluation of large classes of definite integrals is a process well within the capabilities of even the slower computers. However, there are still problems involving quadrature in two or more dimensions which would requiree inordinat amounts of time on even the fastest of present-day computers.

7.2 REVIEW OF THE DEFINITE INTEGRAL

In elementary calculus the definite integral is defined as follows: Let $y = f(x)$ be a function defined (and reasonably well behaved) between $x = a$ and $x = b$. Now divide the interval $a \leqslant x \leqslant b$ into n subintervals by the points $a < x_1 < x_2 < \cdots < x_{i-1} < x_i < \cdots < x_{n-1} < b$. Now let

$$\Delta x_i = x_i - x_{i-1} \qquad (7\text{-}1)$$

and ξ_i be any point between x_{i-1} and x_i, and form the sum

$$\sum_{i=1}^{n} f(\xi_i)\, \Delta x_i \qquad (7\text{-}2)$$

Now let the number of intervals n approach infinity in such a manner that all the lengths of the intervals Δx_i approach zero. Then if the quantity given by expression (7-2) approaches a limit, that limit is called the definite integral of $f(x)$ from a to b and is denoted by the symbol

$$\int_a^b f(x)\, dx \qquad (7\text{-}3)$$

This is the definition of the definite integral. By a most extraordinary circumstance, it happens that there is a function $F(x)$, whose derivative $F'(x)$ is equal to $f(x)$, and the number represented by the definite integral (7-3) is the same number one obtains by computing $F(b) - F(a)$. The fact that

$$\int_a^b f(x)\, dx = F(b) - F(a) \qquad (7\text{-}4)$$

is one of the most remarkable and useful facts in all mathematics. It is called the "fundamental theorem of integral calculus." It can be used to find the value of the definite integral, provided that the function $f(x)$ is sufficiently simple that its antiderivative $F(x)$ [that is, the function which, when differentiated, gives $f(x)$] can be determined. A major part of the usual integral calculus course is devoted to the evaluation of definite integrals in this manner. The hardest part of this problem is the determination of the antiderivative, $F(x)$, and many hours are usually devoted to this problem. When the antiderivative cannot be found, as is quite often the case, it is necessary to employ the trapezoidal rule or Simpson's rule or some other numerical method of approximation. This amounts to computing the definite integral from its definition rather than making use of the fundamental theorem of integral calculus. Such methods, because of the large amounts of calculation involved, ordinarily appear to be undesirable as presented in the elementary calculus course but are quite natural and useful when digital computers are available.

7.21   Geometrical Significance of the Definite Integral

As is well known, the definite integral of $f(x)$ between a and b can be considered to be the area lying between the curve $f(x)$ and the x axis, and between the lines $x = a$ and $x = b$, the area being considered positive if $f(x)$ is above the x axis and negative if it is below. The area marked in Figure 7-1 represents the situation. Because it is so difficult to give a rigorous definition of the geometrical concept of area, it is not uncommon to define area in terms of the definite integral rather than to regard the integral as being represented geometrically by the area. Be that as it may, most of the methods of numerical approximation to the definite integral are easily portrayed in terms of approximation of an area by another area and will be presented from that viewpoint.

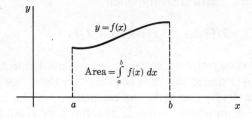

Figure 7-1

7.3 THE TRAPEZOIDAL RULE

A very straightforward approximation to the definite integral is obtained by dividing the interval a to b into n *equal* parts, erecting an ordinate line to the curve at each of the points of division, and connecting the end points of these ordinate lines to form trapezoids, as in Figure 7-2. The sum of the areas of these trapezoids can be seen to approximate the area under the curve. The approximation can be made as close as desired by taking a sufficient number of intervals. Using the notation of Section 7.2, the areas of the trapezoids are

$$A_1 = (\Delta x_1/2)[f(x_1) + f(x_0)]$$
$$A_2 = (\Delta x_2/2)[f(x_2) + f(x_1)]$$
$$\vdots$$
$$A_i = (\Delta x_i/2)[f(x_i) + f(x_{i-1})]$$
$$\vdots$$
$$A_{n-1} = (\Delta x_{n-1}/2)[f(x_{n-1}) + f(x_{n-2})]$$
$$A_n = (\Delta x_n/2)[f(x_n) + f(x_{n-1})]$$

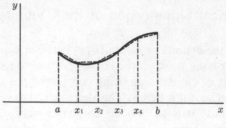

Figure 7-2

Since all the x's are equal,

$$\Delta x_1 = \Delta x_2 = \cdots = \Delta x_i = \cdots = \Delta x_n = \Delta x$$

we may sum up the above areas to obtain

$$A = (\Delta x/2)[f(x_0) + 2f(x_1) + 2f(x_2) + \cdots + 2f(x_i) + \cdots + 2f(x_{n-1}) + f(x_n)]$$

To evaluate this quantity on a computer, one must calculate the function $f(x)$ a total of $n + 1$ times, and then need do very little else. It follows, then, that the time required to perform the calculation will be roughly equal to $n + 1$ times the computation time of the function $f(x)$. Hence the time required is roughly proportional to the number of subdivisions made and thus is closely tied to the question of accuracy. The accuracy consideration will be discussed in a later section. The flow chart given in Figure 7-3 will serve for the application of the trapezoidal rule for any value of $n \geqslant 2$.

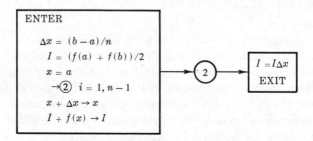

Figure 7-3: Integration by the trapezoidal rule

For remote-terminal use, a program such as the following could be used to evaluate integrals by the trapezoidal rule:

1   GRANF(X) = (insert correct expression for integrand)
2   PRINT, "LOWER AND UPPER LIMITS OF INTEGRA-
       TION"
3   INPUT, A,B

```
 4   1 PRINT, "NUMBER OF SUBDIVISIONS"
 5     INPUT, N
 6     FN=N
 7     DX=(B-A)/FN
 8     FI=(GRANF(A)+GRANF(B))/2.
 9     X=A
10     NN=N-1
11     DO 2 I=1,NN
12     X=X+DX
13   2 FI=FI+GRANF(X)
14     FI=DX*FI
15     PRINT, "INTEGRAL",FI
16     GO TO 1
17     END
```

In this program the statement at line 1 must be completed to be the function which describes the integrand. The program is set up to first ask for A and B, the limits of integration. After these are entered, the program asks for the number of subdivisions to use. After this number is entered, the calculation is performed in accordance with the flow chart, Figure 7-3, and the value of the integral is printed. The program then reaches the GO TO 1 statement at line 16, at which point it returns to statement 1, at line 4, and asks for a new number of subdivisions. The user can enter a new number and run again if he chooses. The relationship among number of subdivisions used, accuracy, and running time is discussed below.

Example 1.   You have previously stored the program given above for integration by the trapezoidal rule under the file name INTEG. You wish to recall it and use it to evaluate

$$\int_{.5}^{1} \cos \sqrt{x} \, dx$$

using 20 subdivisions. List the required typed inputs and the machine responses for accomplishing this action.

The inputs and responses are as listed below, the boldface entries being the machine responses.

LOAD INTEG
READY
1   GRANF(X)=COS(SQRT)(X))
       RUN
       LOWER AND UPPER LIMITS OF INTEGRATION
?   .5,1

NUMBER OF SUBDIVISIONS
?   20
      INTEGRAL = .3242240
      NUMBER OF SUBDIVISIONS
?   S
      STOP

The file was called from storage and the statement at line 1 was rewritten to include the correct integrand. Then the calculation was performed.

7.31   Accuracy for the Trapezoidal Rule

It can be seen that integration by the trapezoidal rule involves computing the value of the integrand for a number of values of the independent variable and very little else in the way of computation. Hence the computer time required to evaluate an integral is just about proportional to the number of times the integrand must be computed, or the number of subdivisions made in the interval of integration. This in turn is determined by the accuracy required in the final answer. Thus a method is needed for estimating the accuracy to be obtained by subdivision into a given number of intervals.

In the trapezoidal rule, the area under a segment of width Δx of the curve $y = f(x)$ is approximated by the area of a trapezoid,

$$A = (\Delta x/2)[f(x) + f(x + \Delta x)]$$

The true value of the area is

$$I = \int_x^{x+\Delta x} f(x)\, dx$$

It can be shown that the difference between these two quantities is given by

$$A - I = \frac{\Delta x^3}{12} f''(\xi_0) \tag{7-5}$$

where ξ_0 is some value between x and $x + \Delta x$.*

If we are integrating from a to b and have applied the trapezoidal rule by dividing this region up into n subintervals, each of width Δx, then the total error is the sum of the errors for the individual subintervals, or

$$E = \frac{\Delta x^3}{12} f''(\xi_1) + \frac{\Delta x^3}{12} f''(\xi_2) + \cdots \frac{\Delta x^3}{12} f''(\xi_n) \tag{7-6}$$

---

* See, for example, B. W. Arden, *An Introduction to Digital Computing*, Addison-Wesley Publishing Company, Inc., Reading, Mass., 1963.

where $\xi_1, \xi_2, \ldots, \xi_n$ are properly selected points in each of the subintervals. This can be written

$$E = \frac{\Delta x^3}{12} [f''(\xi_1) + f''(\xi_2) + \cdots + f''(\xi_n)] \tag{7-7}$$

Now if the function $f''(x)$ is continuous, there is some point ξ in the interval (a, b) at which $f''(\xi)$ is the average of the values at ξ_1, ξ_2, etc.; that is,

$$f(\xi) = \frac{1}{n} [f''(\xi_1) + f''(\xi_2) + \cdots + f''(\xi_n)] \tag{7-8}$$

Thus (7-7) can be written

$$E = \frac{\Delta x^3}{12} n f''(\xi) \tag{7-9}$$

or, since

$$n = (b - a)/\Delta x$$

$$E = \frac{b - a}{12} \Delta x^2 f''(\xi) = \frac{(b - a)^3}{12n^2} f''(\xi) \tag{7-10}$$

where ξ is some point between a and b.

If we let M_2 stand for the maximum absolute value of the second derivative $f''(x)$ anywhere within the interval $a \leqslant x \leqslant b$, we can say that

$$\text{absolute error} \leqslant \frac{(b - a)^3}{12n^2} M_2 \tag{7-11}$$

This relation can be used to determine the number of subintervals required to attain a given accuracy and hence the time requirement for an integration by the trapezoidal rule. For this purpose it is convenient to write the relation in another form. If ΔI is the largest absolute error which can be tolerated, we must choose n so that

$$\text{absolute error} \leqslant \Delta I$$

This will certainly be true if n is chosen so that

$$\frac{M_2(b - a)^3}{12n^2} \leqslant \Delta I \tag{7-12}$$

Solving for n, we must have

$$n \geqslant \sqrt{\frac{M_2(b - a)^3}{12\Delta I}} \qquad\qquad (7\text{-}13)$$

Example 1. The function $y = \cos \sqrt{x}$ is to be integrated from $x = .5$ to $x = 1$ by the trapezoidal rule. How many intervals should be taken to assure that the error in the answer is no worse than $.00001$?

The second derivative is

$$y'' = \frac{1}{4x} \left(\frac{\sin \sqrt{x}}{\sqrt{x}} - \cos \sqrt{x} \right)$$

We wish an upper limit for this quantity over the range $.5$ to 1. As a crude estimate, we note that $\sin u/u$ is never greater than 1. Hence in the range of interest both $\sin \sqrt{x}/\sqrt{x}$ and $\cos \sqrt{x}$ are between zero and 1. Hence the quantity in parentheses is certainly no greater than $1 - 0$, or 1. The quantity $1/4x$ has its maximum when $x = .5$. Its value is then $.5$. Hence we may certainly say that $y'' < .5$ in the range of interest. If we put $M_2 = .5$, $b = 1$, $a = .5$, and $\Delta I = .00001$ in formula (7-13), we obtain $n \geqslant 23$. Thus a subdivision into about 25 subintervals would be adequate. If the machine requires 10 multiplication times to extract a square root and 10 to compute a cosine, about 20 multiplication times would be required to compute the integrand once. Hence computation of the integral would require roughly $25 \times 20 = 500$ multiplication times, or $1/2$ millisecond.

One of the most interesting features of this example is that the determination of the number of subdivisions to use requires much more time and effort than the actual calculation of the integral once the required number of steps has been determined. Even for the relatively simple integrand in the example, determination of the maximum value of the second derivative was somewhat laborious. For more complicated integrands it can be extremely difficult. For this reason, it is advisable to have some other way of estimating the error. One way, extremely simple with the program given above, is to perform the calculation for two different values of n and note how much change takes place in the answer. In Example 1, if we had typed in the value 40 the second-time NUMBER OF SUBDIVISIONS was requested, we would have received the answer $.3243325$. Now, according to equation (7-9), the maximum error is proportional to the cube of the step size. We reduced the step size by the ratio $20/40$, or $1/2$, so we should have reduced the error by $1/8$, or almost a factor of 10. Thus the new answer should be correct to about one more digit than the old one. We are inclined to accept both answers as being correct to the number of places to which they agree, and the second as being accurate to one more place. Note that we have not proved that this is the case. It is possible that the answers obtained by the two calculations may be quite close to each other and still be far from the correct answer. Figure 7-4 illustrates a case in which such an event occurs. In the figure, most of the area occurs inside one of the smallest subintervals shown. Until the interval is

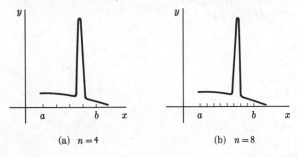

(a) $n = 4$ (b) $n = 8$

Figure 7-4

divided into fine enough zones to sense the abrupt change of the function within that narrow region, an accurate estimate of the value of the integral cannot be obtained. If one were able to obtain an expression for the second derivative of $y = f(x)$ for the curve in Figure 7-4, one would find that the second derivative would be large in those regions where the curve undergoes an abrupt change in direction, so that one would find, on using this upper limit for the number M_2 of relation (7-13), that a large number of terms would be required to achieve any appreciable accuracy. On the other hand, if one merely applies the trapezoidal rule blindly to the problem, using $n = 4$ and $n = 8$, one would miss completely the fact that a sharp peak in the curve exists, and would obtain two estimates of the area with the peak ignored completely. These estimates might be very nearly equal, but both would underestimate the actual area by about a factor of two. Thus the estimation of accuracy based merely on using two different values of n can be trusted only if there is assurance that the curve has no abrupt changes of behavior.

7.32 Automatic Selection of Interval Size

It is easy to write a program which will automatically recompute an integral, using smaller and smaller subdivisions, until the accuracy in the answer is considered sufficient. If the subdivisions are cut exactly in half at each new trial, all the values of the integrand computed in one trial can be reused in the next, so that total running time on the machine is only slightly longer than for a single evaluation of the integral using the number of subdivisions finally arrived at in the program. The flow chart (Figure 7-5) and FORTRAN subroutine below demonstrate the process.

```
         SUBROUTINE  TRAP(A,B,ERR,FI)
         DX =(B − A)/2.
         FI1 =(GRANF(A)+GRANF(B))/2.
         FI2 = GRANF(A + DX)
         FIP=DX*(FI1 + FI2)
         N =1
       1 N =2*N
         TDX=DX
         DX = .5*DX
         X =A + DX
         DO 2 I =1,N
         FI2 =FI2 + GRANF(X)
       2 X =X + TDX
         FI =DX*(FI1 + FI2)
         IF(ABS(FI − FIP) − ERR*ABS(FI))4,4,3
       3 FIP=FI
         GO TO 1
       4 RETURN
         END
```

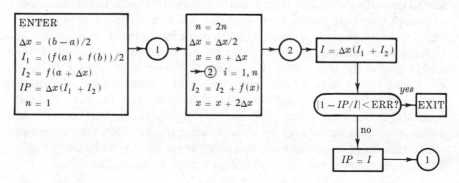

Figure 7-5: Integration by the trapezoidal rule with error control

The subroutine requires A, B, and ERR as inputs and returns the value FI of the integral. It assumes that GRANF(X) is an external function, defined elsewhere in the program as described in Section 4.6.

As the flow chart and subroutine are written, the process will stop when the change in the answer from one trial to the next is less than ERR. This suggests, but does not prove, that the relative error is on the order of ERR. As was pointed out in Section 7.31 and demonstrated in Figure 7-4, this sort of error estimate can be misleading, but it is often the best one available.

Example 1. Subroutine TRAP is stored under the file name ITRAP. Write a program for remote terminal which will call the subroutine and use it to evaluate

$$\int_0^1 \frac{\sin x}{\sqrt{x^3 + 1}}\, dx$$

to an accuracy of .0001.

A suitable program is

```
1 GRANF(X) =SIN(X)/SQRT(X**3+1.)
2 CALL TRAP(0.,1.,.0001,FI)
3 PRINT, "INTEGRAL =",FI
4 STOP
5 END
6 $USE ITRAP
```

The statement at line 1 serves to define the function GRANF. Then the subroutine is called and the answer determined and printed.

Example 2.   An integral has been evaluated using subroutine TRAP, with ERR set at .001. The running time was 1 second. It is to be rerun with ERR set at .00001. How long would you expect this run to take?

It has been indicated that the running time is roughly proportional to the number of subdivisions used. Relation (7-11) indicates that the error should be inversely proportional to the square of the number of subdivisions used. Thus to reduce the error from .001 to .00001, a factor of 100, we should need to increase the number of subdivisions by a factor of 10. Thus the expected running time would be about 10 seconds.

7.33   Effect of Roundoff Error

The accuracy discussion given above assumes that the only source of error was that associated with the approximation of

$$\int_a^b f(x)\, dx$$

by

$$\Delta x[(1/2)f(x_0) + f(x_1) + f(x_2) + \cdots + f(x_{n-1}) + (1/2)f(x_n)] \quad \text{(7-14)}$$

that is, the use of a finite value of Δx and a finite number of subdivisions, n. As was seen in Chapter 6, there will also be computational errors associated with the use of approximate numbers and approximate arithmetic in evaluating expression (7-14). These can become very important, or even limiting, for large values of n.

Let us approach the error problem by first writing an error-bounding version of the FORTRAN program in Section 7.3. Let DGRAN(X) stand for the derivative of GRANF(X); R for the relative error introduced by roundoff; EX, EDX, and EFI for the absolute errors in X, DX, and FI, respectively; and RFI for the relative error in FI. Then after the manner of Section 6.65 we can write the following error-bounding program:

```
 1     GRANF(X) = (insert correct expression for integrand)
 2     DGRAN(X) = (insert expression for derivative of integrand)
 3     PRINT, "LOWER AND UPPER LIMITS OF
            INTEGRATION"
 4     INPUT, A,B
 5     PRINT, "ROUNDOFF ERROR"
 6     INPUT, R
 7   1 PRINT, "NUMBER OF SUBDIVISIONS"
 8     INPUT, N
 9     FN = N
10     EA = R*ABS(A)
11     EB = R*ABS(B)
12     C = B − A
13     EC = EA + EB + R*ABS(C)
14     DX = C/FN
15     EDX = ABS(DX)*(EC/C + R + R)
16     FI = GRANF(A) + GRANF(B)
17     EFI = DGRAN(A)*EA + DGRAN(B)*EB + R*ABS(FI)
18     RFI = EFI/ABS(FI)
19     FI = FI/2,
20     RFI = RFI + R
21     EFI = RFI*ABS(FI)
22     X = A
23     EX = EA
24     NN = N − 1
25     DO 2 I = 1,NN
26     X = X + DX
27     EX = EX + EDX + R*ABS(X)
28     FI = FI + GRANF(X)
29   2 EFI = EFI + ABS(DGRAN(X))*EX + R*ABS(FI)
30     RFI = EFI/ABS(FI)
31     FI = DX*FI
32     RFI = EDX/ABS(DX) + RFI + R
33     PRINT, "INTEGRAL =",FI
34     PRINT, "RELATIVE ERROR BOUNDS",RFI
35     GO TO 1
36     END
```

A few aspects of this program call for some explanation. First, it is assumed that the relative errors in A, B, and FN are R. At lines 10 and 11, the absolute errors EA and EB are found, and these are used at line 13 to find the absolute error in B−A. The initial value of FI is calculated in two steps, at lines 16 and 19. At line 17, the absolute error EFI is first found, then at line 18 it is converted to relative error because relative errors are most convenient to handle in division. At line 20, the relative error in the number two is assumed to be zero, since that number is usually stored exactly in a binary machine. The roundoff error R is added in, but even that is not required in many machines. If the machine happens to store its exponents in binary, it may divide by two merely by reducing the exponent by one, not touching the fractional part and hence not affecting the relative error. If the machine uses hexadecimal exponents, as described in Chapter 2, then the fractional part will be shifted to the right one place, and one binary bit position will be lost, with a 50 : 50 chance of increasing the relative error.

The main chance for error accumulation occurs at lines 25 through 29, where the main loop in the program is performed $N − 1$ times. It would have been simpler, and probably about as effective, to introduce the error bounding into this loop only, ignoring the error accumulation in the very few calculations outside the loop. In the loop itself, it can be seen that EX increases by an amount of at least EDX each time through and that this in turn causes the middle term in the expression at line 29 to tend to increase for each pass. Further, the last term in the expression at 29 ensures that the relative error in FI will increase by at least an amount R at each pass. Then if 1000 subintervals are used, the error bound shown by this program will be at least 1000 times the machine roundoff error. If 10^7 subintervals are used in a machine using seven-decimal-place arithmetic, the error bound will surely indicate that no accuracy remains. As pointed out in Section 3.63, this may be too pessimistic a view, since the roundoff errors should tend to cancel out, but there is a clear indication that machine roundoff can become a problem when the number of subintervals is large.

Note also in the expression at line 29 the effect of the term ABS(DGRAN(X))*EX. If the derivative is large at any point, that is, the curve is steep, this term can contribute a large error. The EX in this term is the error in picking the value of X at which the ordinate is to be taken (see Figure 7-2), and if the curve is steep a small miss in this value can cause a large error in the ordinate.

7.4   OTHER QUADRATURE FORMULAS

For the trapezoidal rule, the inherent error is proportional to the square of the interval size. There are other approximate formulas for the evaluation of definite integrals for which the inherent error is proportional to higher powers of the interval size. For the same accuracy, these formulas tend to require a smaller number of subdivisions. This offers an advantage in that it reduces the number of points at which the integrand must be evaluated and thus the amount of computer time required. This reduction can be quite dramatic. For example, for Simpson's rule, discussed in Section 7.41, the inherent error is proportional to the fourth power of the number of subintervals. In Example 2, Section 7.32, it was shown that with the trapezoidal rule a factor of 10 increase in running time would be expected to produce two additional correct significant digits. With Simpson's rule, only a factor of three increase should be required (since $\sqrt[4]{100} = 3.16 \approx 3$).

One other advantage and one disadvantage are worth mentioning in connection with the higher-order quadrature formulas. The advantage is that the lesser number of subintervals means a lesser number of calculations and thus less opportunity for accumulation of error. This is important only in those uses where the trapezoidal rule requires very large numbers of subdivisions. The disadvantage is that in using a smaller number of subdivisions the higher-order formulas sample the integrand in fewer places, and thus run a bigger risk of making a sizable error by missing some feature of the integrand, as demonstrated in Figure 7-4. The computer calculation cannot provide complete insurance against this risk for any of the integration methods. Only an analytic analysis, as in Example 1, Section 7.31, can provide this insurance.

7.41 Simpson's Rule

The trapezoidal rule was based on passing straight lines through pairs of consecutive points on the graph of $y = f(x)$. Since the equation of a straight line is of the form $y = mx + b$, which is of first degree, or linear, in x, this is equivalent to fitting each pair of points with a first-degree polynomial. In Simpson's rule, a parabola, or second-degree polynomial, is passed through each set of three consecutive points. The formula is obtained by summing the areas under the parabolic arcs. The formula is

$$A = (\Delta x/3)[f(x_0) + 4f(x_1) + 2f(x_2) + 4f(x^3)$$
$$+ 2f(x_4) + \cdots + 4f(x_{n-1}) + f(x_n)] \quad \text{(7-15)}$$

This formula has the same general appearance as that obtained for the trapezoidal rule. Again it involves computing $f(x)$ for $n + 1$ values of x and combining the results, the only difference being that in this case different coefficients are involved. The formula is easily remembered if certain of its features are noted. Inside the brackets, the first and last coefficients are one. The second and next to last are 4. In the middle, they alternate between 4 and 2.

There are several possible ways of programming Simpson's rule for a computer. The flow chart, Figure 7-6, is for using formula (7-15) rearranged as follows:

$$A = (\Delta x/3)\{[f(a) + f(b)] + 4[f(x_1) + \cdots + f(x_{n-1})]$$
$$+ 2[f(x_2) + \cdots + f(x_{n-2})]\} \qquad \textbf{(7-16)}$$

It will work for n any even number.

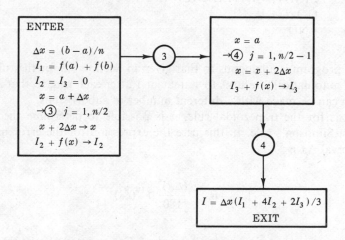

Figure 7-6: Integration by Simpson's rule

The FORTRAN program given below will perform this calculation. Again statement 1 must be completed to describe the integrand completely:

```
1       GRANF(X) = (insert correct expression for integral)
2       PRINT, "LOWER AND UPPER LIMITS OF
            INTEGRATION"
3       INPUT, A,B
4   1   PRINT, "NUMBER OF SUBDIVISION-MUST BE EVEN"
5       INPUT, N
6       FN = N
7       DX = (B−A)/FN
8       TDX = 2.*DX
9       FI1 = GRANF(A) + GRANF(B)
10      FI2 = 0.
11      FI3 = 0.
12      X = A + DX
13      NN = N/2
14      DO 3 J = 1,NN
```

```
15    X = X + TDX
16  3 FI2 = FI2 + GRANF(X)
17    X = A
18    NM = NN − 1
19    DO 4 J = 1,NM
20    X = X + TDX
21  4 FI3 = FI3 + GRANF(X)
22    FI = DX*(FI1 + 4.*FI2 + 2.*FI3)/3.
23    PRINT, "INTEGRAL =",FI
24    GO TO 1
25    END
```

This program is analogous to that given in Section 7.2 for the trapezoidal rule. It automatically returns to statement 1 after each run, so that if desired a rerun can be made with a different number of subdivisions.

Just as for the trapezoidal rule, it is possible to determine the inherent error in Simpson's rule. In this case the expression for the error in a single subinterval Δx is

$$E_0 = \frac{(\Delta x)^5}{180} f^{IV}(\xi_0) \tag{7-17}$$

where ξ_0 is a value between x and $x + \Delta x$.* For an integral over the interval (a, b), the inherent error is

$$E = \frac{b - a}{180} \Delta x^4 f^{IV}(\xi) \tag{7-18}$$

where ξ is some point in the interval (a, b). It is seen that the error depends on the fourth derivative of the function being integrated. This is a surprising fact. It means that, for functions whose fourth derivative is zero, there is no error. Now a third-degree polynomial has a zero fourth derivative. This means that Simpson's rule, which is based on approximating by parabolas, which are second-degree polynomials, is exact for third-degree polynomials as well. This is a bonus in accuracy which accounts for the widespread use of Simpson's rule.

As for the trapezoidal rule, the error relation can be converted into a form for the determination of the number of intervals required for a given maximum error. If ΔI is the maximum allowable error, and if M_4 is the

* For a proof, see, for example, J. B. Scarborough, *Numerical Mathematical Analysis*, 5th ed., The Johns Hopkins Press, Baltimore, 1962.

maximum value of the fourth derivative of $f(x)$ in the range of integration, then the number of subdivisions n required to make the error less than ΔI is

$$n \geqslant \sqrt[4]{\frac{M_4(b-a)^5}{180\Delta I}} \qquad (7\text{-}19)$$

As demonstrated in Example 1, Section 7.31, for the trapezoidal rule, this relation can be used to determine the number of subdivisions sufficient to produce any given accuracy. Note, however, that application requires determination of an upper limit for the fourth derivative of the integrand, frequently a forbidding task. The substitute method, varying the interval size and inferring the error from the change in answer, can be used, with the risk already pointed out. The flow chart, Figure 7-7, and the FORTRAN

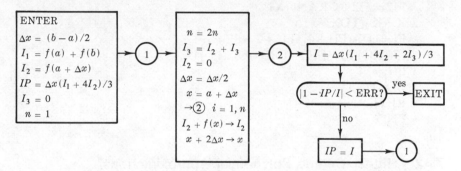

Figure 7-7: Integration by Simpson's rule with error control

subroutine given below demonstrate the process for Simpson's rule. In the flow chart, the sum of the two end ordinates is denoted by I_1, the sum of all ordinates which must be multiplied by the coefficient 4 in Simpson's rule is denoted by I_2, and the sum of all ordinates which must be multiplied by 2 is I_3. The pattern of the coefficients in two successive applications of Simpson's rule is

$$1 \quad 4 \quad 2 \quad 4 \quad 2 \quad 4 \quad 1$$
$$1 \ 4 \ 2 \ 4 \ 2 \ 4 \ 2 \ 4 \ 2 \ 4 \ 2 \ 4 \ 1$$

so that except for the first and last ordinates, all ordinates computed at one trial will have the coefficient 2 at the next trial. Hence the I_3 for a new trial is simply the sum of $I_2 + I_3$ for the preceding trial.

Again, the subroutine has four parameters. Inputs are A and B, the limits of integration, and ERR, a representation for the allowed relative error. The routine returns FI, the value of the integral.

```
        SUBROUTINE SIMP(A,B,ERR,FI)
        DX =(B−A)/2.
        FI1 =GRANF(A)+GRANF(B)
        FI2=GRANF(A+DX)
        FI3 =0.
        FIP=DX*(FI1+4.*FI2)/3.
        N=1
      1 N=2*N
        FI3 =FI2+FI3
        FI2=0.
        TDX=DX
        DX=.5*DX
        X=A+DX
        DO 2 I=1,N
        FI2=FI2+GRANF(X)
      2 X=X+TDX
        FI =DX*(FI1+4.*FI2+2.*FI3)/3.
        IF(ABS)(FI−FIP)−ERR*ABS(FI))4,4,3
      3 FIP=FI
        GO TO 1
      4 RETURN
        END
```

7.42   Higher-Degree Polynomial Approximations

It is possible to obtain other approximations by passing higher-degree
polynomials through consecutive points. In general, if the nth-degree polynomial

$$y = a_1 x^n + a_2 x^{n-1} + \cdots + a_n x + a_{n+1}$$

is passed through $n + 1$ points equally spaced in the x direction, the area
under the curve can be expressed in the form

$$\int_{x_0}^{x_n} y \, dx = \Delta x[c_0 y_0 + c_1 y_1 + \cdots + c_n y_n]$$

where the c's are constants which must be determined for the degree of
polynomial used.

One can easily write FORTRAN programs similar to the ones given for
the trapezoidal rule and Simpson's rule, for polynomials of any desired
degree. In spite of the improved accuracy associated with these formulas, they
are no longer widely used, because some other formulas discussed below are
generally superior.

7.43 Romberg Integration

In using the trapezoidal rule with error control we computed values of the integral several times, each time with subdivision size cut in half from the previous time. In doing so, we discarded information that could have been used to improve the accuracy of the result. Let

$$T(\Delta x) = (\Delta x/2)[f(a) + 2f(a + \Delta x) + 2f(a + 2\,\Delta x) + \cdots + f(b)] \quad \textbf{(7-20)}$$

That is, $T(\Delta x)$ is the value given by the trapezoidal rule when step size Δx is used. We can write

$$\int_a^b f(x)\,dx = T(\Delta x) + E(\Delta x) \quad \textbf{(7-21)}$$

where $E(\Delta x)$ is the error incurred when step size Δx is used. By a very involved method which will not be reproduced here, $E(\Delta x)$ can be expanded in a power series in Δx, giving an expression of the form

$$E(\Delta x) = a_2\,\Delta x^2 + a_4\,\Delta x^4 + a_6\,\Delta x^6 + \cdots \quad \textbf{(7-22)}$$

Only even powers of Δx appear in the expansion. The constants a_i are of known value, but the values are not required for our present purposes. We can write equation (7-21) as

$$\int_a^b f(x)\,dx = T(\Delta x) + a_2\,\Delta x^2 + a_4\,\Delta x^4 + a_6\,\Delta x^6 + \cdots \quad \textbf{(7-23)}$$

Now note that in the calculation of $T(\Delta x)$ in the subroutine TRAP, we also calculated $T(2\,\Delta x)$, which will satisfy the relation

$$\int_a^b f(x)\,dx = T(2\,\Delta x) + a_2(2\,\Delta x)^2 + a_4(2\,\Delta x)^4 + a_6(2\,\Delta x)^6 + \cdots \quad \textbf{(7-24)}$$

We can eliminate the Δx^2 term between these two equations by multiplying (7-23) by four and subtracting (7-24) from it, giving

$$3\int_a^b f(x)\,dx = 4T(\Delta x) - T(2\,\Delta x) - 12a_4\,\Delta x^4 - 60a_6\,\Delta x^6 + \cdots \quad \textbf{(7-25)}$$

or

$$\int_a^b f(x)\,dx = \frac{4T(\Delta x) - T(2\,\Delta x)}{3} - 4a_4\,\Delta x^4 - 20a_6\,\Delta x^6 + \cdots \quad \textbf{(7-26)}$$

This last relation says that $(4T(\Delta x) - T(2\,\Delta x))/3$ should be a better approximation to the integral than either $T(\Delta x)$ or $T(2\,\Delta x)$, since the error goes as Δx^4 rather than Δx^2. The process which led to this improvement can be carried further. In getting to step size Δx in subroutine TRAP, we also need step sizes of $4\,\Delta x$, $8\,\Delta x$, $16\,\Delta x$, etc., all the way to a step size of $(b - a)/2$. Thus we can write

$$\int_a^b f(x)\,dx = T(4\,\Delta x) + a_2(4\,\Delta x)^2 + a_4(4\,\Delta x)^4 + a_6(4\,\Delta x)^6 + \cdots \qquad \text{(7-27)}$$

$$\int_a^b f(x)\,dx = T(8\,\Delta x) + a_2(8\,\Delta x)^2 + a_4(8\,\Delta x)^4 + a_6(8\,\Delta x)^6 + \cdots \qquad \text{(7-28)}$$

$$\int_a^b f(x)\,dx = T(16\,\Delta x) + a_2(16\,\Delta x)^2 + a_4(16\,\Delta x)^4 + a_6(16\,\Delta x)^6 + \cdots \qquad \text{(7-29)}$$

etc., where the values of $T(4\,\Delta x)$, $T(8\,\Delta x)$, $T(16\,\Delta x)$, etc., are known. We can combine these equations by pairs so as to eliminate the Δx^2 terms, then combine the resulting equations to eliminate the Δx^4 terms, then the Δx^6 terms, and so on. If, for example, we had subdivided five times in all, we could combine equations to eliminate four terms, and eliminate the second, fourth, sixth, and eighth powers of Δx, ending up with a final expression having error proportional to Δx^{10}. This expression should represent the most accurate estimate available from five repeated applications of the trapezoidal rule.

In order to develop a flow chart and FORTRAN subroutine for the Romberg method, assume that we have applied the trapezoidal rule K times, first with subinterval size $(b - a)/2$, the second with half that size, the third with half that, and so on. Denote the results by $TR(1)$, $TR(2)$, ..., $TR(K)$. In order to eliminate the Δx^2 error between two adjacent values, we multiply the second by four, subtract the first and divide by three, as was done in equation (7-25). In order to eliminate the Δx^4 term we multiply the second by 4^2, subtract the first and divide by $4^2 - 1$. In order to eliminate the Δx^{2m} term, we would multiply the second quantity by 4^m, subtract the first, and divide by $4^m - 1$. Figure 7-8 shows schematically how this process would be applied to eliminate all powers up to Δx^{2k}. Each operation involves an adjacent pair of the numbers, denoted as x for the first number, y for the second. In the figure, the names $TR(1)$, $TR(2)$, etc., have been given to each new set of values. The old values at each stage are not needed, and this renaming will save storage. The first step is to start at the top of the figure, do the operation $(4y - x)/3$ on each pair of numbers, working from left to right across the row. Then do the operation $(4^2 y - x)/(4^2 - 1)$ from left to right across the new row. Continue in this manner, finally doing the operation $(4^{k-1} y - x)/(4^{k-1} - 1)$ to the last $TR(1)$ and $TR(2)$ values to obtain a final $TR(1)$. This $TR(1)$ is the best current estimate of the integral. Now suppose we want to do one more subdivision and incorporate that result in a similar fashion.

Operation

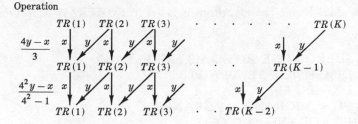

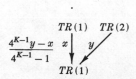

Figure 7-8: Diagram to illustrate Romberg integration

We calculate $TR(K + 1)$ by the trapezoidal rule. It belongs at the right-hand end of the first row in Figure 7-8. We can now combine it with $TR(K)$ according to the operation $(4y - x)/3$, to get a $TR(K)$ for the second row, then combine that with the $TR(K - 1)$ in the second row according to the operation $(4^2 y - x)/(4^2 - 1)$ to get a $TR(K - 2)$ for the third row, and so on, continuing in this manner until we finally arrive at a new value for $TR(1)$. It can be compared to the old value to estimate the error and determine if still further subdivision should be made. Figure 7-9 reduces the above description

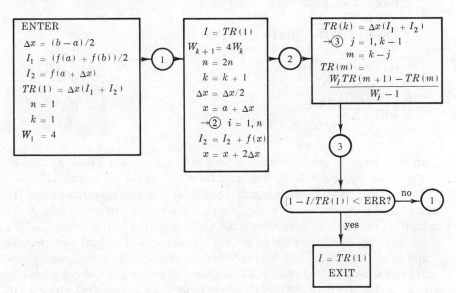

Figure 7-9: Integration by the Romberg method

to flow-chart form. The algorithm is given in FORTRAN subroutine form
as follows:

```
      SUBROUTINE ROMB(A,B,ERR,FI,FAIL)
      DIMENSION TR(20),W(20)
      DX=(B-A)/2.
      FI1=.5*(GRANF(A)+GRANF(B))
      FI2=GRANF(A+DX)
      TR(1)=DX*(FI1+FI2)
      N=1
      K=1
      W(1)=4.
    1 FI=TR(1)
      W(K+1)=4.*W(K)
      N=2*N
      K=K+1
      TDX=DX
      DX=.5*DX
      X=A+DX
      DO 2 I=1,N
      FI2=FI2+GRANF(X)
    2 X=X+TDX
      TR(K)=DX*(FI1+FI2)
      KK=K-1
      DO 3 J=1,KK
      M=K-J
    3 TR(M)=(W(J)*TR(M+1)-TR(M))/(W(J)-1.)
      FAIL=ABS(TR(1)-FI)/ABS(TR(1))
      IF(FAIL-ERR)6,6,4
    4 IF(K-20)1,6,6
    6 FI=TR(1)
      RETURN
      END
```

In most respects this subroutine is similar to subroutines TRAP and SIMP
given in earlier sections. A few new features should be noted, however.
First a new output parameter has been added, a quantity named FAIL,
computed by the statement immediately following statement 3. FAIL is the
calculated estimate of the relative error. Ordinarily, if FAIL is larger than the
input quantity ERR, the allowed error, the program will go back to statement
1 and subdivide again. However, each new subdivision process requires a new
entry in the array of values $TR(K)$. Since the dimension statement allows 20
entries in this array, an attempt to repeat more than 20 times the process will
overrun this storage space, causing the program to fail. Therefore the IF

statement 4 is used to cause an exit if the storage has been used up, regardless of the size of the error. If the user wants to check for this condition after the subroutine has been executed he can compare FAIL to ERR. If FAIL is smaller than ERR, the subroutine performed normally. If FAIL is the larger, the desired accuracy was not produced, and the user must modify the program or use another integration method.

A second subscripted variable has been introduced in the subroutine, W(J). This array is used to store values of 4.**J needed repeatedly, to avoid unnecessary recomputation.

The inherent error in Romberg integration is given by*

$$E = \frac{(b-a)B_{2m}}{(2m)!\,2^{(m-1)m}}\,\Delta x^{2m} f^{(2m)}(\xi) \tag{7-30}$$

where m is the number of times the trapezoidal rule has been applied and Δx is the step size in the final, smallest subdivision, that is,

$$\Delta x = \frac{b-a}{2^m}$$

The quantity B_{2m} is the $(2m)$th Bernoulli number. Bernoulli numbers are a sequence of numbers first defined by Jacques Bernoulli in connection with evaluating sums of powers of the integers. They can be obtained by expanding the function $x/(e^x - 1)$ in a Maclaurin series. The nth Bernoulli number is $n!$ times the coefficient of x^n in this expansion, that is,

$$\frac{x}{e^x - 1} = \sum_{n=0}^{\infty} \frac{B_n}{n!}\,x^n \tag{7-31}$$

For example, $B_0 = 1$, $B_1 = -1/2$, $B_2 = 1/6$, etc.

7.44   Gauss's Formula

In the quadrature formulas so far discussed, values of the function at equally spaced intervals are required. Gauss investigated the possibility of obtaining greater accuracy with the same number of values of $f(x)$ by using a different spacing. The problem which he undertook to solve is the following: If $\int_b^a f(x)\,dx$ is to be computed from a given number of values of $f(x)$, what

---

* For a proof, see, for example, H. S. Wilf, Advances in Numerical Quadrature, in *Mathematical Methods for Digital Computers* (Anthony Ralston and H. S. Wilf, eds.), Vol. II, John Wiley & Sons, Inc., New York, 1967.

selection of abscissas will in general give the most accurate result? He found that the points of subdivision should not be equally spaced, but should be symmetrically placed with respect to the midpoint of the interval of integration.

Gauss's formula is usually expressed in the form

$$\int_{-1/2}^{1/2} \phi(u) \, du = R_1\phi(u_1) + R_2\,\phi(u_2) + \cdots + R_n\,\phi(u_n) \qquad (7\text{-}32)$$

that is, for an integral over the interval from $-1/2$ to $1/2$.

The values of R_i and u_i to be used in this formula have been tabulated for various values of n, for example, for $n = 2$, $R_1 = R_2 = 1/2$ and $-u_1 = u_2 = .2886751346$. For this case then, the formula is

$$\int_{-1/2}^{1/2} \phi(u) \, dn = (1/2)\phi(- .2886751346) + (1/2)\phi(.2886751346)$$

While Gauss's formula is given in terms of integration over the interval from $-1/2$ to $1/2$, a change of variable can be employed to make it applicable to any interval. If we wish to evaluate

$$I = \int_a^b f(x) \, dx \qquad (7\text{-}33)$$

we make the substitution

$$x = (b - a)u + \frac{a + b}{2}$$

Then when $u = -1/2$, $x = a$, and when $u = 1/2$, $x = b$. If we then write

$$f(x) = f\left[(b - a)u + \frac{a + b}{2}\right] = \phi(u), \text{ say}$$

then, since

$$dx = (b - a) \, du$$

the integral (7-33) becomes

$$I = (b - a) \int_{-1/2}^{1/2} \phi(u) \, du$$

and Gauss's formula can be applied.

The problem with the Gaussian quadrature formula comes in finding the R's and u's for equation (7-32). This task is quite difficult, so the results are usually tabulated for some values of n, and Gauss's formula used only for those values of n where the tabulated values are available. Some sample values are given in Table 1.

TABLE 1
Coefficients and Abscissas for Gaussian Quadrature

n	R		u	
2	R_1, R_2:	.5	$-u_1, u_2$:	.2886751346
5	R_3:	.2844444444	u_3:	0
	R_2, R_4:	12393143352	$-u_2, u_4$:	.2692346551
	R_1, R_5:	.1184634425	$-u_1, u_5$:	.4530899230
10	R_5, R_6:	.1477621124	$-u_5, u_6$:	.0744371695
	R_4, R_7:	.1346333597	$-u_4, u_7$:	.2166976971
	R_3, R_8:	.1095431813	$-u_3, u_8$:	.3397047841
	R_2, R_9:	.07472567458	$-u_2, u_9$:	.4325316833
	R_1, R_{10}:	.03333567215	$-u_1, u_{10}$:	.4869532643

The flow chart, Figure 7-10, and the FORTRAN function subprogram GINT given below show how Gaussian integration with n ordinates can be applied to a function $f(x)$.

```
FUNCTION GINT(A,B,N)
COMMON R(40),U(40)
FI=0.
NN=N/2
FN=N-2*NN
A1=B-A
A2=(A+B)/2.
DO 1 I=1,NN
X1=A2+A1*U(I)
X2=A2-A1*U(I)
1 FI=FI+R(I)*(GRANF(X1)+GRANF(X2))
GINT=A1*(FI+FN*R(NN+1)*GRANF(A2))
RETURN
END
```

In the subprogram it is assumed that the R's and u's are already stored in common before the subroutine is called. Only the values of R_i and n_i through $n/2+1$ need be stored because the routine takes advantage of the fact that $R_1 = R_n$, $R_2 = R_{n-1}$, $u_1 = -u_n$, $u_2 = -u_{n-1}$, etc., by adding on new terms

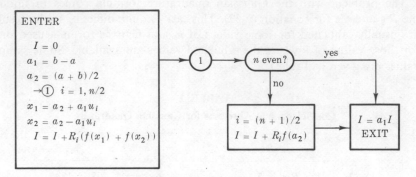

Figure 7-10: Gaussian Integration

in pairs. Some trickery is required to determine if n is odd, and add in the center ordinate if necessary. This is done by the computation of the quantity, FN, which will be zero if n is even and one if n is odd. Then the statement after statement 1 will provide the value at the center ordinate if required.

In using Gaussian quadrature, one may wish to exercise some sort of error control by performing the integration with two different step sizes and comparing answers. Since the coefficients and abscissas for subdivision into large numbers of subintervals are not easily obtained, the simplest way to accomplish this is to pick a number of subintervals for Gauss's formula, say 10, and proceed as follows. First perform Gaussian integration with 10 subintervals on the entire region of integration. Then split this region into two halves and use Gaussian integration with 10 subintervals on each of these. If the sum of the integrals for these two halves is not sufficiently close to the value obtained for the entire region, subdivide again in the same manner. At each subdivision the integrand has to be evaluated at a completely new set of abscissas, and the old values are discarded. This wasted effort is more than compensated for by the fact that the integrand is being represented accurately up to the twentieth derivative (see Section 7.45), and so not many subdivisions should be required to perform the process. Figure 7-11 is a flow chart for the calculation, and the algorithm is given in FORTRAN subroutine form below. This subroutine merely controls the selection of interval size, and calls function GINT to do the actual calculation.

```
        SUBROUTINE GAUSIN(A,B,ERR,FI,N)
        DX=B-A
        FIP=GINT(A,B,N)
        M=1
     1  M=2*M
        DX=.5*DX
        FI=0.
```

```
      X2=A
      DO 2 I=1,M
      X1=X2
      X2=X1+DX
    2 FI=FI+GINT(X1,X2,N)
      IF(ABS(FI−FIP)−ERR*ABS(FI))4,4,3
    3 FIP=FI
      GO TO 2
      RETURN
      END
```

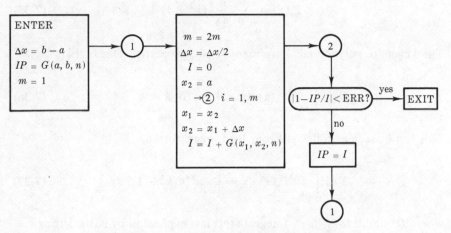

Figure 7-11: Gaussian integration with error control

7.45 Abscissas and Coefficients in Gaussian Quadrature

As remarked in Section 7.44, the abscissas u_i and coefficients R_i for the Gaussian quadrature formula (7-32) are somewhat difficult to obtain. Equations for determining these coefficients will be derived below but will not be solved. FORTRAN programs which can be used to solve these equations are given in later sections of the book.

Define the set of polynomials

$$P_0(x) = 1$$

$$P_1(x) = x \tag{7-34}$$

$$nP_n(x) = (2n - 1)xP_{n-1}(x) - (n - 1)P_{n-2}(x) \qquad \text{for } n \geqslant 2$$

With these definitions we have

$$P_2(x) = (1/2)(3x^2 - 1)$$

$$P_3(x) = (1/2)(5x^3 - 3x) \qquad \text{etc.}$$

$P_n(x)$ is a polynomial in x of degree n and is known as the Legendre polynomial of degree n. $P_n(x)$ can also be described by Rodriguez' formula, which states that

$$P_n(x) = \frac{1}{2^n n!} \frac{d^n}{dx^n} (x^2 - 1)^n \qquad \text{(7-35)}$$

The Legendre polynomials have many useful properties. One is that

$$\int_{-1}^{1} x^k P_n(x)\, dx = 0 \qquad \text{for } k < n \qquad \text{(7-36)}$$

This can be seen as follows. From Rodriquez' formula,

$$2^n n! \int_{-1}^{1} x^k P_n(x)\, dx = \int_{-1}^{1} x^k D^n (x^2 - 1)^n\, dx \qquad \text{(7-37)}$$

where D^n stands for d^n/dx^n. Integrate this last expression by parts, letting

$$u = x^k \qquad\qquad dv = D^n(x^2 - 1)^n\, dx$$

$$du = kx^{k-1}\, dx \qquad v = D^{n-1}(x^2 - 1)^n$$

Then

$$\int_{-1}^{1} x^k D^n(x^2 - 1)^n\, dx = x^k\, D^{n-1}(x^2 - 1)^n \Big|_{-1}^{1}$$

$$- k \int_{-1}^{1} x^{k-1} D^{n-1}(x^2 - 1)^n\, dx$$

Since the $(n-1)$st derivative of $(x^2 - 1)^n$ will have $x^2 - 1$ as a factor, the first term will be zero at $x = 1$ and $x = -1$. Hence we have

$$\int_{-1}^{1} x^k D^n(x^2 - 1)^n\, dx = -k \int_{-1}^{1} x^{k-1} D^{n-1}(x^2 - 1)^n\, dx$$

If we integrate by parts repeatedly, on the right-hand side the power of x will decrease by one at each step until finally we have the first power, and

$$\int_{-1}^{1} xD^{n-k+1}(x^2 - 1)^n \, dx = \frac{1}{2} D^{n-k}(x^2 - 1)^n \Big|_{-1}^{1}$$

For $k < n$, $D^{n-k}(x^2 - 1)^n$ has a factor $x^2 - 1$, and so vanishes for $x = 1$ and $x = -1$, and (7-36) is proved.

Now let $u = x/2$, and form the integrals

$$\int_{-1/2}^{1/2} P_n(u) \, du, \quad \int_{-1/2}^{1/2} uP_n(u) \, du, \ldots, \int_{-1/2}^{1/2} u^{n-1}P_n(u) \, du$$

By (7-36) these are all zero. If they are evaluated by Gaussian quadrature with n ordinates, we should have

$$\int_{-1/2}^{1/2} P_n(u) \, du = R_1 P_n(u_1) + R_2 P_n(u_2) + \cdots + R_n P_n(u_n) = 0$$

$$\int_{-1/2}^{1/2} uP_n(u) \, du = R_1 u_1 P_n(u_1) + R_2 u_2 P_n(u_2) + \cdots + R_n u_n P_n(u_n) = 0$$

$$\vdots$$

$$\int_{-1/2}^{1/2} u^{n-1}P_n(u) \, du = R_1 u_1^{n-1} P_n(u_1) + R_2 u_2^{n-1} P_n(u_2) + \cdots$$

$$+ R_n u_n^{n-1} P_n(u_n) = 0$$

If we choose $u_1, u_2, \ldots, u_n$ such that

$$P_n(u_1) = P_n(u_2) = P_n(u_3) = \cdots P_n(u_n) = 0 \tag{7-38}$$

then these equations are all satisfied. Since $P_n(u)$ is a polynomial of degree n, it has n roots, and these are the values we are choosing for u_1, u_2, etc. Now let $\phi(u)$ be a polynomial of degree $2n - 1$ or less. Then $\phi(u)$ can be written

$$\phi(u) = g(u)P_n(u) + h(u) \tag{7-39}$$

where $g(u)$ and $h(u)$ are of degree $n - 1$ or less. This is accomplished by merely dividing $\phi(u)$ by $P_n(u)$, and $g(u)$ will be the quotient and $h(u)$ the remainder. Now

$$\int_{-1/2}^{1/2} \phi(u) \, dn = \int_{-1/2}^{1/2} g(u)P_n(u) \, dn + \int_{-1/2}^{1/2} h(u) \, dn \tag{7-40}$$

Since $g(u)$ is a sum of terms of the form u^K, with $K < n$, the first term vanishes by (7-36). Hence

$$\int_{-1/2}^{1/2} \phi(u) \, dn = \int_{-1/2}^{1/2} h(u) \, dn \tag{7-41}$$

Let

$$h(u) = c_1 + c_2 u + c_3 u^2 + \cdots + c_n u^{n-1} \tag{7-42}$$

Then

$$\int_{-1/2}^{1/2} h(u) \, dn = c_1 + c_3(1/2)^2(1/3) + c_5(1/2)^4(1/5) + \cdots \tag{7-43}$$

If Gaussian quadrature is applied to $h(u)$, we have

$$\int_{-1/2}^{1/2} h(u) \, dn = R_1 h(u_1) + R_2 h(u_2) + \cdots + R_n h(u_n) \tag{7-44}$$

From (7-41) we also have

$$h(u_1) = c_1 + c_2 u_1 + c_3 u_1^2 + \cdots + c_n u_1^{n-1}$$

$$h(u_2) = c_1 + c_2 u_2 + c_3 u_2^2 + \cdots + c_n u_2^{n-1} \qquad \text{etc.}$$

Substituting these relations in (7-44) we have

$$\int_{-1/2}^{1/2} h(u) \, dn = R_1(c_1 + c_2 u_1 + c_3 u_1^2 + \cdots + c_n u_1^{n-1}$$

$$+ R_2(c_1 + c_2 u_2 + c_3 u_2^2 + \cdots + c_n u_2^{n-1}$$

$$+ \cdots$$

$$+ R_n(c_1 + c_2 u_n + c_3 u_n^2 + \cdots + c_n u_n^{n-1})$$

or, rearranging the terms,

$$\int_{-1/2}^{1/2} h(u) \, dn = c_1(R_1 + R_2 + \cdots + R_n)$$

$$+ c_2(R_1 u_1 + R_2 u_2 + \cdots + R_n u_n)$$

$$+ \cdots$$

$$+ c_n(R_1 u_1^{n-1} + R_2 u_2^{n-1} + \cdots + R_n u_n^{n-1}) \tag{7-45}$$

The value of this integral is now given in two different ways in expressions (7-43) and (7-45). If these are to be the same, the coefficients of each c_i must be the same in the two expressions. Thus we have

from c_1: $R_1 + R_2 + R_3 + \cdots + R_n = 1$

from c_2: $R_1 u_1 + R_2 u_2 + R_3 u_3 + \cdots + R_n u_n = 0$

from c_3: $R_1 u_1^2 + R_2 u_2^2 + R_3 u_3^2 + \cdots + R_n u_n^2 = (1/2)^2 \cdot 1/3$ **(7-46)**

$$\vdots$$

from c_{2i-1}: $R_1 u_1^{2i-2} + R_2 u_2^{2i-2} + R_3 u_3^{2i-2} + \cdots + R_n u_n^{2i-2}$

$$= (1/2)^{2i-2} \cdot \frac{1}{2i-1}$$

from c_{2i}: $R_1 u_1^{2i-1} + R_2 u_2^{2i-1} + R_3 u_3^{2i-1} + \cdots + R_n u_n^{2i-1} = 0$
etc.

There are n equations in the n unknowns $R_1, R_2, \ldots, R_n$. If they are solved for these values, using the $u_1, u_2, \ldots, u_n$ from equation (7-38), these R's and u's when used in the Gaussian quadrature formula (7-32) will give an exact value for the integral for $\phi(x)$ any polynomial of degree $2n - 1$ or less. Other functions $\phi(u)$, can be represented by Maclauren's formula as

$$\phi(u) = \phi(0) + \phi'(0)u + \phi''(0)\frac{u^2}{2!} + \cdots + \phi^{(2n-1)}(0)\frac{u^{2n-1}}{(2n-1)!}$$

$$+ \phi^{(2n)}(\theta u)\frac{u^{2n}}{(2n)!} \tag{7-47}$$

so that

$$\int_{-1/2}^{1/2} \phi(u)\, dn = R_1\phi(u_1) + R_2\phi(u_2) + \cdots + R_n\phi(u_n)$$

$$+ \int_{-1/2}^{1/2} \phi^{(2n)}(\theta u)\frac{u^{2n}}{(2n)!}\, du \tag{7-48}$$

The last term is then the error term for Gaussian quadrature. It can be reduced to a simpler form by use of the mean value theorem for integrals, which states that if $f(x)$ is continuous on the closed interval (a, b), then there is some number X, $a \leqslant X \leqslant b$, such that

$$\int_a^b f(x)\, dx = (b - a)f(X)$$

Applying this to (7-48), we may write

$$\int_{-1/2}^{1/2} \phi(u)\, dn = R_1\phi(u_1) + R_2\phi(u_2) + \cdots + R_n\phi(u_n) + \phi^{(2n)}(\theta\xi) \frac{\xi^{2n}}{(2n)!} \quad (7\text{-}49)$$

where $-1/2 \leqslant \xi \leqslant 1/2$.

This indicates that the error behavior for Gaussian integration should be far better than that for any of the methods previously discussed. By evaluating the integrand at n ordinates, we have an error proportional to the $(2n)$th derivative of $\phi(u)$. This is a far better result than with any of the methods discussed previously. For Romberg integration, the best discussed so far, according to equation (7-30) we would have to subdivide n times to have this order of accuracy, that is, use 2^n ordinates.

The procedures described above can be used to calculate the u_i's and R_i's for any value of n. The procedure can be summarized in the following three steps.

(1) Write the nth-degree Legendre polynomial $P_n(x)$. This can be done from the recursion relations (7-34).

(2) Find the n roots of $P_n(x)$, $x_1, x_2, \ldots, x_n$. Subroutines for doing this are given in Chapter 9. Then the u_i's are given by $u_i = x_i/2$.

(3) Solve the set of linear equations (7-46) for the R_i's. Subroutines for doing this are given in Chapter 10. The equations (7-46) are a little more convenient if written in terms of the x_i's. Since $x_i = 2u_i$, they become

$$R_1 + R_2 + R_3 + \cdots + R_n = 1$$

$$R_1 x_1 + R_2 x_2 + R_3 x_3 + \cdots + R_n x_n = 0$$

$$R_1 x_1^2 + R_2 x_2^2 + R_3 x_3^2 + \cdots + R_n x_n^2 = 1/3$$

$$R_1 x_1^3 + R_2 x_2^3 + R_3 x_3^3 + \cdots + R_n x_n^3 = 0$$

$$R_1 x_1^4 + R_2 x_2^4 + R_3 x_3^4 + \cdots + R_n x_n^4 = 1/5 \qquad \text{etc.}$$

EXERCISE 22

1. Assume that you have the following subprograms from this chapter stored in permanent files at the computer and have access from a remote terminal.

Subprogram Name	File Name
TRAP	ITRAP
SIMP	ISIMP
ROMB	IROMB
GINT	IGINT
GAUSIN	IGAUS

a.   Write a remote terminal program which will use subroutine TRAP for
$\int_0^1 e^{\sin x} dx$. Run the program with ERR $= .1$ and ERR $= .01$. What
would you expect the ratio of running times to be?

b.   Do part a for subroutine SIMP.

c.   Do part a for subroutine ROMB.

d.   Write a subroutine GCON which will store the Gaussian quadrature
constants R_i and u_i for $n = 10$ into COMMON for use with subroutine
GINT. Write the remote-terminal commands to store this subroutine
under file name IGCON. Do part a for subroutine GAUSIN.

2.   Using the programs under problem 1, evaluate the following integrals, to an
accuracy of $.01$.

a. $\int_0^2 \ln\sqrt{1 + x}\, dx$ b. $\int_1^2 \frac{dx}{\sqrt{x^2 - .9}}$

3.   The function $f(x) = \log(1 + \tan^{-1} x)$ is to be expanded in a Chebyshev series.
Using subroutine ROMB, write a remote terminal program which will ask
for the relative error ERR as in input, then ask for the term number, N, as
an input, compute and print the coefficient

$$C_n = \int_0^\pi f(\cos \theta) \cos n\theta\, d\theta$$

and then request a new value of N. Use the program to find coefficients and
write a Chebyshev series which should give $f(x)$ to four-place accuracy.

4.   Write an error-bounding version of subroutine ROMB. Use this version to
redo problem 3.

5.   a.   Write a remote-terminal program which will compute and print $-\log a$,
and run for $a = .1$.

b.   Use the remote-terminal program given in Section 7.3 for the trapezoidal
rule to evaluate the integral

$$\int_a^b \frac{dx}{x}$$

c.   Run with $a = .1$, $b = 1$, and $n = 2, 4$, and 8.

d.   Compare the result with $-\log(.1)$, obtained in part a. How does the
relative error change as n is increased? What value of n do you estimate
to be required to give a relative error less than $.0001$?

e.   Revise your program so that it will ask for a and n as inputs and automatically compute and print

$$\int_a^1 \frac{dx}{x} \qquad \text{and} \qquad \frac{\int_a^1 dx/x}{-\log a} - 1$$

f.   Run this program for $a = .1$ and $n = 8$ to confirm your previous hand
calculation of relative error.

g. Run the program for $a = .01$ and $n = 4$, 8, and 16. What value of n do you estimate to be required to give a relative error less than .0001?

6. Using equations (7-11) and (7-19), estimate the number of subdivisions required to evaluate $\int_{.01}^{1} dx/x$ by the trapezoidal rule and by Simpson's rule.

7. According to equation (7.31), the Bernoulli numbers B_n are the coefficients in the Maclauren expansion

$$\frac{x}{e^x - 1} = \sum_{n=0}^{n} \frac{B_n}{n!} x^n$$

a. Prove that the Maclauren expansion for $\dfrac{e^x - 1}{x}$ is

$$\frac{e^x - 1}{x} = 1 + \frac{x}{2!} + \frac{x^2}{3!} + \cdots + \frac{x^{n-1}}{n!} + \cdots$$

Hence

$$\left(B_0 + B_1 x + B_2 \frac{x^2}{2!} + B_3 \frac{x^3}{3!} + \cdots \right) \left(1 + \frac{x}{2!} + \frac{x^2}{3!} + \cdots \right) = 1$$

b. Use this relation to find B_0, B_1, B_2, B_3, B_4.

c. Show that

$$\frac{B_n}{n!1!} + \frac{B_{n-1}}{(n-1)!(2!)} + \frac{B_{n-2}}{(n-2)!(3!)} + \cdots + \frac{B_1}{(1!)(n!)} + \frac{B_0}{(n+1)!} = 0$$

or

$$\binom{n+1}{1} B_n + \binom{n+1}{2} B_{n-1} + \binom{n+1}{3} B_{n-2} + \cdots + \binom{n+1}{n} B_1 + \binom{n+1}{n} B_0 = 0$$

where $\binom{j}{i}$ is the coefficient of x^i in $(1 + x)^j$, that is, a binomial coefficient.

d. From Pascal's triangle for the binomial coefficients, show that the equations for the B_i are

$$2B_1 + B_0 = 0$$
$$3B_2 + 3B_1 + B_0 = 0$$
$$4B_3 + 6B_2 + 4B_1 + B_0 = 0 \qquad \text{etc.}$$

e. By comparing coefficients of x^i in

$$(1 + x)^{n+1} \qquad \text{and} \qquad (1 + x)(1 + x)^n$$

show that

$$\binom{n+1}{i} = \binom{n}{i-1} + \binom{n}{i} \qquad \text{for} \quad 0 < i < n+1$$

f. Using the results of parts e and c, write a FORTRAN program that will compute the Bernoulli number B, given that $B_0 = 1$.

8. Writing the nth-degree Legendre polynomial as

$$P_n(x) = a_1(n)x^n + a_2(n)x^{n-1} + \cdots + a_n(n)x + a_{n+1}(n)$$

use (7.34) to derive a recursion relation for $a_i(n)$. Using this relation, write a FORTRAN subroutine which, given n, will compute the coefficients of $P_n(x)$.

9. Using (7-36), show that

$$\int_{-1}^{1} P_n(x)P_m(x)\,dx = 0 \qquad \text{for} \quad n \neq m$$

That is, the Legendre polynomials are orthogonal over the interval $(-1, 1)$.

10. Using Rodriguez' formula and integration by parts, show that

$$\int_{-1}^{1} x^n P_n(x)\,dx = \frac{1}{2^n} \int_{-1}^{1} (1-x^2)^n\,dx = \frac{1}{2^n} \int_{-\pi/2}^{\pi/2} \cos^{2n+1}\theta\,d\theta$$

11. Given that

$$\int_{-1}^{1} x^n P_n(x)\,dx = \frac{2^{n+1}(n!)^2}{(2n+1)!}$$

show that

$$\int_{-1}^{1} [P_n(x)]^2\,dx = \frac{2}{2n+1}$$

12. Show that if $f(x)$ can be expanded in the series

$$f(x) = \sum_{n=0}^{\infty} b_n P_n(x)$$

then

$$b_n = \frac{2n+1}{2} \int_{-1}^{1} f(x)P_n(x)\,dx$$

13. Given $f(x)$ defined by the following table of values:

x	$f(x)$	x	$f(x)$	x	$f(x)$
-1	.0793	$-.3$	.3159	.4	.4452
$-.9$	.1179	$-.2$	.3413	.5	.4554
$-.8$	.1554	$-.1$	.3643	.6	.4641
$-.7$	.1915	0	.3849	.7	.4713
$-.6$	.2258	.1	.4032	.8	.4773
$-.5$	.2580	.2	.4192	.9	.4821
$-.4$	.2881	.3	.4332	1.	.4861

a. Write a FORTRAN program which will compute

$$\int_{-1}^{1} f(x)\,dx$$

by Simpson's rule.

b. Write a FORTRAN program which will compute

$$b_n = \frac{2n+1}{2} \int_{-1}^{1} f(x)P_n(x)\, dx$$

for input values of n.

c. Use the above program to generate the first six coefficients of an expansion of $f(x)$ in Legendre polynomials.

d. Explain the difficulty in using this same technique to perform an expansion in Chebyshev polynomials for the above table.

14. Write a FORTRAN subroutine which will convert the coefficients of a Legendre expansion into coefficients of a polynomial in x.

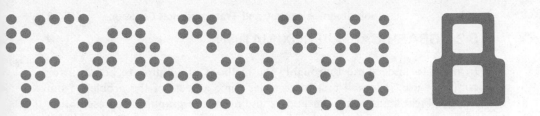

Solution of Algebraic and Transcendental Equations

8.1 INTRODUCTION

In the preceding chapters methods have been described by which functions of many different types can be evaluated; that is, given a value of x, we can find $y = f(x)$. The reverse problem is frequently of interest; that is, given a value of y, find x such that $f(x) = y$. Of particular interest are the values of x which make $y = 0$. These are called the zeros of the function $f(x)$ or the roots or solutions of the equation $f(x) = 0$. If $f(x)$ is a quadratic expression, the roots are given by the quadratic formula of algebra. Algebraic formulas are also available for the solution of cubic or biquadratic equations. For higher-degree polynomials, however, formulas do not exist, and numerical methods must be employed to find the roots. For equations such as $ae^x + b \cos x = 0$ or $ax + b \tan x = 0$, algebraic methods are not available. These are transcendental equations, and there is no general method of stating their roots in terms of their coefficients. This means that the solutions to such equations are quite difficult to tabulate, and so, prior to the advent of computers, such solutions have been considered relatively inaccessible, and obtainable only by means of a considerable expenditure of effort. The capability for numerical calculations at high speed provided by the computer has changed this picture considerably. Although the problem of finding the roots of an equation on a computer is by no means trivial, it is a problem solvable in the majority of cases by straightforward application of simple procedures. The object of the present chapter is to set forth the most useful methods for performing this task.

8.2   GRAPHICAL APPROXIMATION

In order to understand thoroughly the methods of solution to be given for computer use, it is well first to consider some aspects of the problem from the graphical standpoint. An understanding of the graphical representation of functions and of the graphical analogues of the methods of solution will be quite helpful in following the methods themselves at a later stage. Let

$$f(x) = 0 \tag{8-1}$$

be the equation whose roots are to be found. Then if we take a set of rectangular coordinates axes and plot the graph of

$$y = f(x) \tag{8-2}$$

it is evident that the abscissas of the points where the graph crosses the x axis are the real roots of the given equation, for at these points y is zero and therefore (8-1) is satisfied. Thus a graph of the function $y = f(x)$ displays the roots of the equation $f(x) = 0$. One method, then, of determining the roots is to compute y for a sequence of values of x and plot the results, then use the graph to read the roots. The accuracy of the determination will depend upon the number of values of x for which y is computed and upon the care used in constructing the graph. The first attempt by this method ordinarily gives only rough values for the roots. It is possible to improve the accuracy by replotting the curve in the vicinity of the roots on a graph of larger scale, but such a procedure is not particularly adaptable to computer use, so it will not be considered here. The discussion of the graphical approach will be limited to two topics: obtaining a rough estimate of the roots, and clarification of the computer methods for obtaining precise values, which will be described later.

8.21   Isolation of the Roots

In many respects, the problem of obtaining a first rough estimate of the roots of an equation is the most difficult part of the problem of determination of these roots. Once a reasonably decent first estimate is available, the methods that will be described later will ordinarily furnish an accurate value of the root. An important first step will have been made when the roots of $f(x)$ have been isolated, that is, when the range of possible values of x has been divided into intervals, each of which contains only one root of $f(x)$. This process is not an easy one but is necessary for some of the methods of finding roots. Some theorems which are useful in this connection will be given without proof.

Theorem 1.   *If $f(x)$ is continuous from $x = a$ to $x = b$ and if $f(a)$ and $f(b)$ have opposite signs, then there is at least one real root of $f(x) = 0$ between a and b.*

Stated in terms of the graph, if $f(x)$ is on one side of the x axis at $x = a$ and the other side at $x = b$, and if $f(x)$ is continuous from $x = a$ to $x = b$, the graph must cross the x axis at least once between a and b. Figure 8-1 shows

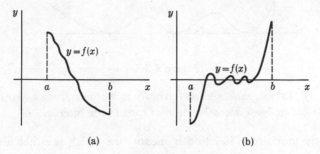

(a)   (b)

Figure 8-1

some of the possible configurations for which this theorem applies. In Figure 8-1(a), the root between a and b is isolated; that is, it is the only root in the interval. In Figure 8-1(b), on the other hand, there are several roots between a and b, so the roots have not yet been isolated. To complete the task of isolation, other values of x between a and b would have to be found for which the sign of $f(x)$ changes. Figure 8-2 shows a case in which the theorem does not apply because of a discontinuity between $x = a$ and $x = b$. $f(a)$ is positive and $f(b)$ is negative, yet there is no root between a and b.

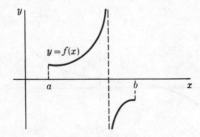

Figure 8-2

This theorem is of great value in isolating the roots of $f(x) = 0$, but it is not in itself sufficient. For example, in Figure 8-1(b), there are many roots between a and b. On the other hand, Figure 8-3 illustrates a case in which $f(x)$ never crosses the x axis, yet $f(x)$ is equal to zero at several points. The roots of a function such as this one can be quite difficult to locate. It may be difficult to determine that the function even has real zeros. The following theorem is of some assistance in this regard.

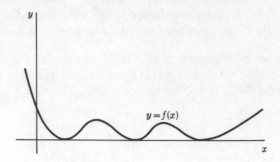

Figure 8-3

Theorem 2.   *If $f(x)$ is continuous and strictly monotonic from $x = a$ to $x = b$, then $f(x) = 0$ has at most one discrete real root in the interval.*

By a strictly monotonic function is meant one which is either always increasing as x increases or always decreasing as x increases. It is characterized by a derivative which does not change sign or become zero. For example, Figure 8-4 illustrates functions that are strictly monotonic for the range of

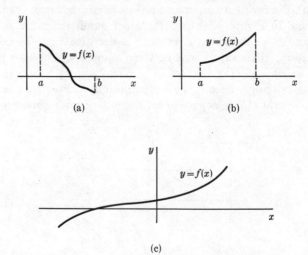

Figure 8-4

values of x indicated. In Figure 8-4(c) the interval is intended to be $-\infty$ to $+\infty$. Figure 8-5 represents a case in which the theorem does not apply because the function is discontinuous between a and b. Although it is strictly monotonic except at the point of discontinuity, it has two roots between a and b. Fortunately, in many cases of practical interest corresponding to physical problems, there is only one root. Knowledge of this fact sometimes simplifies the problem of finding the root.

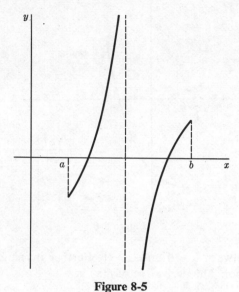

Figure 8-5

8.22   Graphical Representation as the Intersection of Two Curves

The representation of the roots of $f(x) = 0$ as points where the function $y = f(x)$ crosses the x axis is not the only possible graphical representation, nor is it necessarily the most useful one in all cases. If the equation $f(x) = 0$ can be written in the form $f_1(x) = f_2(x)$, then the roots of the equation $f(x) = 0$ are those values of x for which the curves $y_1 = f_1(x)$ and $y_2 = f_2(x)$ intersect.

Example.   Find approximate values of the real roots of $xe^x = 2$.

We first write the equation as

$$xe^x - 2 = 0$$

It is now in the form $f(x) = 0$, where

$$f(x) = xe^x - 2$$

Figure 8-6 shows a graph of $y = f(x)$ and the table of values used to plot the graph. Since y is negative for $x = 0$ and positive for $x = 1$, there is at

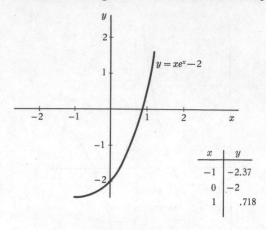

Figure 8-6

least one root between $x = 0$ and $x = 1$. Further investigation shows that this is the only root. The derivative of $f(x)$ is

$$f'(x) = (x + 1)e^x$$

which is positive as long as x is greater than -1 and negative as long as x is less than -1. Hence $f(x)$ is monotonically decreasing in the interval from $-\infty$ to -1 and monotonically increasing in the interval from -1 to $+\infty$. By Theorem 2, then, it has at most one discrete root in each of these intervals. Now when x is negative, the quantity xe^x is also negative, so that $f(x)$ must be less than -2. Hence there can be no root in the interval $-\infty$ to -1, so that the root between $x = 0$ and $x = 1$ is the only real root.

Figure 8-7 shows a different way in which the root can be displayed graphically. The equation can be written as $e^x = 2/x$ (this representation is not valid at $x = 0$, since division by zero is not defined in the laws of arithmetic). The root is then the abscissa of the point of intersection of the curves $y_1 = e^x$ and $y_2 = 2/x$.

Example 2.   Find the approximate values of the real roots of

$$2x - \cos x - 1 = 0$$

This equation can be written in the form

$$2x - 1 = \cos x$$

The two curves $y_1 = 2x - 1$ and $y_2 = \cos x$ are plotted in Figure 8-8. It is evident from the figure that there is only one point of intersection and hence only one real root.

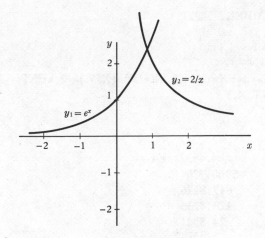

Figure 8-7

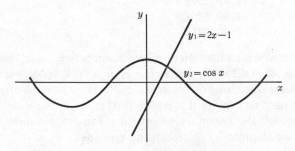

Figure 8-8

8.23   Computer Assistance in Graphical Approximation

At a remote terminal with direct computer access, the graphical method can be employed quite handily, using the computer to provide values of y for given values of x as required.

Example 1.   Find a root of $y = x^2 e^x - 5x \cos(x/5) + 2e^{-x}$ by graphical approximation, using a remote-terminal program to generate values of y as needed.

To get an idea of how the curve looks, we might first construct and run a program to generate a series of points from $x = -10$ to $x = 10$ by twos.

The communication might be as follows:

```
USER NUMBER? 54321
READY
1     DO 5 N=1,11
2     X=2*N-12
3     Y=X**2*EXP(X)-5.*X*COS(X/5.)+2.*EXP(-X)
4   5 PRINT X,Y
5     STOP
6     END
RUN
       -10.        4.403213E4
        -8.        5960.770
        -6.        817.8176
        -4.        123.4235
        -2.        24.5301
         0.        2.0
         2.        20.6163
         4.        859.6729
         6.        1.451257E4
         8.        1.907825E5
        10.        2.20267E6
```

There is no point at which the function became negative, so at this point no root has been isolated. Clearly, if there is a root it must be between $x = -2$ and $x = 2$, because the function becomes quite large and positive outside this region. We need to explore this region further.

In order to utilize the computer to obtain y for other values of x, the program might be changed by the following type-ins.

```
1     1 PRINT, "X"
2       INPUT, X
4.5     GO TO 1
```

To see how the program now looks, we can type

```
LIST
```

and obtain the response

```
1     1 PRINT, "X"
2       INPUT, X
3       Y=X**2*EXP(X)-5.*X*COS(X/5.)+2.*EXP(-X)
4     5 PRINT, X,Y
4.5     GO TO 1
5       END
READY
```

The commands and responses for finding other points on the curve might be as follows:

```
RUN
  X
? 1.
  1.00      -1.496292
  X
?
```

Since y is negative for $x = 1$, we have located two roots, one in the range $0 < x < 1$ and one in the range $1 < x < 2$. Let us pursue the root between zero and one. The commands and responses might go as follows. Continuing from the question mark above, we have

```
?    .5
           .5     - .8623316
  X
?    .3
           .3      .1058229
  X
?    .32
           .32    - .003480834
  X
?    .319
           .319    .00119402
  X
?    .3193
           .3193   .00037265
  X
?    .3194
           .3194  - .00016769
```

The root is now known to lie between .3193 and .3194. The process can be continued until the root is known to desired accuracy or until the accuracy of the calculation becomes a problem. As was shown in Chapter 5, the error in calculating some function $y = f(a_1, a_2, \ldots, a_n, x)$, a function of x and also of some input numbers, a_1, a_2, etc., is approximately

$$\Delta y = \frac{\partial f}{\partial a_1} \Delta a_1 + \frac{\partial f}{\partial a_2} \Delta a_2 + \cdots + \frac{\partial f}{\partial a_n} \Delta a_n + \frac{\partial f}{\partial x} \Delta x$$

When we are very near a root, that is, y is very nearly zero, this quantity can be large enough to make the calculated value of y be negative when it should be positive, or vice versa. When this point has been reached, there is no point in trying to improve the value of x any further.

A remark about the guesses for values of x in the above calculation might be in order. They appear to be remarkably lucky choices, and lead to an accurate result very quickly. Actually they are a reasonable set of choices, based on crude interpolation from previous values, an ancient method of root finding usually known as the method of false position. For $x = 0$, we had $y = 2$, and for $x = .5$, we had $y = -.9$. This suggests a value closer to .5 than to zero, hence the choice $x = .3$. At this x value, $y \approx .1$. Hence y should be zero for x about one-tenth of the way from .3 to .5, or $x = .32$. The value $y \approx -.003$ at $x = .32$ compared to $y = .1$ at $x = .3$ suggests that y should be zero about one-thirtieth of the way from .32 to .3. The value tried was .319, which is about one-twentieth of the way. The next step, .3193, is really one-thirtieth of the way from .32 to .30, so that guess could have been made directly. The computer is supposed to be saving mental labor, so performing this guess with care as a hand calculation on the side seems to be defeating that purpose. A more attractive approach is to change the program to have the computer provide a recommended next value. This is included in Exercise 23 as a problem for the reader.

8.3   THE BISECTION METHOD

The first method of solving an equation is based on Theorem 1 of Section 8.21. Suppose a continuous function $f(x)$ is negative at $x = a$ and positive at $x = b$. Then there is at least one root between a and b. Let us calculate $f[(a + b)/2]$, the value of the function halfway between a and b. If this is zero, we have the root. If it is negative, the root is between that point and b. If it is positive, the root is between that point and a. Thus either we have the root or we have it bracketed within an interval half as large as the previous one. This process can be continued, each time bisecting the interval. It can be continued until the root is known to the desired accuracy.

8.31   Flow Chart and Program for the Bisection Method

A flow chart describing the process for finding a root by bisection when the function is known to be negative at $x = A$ and positive at $x = B$ is given in Figure 8-9. The symbol d is used in the chart to represent the prescribed error limit on x which will be accepted. A remote-terminal FORTRAN program generally following this flow chart is given below.

```
1   1 PRINT, "A,B,D"
2     INPUT, A,B,D,
3   2 X=(A+B)/2.
```

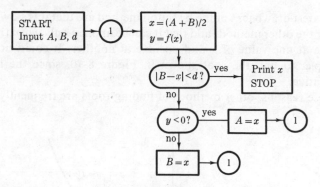

Figure 8-9

```
 4      Y = (insert given function of x)
 5      IF(ABS(B − X) − D)3,3,4
 6   3  PRINT, X
 7      STOP
 8   4  IF(Y)5,3,6
 9   5  A = X
10      GO TO 2
11   6  B = X
12      GO TO 2
13      END
```

8.32 Comments on the Bisection Method

The method described above for finding the value of the root is simple, but it does have several virtues. As is seen, the program is rather easy to write. On a computer, the operation is fairly rapid. The main calculation is the determination of $f(x)$, which must be done once each time the loop is repeated. Each repetition reduces the maximum error by a factor of two, so three repetitions produce roughly an order of magnitude improvement. If the root were initially known to within about 10 units, it would be determined by the program to 10^{-4} unit, an improvement of five orders of magnitude in about 15 iterations.

The greatest virtue of the bisection method is that it is virtually assured to converge to a root. It can fail to do so under the unusual circumstance that an accumulation of errors would cause y at some step to be calculated, say, as a small negative value when actually it should have a small positive value. The machine could then be halving the wrong interval from then on. If proper precautions have been taken concerning accuracy, this should not occur. It will be seen later that this assurance of convergence is not a property of many other methods of finding a root.

The greatest drawbacks of the bisection method are that it is slow compared to some of the other methods and that it can be applied only when the function is negative at one value of x and positive at another. It could not be used, for example, to find the root shown in Figure 8-10, since the function is never negative.

For these reasons, other methods of finding roots are frequently used.

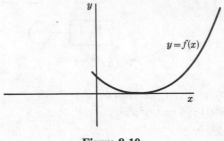

Figure 8-10

8.4   THE METHOD OF ITERATION

When an equation $f(x) = 0$ can be written in the form

$$x = \phi(x) \tag{8-3}$$

accurate values for the real roots can usually be found by the process of iteration. The process is as follows: By some means we obtain a rough approximation x_0 of the desired root. We then substitute it into the right-hand side of (8-3) to obtain a better approximation $x^{(1)}$, given by

$$x^{(1)} = \phi(x_0)$$

This new value is again substituted into the right-hand side of (8-3) to obtain a still better approximation $x^{(2)}$, given by

$$x^{(2)} = \phi(x^{(1)})$$

The process is then repeated (hence the name "iteration"), to give successive approximations:

$$x^{(3)} = \phi(x^{(2)})$$

$$x^{(4)} = \phi(x^{(3)}) \qquad \text{etc.}$$

until a sufficiently close approximation to the true root is obtained. This process is quite readily adaptable to computer use, since the program merely involves the repeated calculation of the same function $\phi(x)$. The flow chart, Figure 8-11, describes the process. In this chart, the number x_0 is the starting value for x, and the number a is a quantity used to cause the calculation to

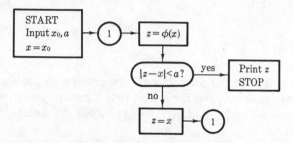

Figure 8-11: Iteration method for finding root

stop when the answer has been obtained with sufficient accuracy. For example, if it is desired that the value of x should be accurate to .01, then $a = .01$ would be used. If a value of x accurate to .0001 is desired, then $a = .0001$ would be used. (This statement is not completely accurate, as will be seen in Section 8.41.) Following through the steps of the flow chart, we find that the first value of x used is $x = x_0$. Hence the first value of z obtained is $z = \phi(x_0)$. This is the quantity denoted by $x^{(1)}$ in the description given above. Hence the first time through, the quantity called $|z - x|$ in the flow chart is actually $|x^{(1)} - x_0|$. In the test we determine if $x^{(1)}$ and x_0 are sufficiently close together that we are willing to accept the value $x^{(1)}$ as the true value x. If the answer is yes, we print $x^{(1)}$ and stop. If the answer is no, we replace x_0 by $x^{(1)}$ and again return to the assertion box $z = \phi(x)$, this time with $x = x^{(1)}$, and compute a new value of z. This time the value obtained for z will be the quantity denoted by $x^{(2)}$ in the previous discussion. This time, then, the quantity called $|z - x|$ in the flow chart will actually be $|x^{(2)} - x^{(1)}|$. If this quantity is small enough, the value of $x^{(2)}$ will be printed. If not, the process will be repeated (or iterated) with $x = x^{(2)}$. After n iterations, we will have $z = x^{(n)}$ and $x = x^{(n-1)}$. If $|x^{(n)} - x^{(n-1)}|$ is sufficiently small, the process will stop and $x^{(n)}$ will be printed.

Example 1. Write a remote-terminal FORTRAN program for finding the root of $xe^x = 1$ to an accuracy of .0001.

This equation can be written as $x = e^{-x}$, which is then in the form

$$x = \phi(x), \quad \text{with } \phi(x) = e^{-x}$$

A suitable remote-terminal program is

```
1     X=1.
2  1  Z=EXP(−X)
3     IF(ABS(Z−X)−.0001)3,3,2
4  2  X=Z
5     GO TO 1
6  3  PRINT, Z
7     STOP
8     END
```

This program will print out nothing until the final result is obtained, so one may sit at the remote terminal for a long time wondering what is happening at the computer. By simply adding the line number 4.5 with the statement

4.5 PRINT, X

we can obtain a printout at each iteration. The successive values of x that would be obtained are

$$x^{(1)} = .36788$$
$$x^{(2)} = .69215$$
$$x^{(3)} = .50042$$
$$x^{(4)} = .60620$$
$$x^{(5)} = .54534$$
$$x^{(6)} = .57959$$
$$x^{(7)} = .56005$$
$$x^{(8)} = .57111$$
$$x^{(9)} = .56484$$
$$x^{(10)} = .56839$$
$$x^{(11)} = .56637$$
$$x^{(12)} = .56751$$
$$x^{(13)} = .56686$$
$$x^{(14)} = .56720$$
$$x^{(15)} = .56704$$
$$x^{(16)} = .56712$$

Sixteen iterations would be performed before $|x^{(n)} - x^{(n-1)}| < .0001$.

Example 2. Modify the program of Example 1 for finding the root of $2x - 1 - 2 \sin x = 0$ to an accuracy of .0001.

This equation can be written as $x = \sin x + .5$, which is then of the form

$$x = \phi(x), \quad \text{with } \phi(x) = \sin x + .5$$

The remote-terminal program of Example 1 can be modified to do this problem by typing in a new statement at line 2,

2   1 Z=SIN(X)+.5

The successive values of x that would be obtained by this routine are

$$x^{(1)} = 1.34147$$
$$x^{(2)} = 1.47381$$
$$x^{(3)} = 1.49528$$
$$x^{(4)} = 1.49713$$
$$x^{(5)} = 1.49727$$
$$x^{(6)} = 1.49729$$

In this case only six iterations are required to obtain a four-decimal-digit accuracy.

Example 3.   Write a program for finding the positive root of $x^2 = 3$ to an accuracy of .0001.

This equation can be written as $x = 3/x$, which is then of the form

$$x = \phi(x), \qquad \text{with } \phi(x) = 3/x$$

The remote-terminal program of Example 1 can be modified to do this problem simply by typing in the statement

2   1 Z=3./X

If the extra print statement inserted for Example 1 were omitted, the program would read

```
1     X=1.
2   1 Z=3./X
3     IF(ABS(Z−Z)− .0001)3,3,2
4   2 X=Z
5     GO TO 1
6   3 PRINT, Z
7     STOP
8     END
```

If we attempt to utilize this program on a computer, we will find that it will never print an answer! It will never perform the transfer to statement 3 but will loop through statements 1 and 2 until the machine is stopped.

The reason for this is seen if the first few values for x are computed by hand. We have

$$x_0 = 1$$
$$x^{(1)} = 3/x_0 = 3$$
$$x^{(2)} = 3/x^{(1)} = 1$$
$$x^{(3)} = 3/x^{(2)} = 3$$
$$x^{(4)} = 3/x^{(3)} = 1 \qquad \text{etc.}$$

The value of x alternates between the numbers 1 and 3 and never converges toward a limit. Thus $|z - x|$ will never be less than .0001.

Before considering the problem of convergence or lack of convergence, let us point out how, in a program such as the one above, we can prevent the machine from hanging up in a loop and wasting valuable time by performing it millions of times. The FORTRAN program below provides this kind of protection:

```
1     X=1.
2     I=1
3   1 Z=3./X
4     IF(ABS(Z−X)−.0001)3,3,2
5   2 X=Z
6     I=I+1
7     IF(I−100)1,4,4
8   3 PRINT, Z
9     STOP
10  4 PRINT, "NOT CONVERGING"
11    STOP
12    END
```

In this program we have included a counter, which only allows the loop to be performed a maximum of 100 times and then skips to statement 4, which prints out "NOT CONVERGING" and stops the problem. If the answer had been obtained in less than 100 steps, it would have been printed out in the normal fashion.

8.41 Convergence of the Iteration Process

From the three examples of Section 8.4 we see that the iteration process may converge slowly, as in the first example, or rapidly, as in the second example, or it may not converge at all, as in the third example. We shall look at the iteration process in more detail in order to determine the conditions under which it does converge. Let x stand for the true value of the root, which would exactly satisfy the equation

$$x = \phi(x) \qquad (8\text{-}4)$$

The first approximation in the iteration process is

$$x^{(1)} = \phi(x_0) \qquad (8\text{-}5)$$

Subtracting (8-5) from (8-4), we have

$$x - x^{(1)} = \phi(x) - \phi(x_0) \qquad (8\text{-}6)$$

By the mean value theorem of differential calculus, the right-hand member of (8-6) can be written

$$\phi(x) - \phi(x_0) = (x - x_0)\phi'(\xi_0) \qquad (8\text{-}7)$$

where ξ_0 is a value between x_0 and x. Hence (8-6) can be written

$$x - x^{(1)} = (x - x_0)\phi'(\xi_0) \qquad (8\text{-}8)$$

In like manner, the second approximation is

$$x^{(2)} = \phi(x^{(1)}) \qquad (8\text{-}9)$$

Subtracting (8-9) from (8-4), we have

$$x - x^{(2)} = \phi(x) - \phi(x^{(1)}) \qquad (8\text{-}10)$$

Again by the mean value theorem

$$\phi(x) - \phi(x^{(1)}) = (x - x^{(1)})\phi'(\xi_1) \qquad (8\text{-}11)$$

where ξ_1 is a value between $x^{(1)}$ and x. Substituting in (8-10), we have

$$x - x^{(2)} = (x - x^{(1)})\phi'(\xi_1) \qquad (8\text{-}12)$$

A similar equation can be obtained for each of the approximations, so that

$$x - x^{(3)} = (x - x^{(2)})\phi'(\xi_2)$$
$$x - x^{(4)} = (x - x^{(3)})\phi'(\xi_3)$$
$$\vdots$$
$$x - x^{(n)} = (x - x^{(n-1)})\phi'(\xi_{n-1})$$

Multiplying together all these equations,

$$(x - x^{(1)})(x - x^{(2)}) \cdots (x - x^{(n)})$$
$$= (x - x_0)(x - x^{(1)}) \cdots (x - x^{(n-1)})\phi'(\xi_0)\phi'(\xi_1) \cdots \phi'(\xi_{n-1}) \quad \text{(8-13)}$$

Dividing the result by the common factors

$$x - x^{(1)}, x - x^{(2)}, \ldots, x - x^{(n-1)}$$

we obtain

$$x - x^{(n)} = (x - x_0)\phi'(\xi_0)\phi'(\xi_1) \cdots \phi'(\xi_{n-1}) \quad \text{(8-14)}$$

The iteration process converges if $x^{(n)}$ becomes close to x as n becomes large; that is, $|x - x^{(n)}|$ becomes small. It can be seen that this will happen if the quantities $\phi'(\xi_1)$ are less than one, for in this case each iteration will multiply the right-hand side of (8-14) by a number less than one and thus reduce the error. Continuing the iteration another step reduces the error by the amount $\phi'(\xi_n)$, where ξ_n is some value of x between $x^{(n)}$ and x. If in equation (8-14) each of the quantities $\phi'(\xi_0)$, $\phi'(\xi_1)$, etc., is less than or equal to some number m less than one, then

$$|x - x^{(n)}| \leqslant |x - x_0|m^n \quad \text{(8-15)}$$

The right-hand member of (8-15) approaches zero as n becomes large, so that the absolute error $|x - x^{(n)}|$ can be made as small as we please by repeating the iteration process a sufficient number of times.

It can be seen from equations (8-14) and (8-15) that the number of iterations required to attain a given accuracy depends on two things: the value of $|x - x_0|$, and the value of $\phi'(x)$ in the neighborhood of the root. If the initial estimate x_0 is well chosen, so that $|x - x_0|$ is small, then fewer iterations will suffice to make $|x - x^{(n)}|$ sufficiently small. If $\phi'(x)$ is small, then the convergence will be rapid, so that few iterations will be required. In Example 1, Section 8.4, the initial estimate x_0 was 1 and the final value was .5671. Hence $|x - x_0|$ was about .43. The value of $\phi'(x) = -e^{-x}$ was, in this case, about $-e^{-.56}$ or $-.57$. Thus each iteration should have reduced the error by about .6. Hence, after 16 iterations, the error should have been about

$$(.43)(.6)^{15} \approx 10^{-4}$$

This was found to be the case. (It might be noted that the bisection method, which reduces the error by a factor of .5 at each step, would have been a little faster in this case.) In Example 3, on the other hand, $\phi'(x) = -3/x^2$, which is equal to -1 at the root $x = \sqrt{3}$, and is greater than 1 in absolute value for $x < \sqrt{3}$. Hence the iteration process does not converge in this case.

8.42   Geometry of the Iteration Process

It is instructive to look at a graphical representation of the iteration process. In Figure 8-12, two curves are drawn, one corresponding to $y_1 = x$ and the other $y_2 = \phi(x)$. The root of $x = \phi(x)$ is the abscissa of the point of intersection. The starting value x_0 is depicted as being to the left of that point.

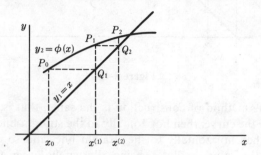

Figure 8-12

Since the absolute value of $\phi'(x)$ must be less than one for convergence, the inclination of the curve $y_2 = \phi(x)$ is depicted as being less than 45° in the vicinity of the root. The iteration process can be traced as follows: At the abscissa x_0, draw a vertical line intersecting the curve $y_2 = \phi(x)$ at P_0. The ordinate at P_0 is then $\phi(x_0)$. From this point draw a horizontal line intersecting the line $y_1 = x$ at the point Q_1. The ordinate at Q_1 is then $\phi(x_0)$, and so, since ordinate equals abscissa along this line, the abscissa is also equal to $\phi(x_0)$. Let us denote this abscissa by $x^{(1)}$. From the point Q_1, draw a vertical line, intersecting the curve $y_2 = \phi(x)$ at P_1. The ordinate at P_1 is then $\phi(x^{(1)})$. From P_1 draw a horizontal line intersecting the line $y_1 = x$ at Q_2. The ordinate at Q_2 is then $\phi(x^{(1)})$, so the abscissa is also. Denote this abscissa by $x^{(2)}$. So far we have

$$\text{at } Q_1, \qquad x^{(1)} = \phi(x_0)$$

$$\text{at } Q_2, \qquad x^{(2)} = \phi(x^{(1)})$$

We can continue this process, obtaining

$$Q_3, \qquad \text{where } x^{(3)} = \phi(x^{(2)})$$
$$\vdots$$
$$Q_n, \qquad \text{where } x^{(n)} = \phi(x^{(n-1)})$$

The points Q_1, Q_2, etc., converge toward the intersection of the two curves.

In Figure 8-12, $\phi'(x)$ is positive, and the convergence is from one side; that is, $x_0 < x_n^{(1)} < x^{(2)} < \cdots < x$. Figure 8-13 shows a case in which $\phi'(x)$

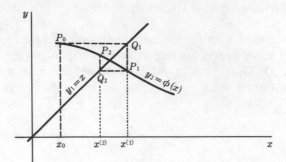

Figure 8-13

is negative. The method of construction is the same, that is, start at x_0 and go vertically to the curve, then horizontally to the straight line, then vertically to the curve, then horizontally to the straight line, and so on. In this case, it is seen that the value oscillates, being alternately larger and then smaller than the true value of the root.

Figure 8-14 depicts a case in which the derivative $\phi'(x)$ is bigger than one in absolute value, so that the process diverges instead of converging.

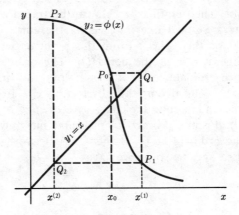

Figure 8-14

8.43   Methods of Inducing or Speeding Convergence

There are many ways of modifying the use of the iteration method to induce or speed convergence. A few of these will be discussed in this section.

One point worthy of note is that if the iteration process starts a wrap-around pattern as in Figures 8-13 and 8-14, then whether this pattern converges or not, a root has been isolated and can be found by the bisection method.

In both figures, at point P_0 the quantity $x - \phi(x)$ is negative and at P_1 it is positive and hence zero at some point in between, and there is a root in between. Hence in order to pick up this sort of behavior, one need only put instructions in the program to monitor the sign of $x - \phi(x)$ and transfer to a different set of instructions for the bisection method if the sign changes. The construction of such a program is left as an exercise for the reader.

There are many possible ways of rewriting an equation $f(x) = 0$ in the form $x = \phi(x)$. It is necessary to choose one for which $\phi'(x)$ is less than one in absolute value, so that the iteration process will converge. It is highly desirable to choose one for which $\phi'(x)$ will be small, so that the convergence will be rapid and not many iterations will be required on the machine. [At the same time it is desirable that $\phi(x)$ itself be kept to a reasonably simple function, so that it may be computed without an inordinate amount of machine time or programming effort.]

Example 1. Write an iteration scheme for the solution of $2 \tan x - x - 1 = 0$.

If we write the equation as

$$x = 2 \tan x - 1$$

then

$$\phi(x) = 2 \tan x - 1$$

and

$$\phi'(x) = 2 \sec^2 x$$

which is greater than one for all values of x. Hence the iteration scheme for this system will diverge. The equation can also be written

$$x = \tan^{-1}\left(\frac{x + 1}{2}\right)$$

In this form

$$\phi(x) = \tan^{-1}\left(\frac{x + 1}{2}\right)$$

and

$$\phi'(x) = \frac{1}{2}\frac{1}{1 + \left(\dfrac{x + 1}{4}\right)^2}$$

which is less than one for all values of x. Hence the iteration scheme based on that formulation will converge. The graphs for the two cases are shown in Figure 8-15(a) and (b). (There are actually an infinite number of solutions in this case, one on each of the branches of the tangent function. If a standard arctangent subroutine is used, it will give a value between $-\pi/2$ and $\pi/2$, so that the intersection indicated by the solid lines will be obtained from the iteration process.) Figure 8-15(b) was obtained by reflecting Figure 8-15(a) about the line $y = x$. This process changed the curve whose slope is greater than one into a curve whose slope is less than one, and hence produced a situation wherein the iteration process would be convergent.

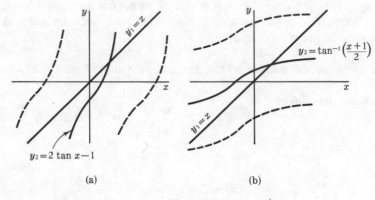

(a)   (b)

Figure 8-15

8.5   THE NEWTON-RAPHSON METHOD

When the derivative of $f(x)$ is a simple expression and easily found, the real roots of $f(x) = 0$ can sometimes be computed by the Newton-Raphson method. Let x_0 denote an approximate value for the root and h denote the correction that must be applied to give the true value of the root, which we shall denote by x. Then

$$x = x_0 + h$$

The equation $f(x) = 0$ then becomes

$$f(x_0 + h) = 0$$

Expanding this by Taylor's formula, we obtain

$$f(x_0 + h) = f(x_0) + hf'(x_0) + (h^2/2)f''(x_0 + \theta h), \qquad \text{where } 0 < \theta < 1$$

Hence

$$f(x_0) + hf'(x_0) + (h^2/2)f''(x_0 + \theta h) = 0$$

If h is sufficiently small, we may ignore the term containing h^2 and write

$$f(x_0) + hf'(x_0) \simeq 0$$

or

$$h \simeq -\frac{f(x_0)}{f'(x_0)}$$

Denoting this approximate value of h by h_1, we have

$$h_1 = -\frac{f(x_0)}{f'(x_0)}$$

and the quantity $x^{(1)}$ given by

$$x^{(1)} = x_0 + h_1 = x_0 - \frac{f(x_0)}{f'(x_0)} \tag{8-16}$$

should be a better estimate of the root than was the quantity x_0. In like manner, better approximations can be obtained from continued application of the process, thus

$$x^{(2)} = x^{(1)} - \frac{f(x^{(1)})}{f'(x^{(1)})}$$

$$x^{(3)} = x^{(2)} - \frac{f(x^{(2)})}{f'(x^{(2)})} \tag{8-17}$$

$$\vdots$$

$$x^{(n)} = x^{(n-1)} - \frac{f(x^{(n-1)})}{f'(x^{(n-1)})}$$

Thus we have an iteration scheme which, like the one discussed in Section 8-4, can be continued until the root is known to the desired accuracy.

8.51   The Square-Root Subroutine

One application of the Newton-Raphson method in computer programming is as a subroutine for computing the square root of a number. The Taylor series is not very good for this purpose because of convergence problems. The Newton-Raphson method is applied as follows: Finding the square root of a number A is equivalent to finding the positive root of the equation

$$x^2 = A$$

or

$$x^2 - A = 0$$

Hence we take

$$f(x) = x^2 - A$$

so that

$$f'(x) = 2x$$

The formula (8-16) can then be written

$$x^{(1)} = x_0 - \frac{x_0{}^2 - A}{2x_0} = \frac{1}{2}\left(x_0 + \frac{A}{x_0}\right)$$

or, for the nth iteration,

$$x^{(n)} = \frac{1}{2}\left(x^{(n-1)} + \frac{A}{x^{(n-1)}}\right)$$

Example 1.   Compute the square root of 10 to seven significant figures.

Taking $x_0 = 1$, we find

$$x^{(1)} = \frac{1}{2}\left(1 + \frac{10}{1}\right) = 5.5$$

$$x^{(2)} = \frac{1}{2}\left(5.5 + \frac{10}{5.5}\right) = 3.66$$

$$x^{(3)} = \frac{1}{2}\left(3.66 + \frac{10}{3.66}\right) = 3.181$$

$$x^{(4)} = \frac{1}{2}\left(3.181 + \frac{10}{3.181}\right) = 3.1623$$

$$x^{(5)} = \frac{1}{2}\left(3.1623 + \frac{10}{3.1623}\right) = 3.162278$$

Five iterations are required to obtain seven-digit accuracy.

It is seen that this method furnishes a convenient and rapidly convergent method for finding the square root. For standard square-root subroutines subject to repeated use, however, still more rapidly convergent processes using rational functions are ordinarily employed.

8.52 Graphical Representation of the Newton-Raphson Method

Figure 8-16 gives a graphical representation of the Newton-Raphson method. The graph of $y = f(x)$ near the root is shown. The steps of the process are as follows: At x_0 draw a vertical line intersecting the curve at P_0. At P_0 draw the tangent line. This will intersect the x axis at $x^{(1)}$. At $x^{(1)}$ draw a vertical

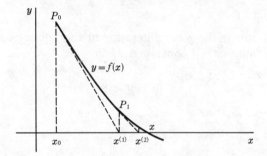

Figure 8-16

line intersecting the curve at P_1. At P_1 draw the tangent line. This will intersect the x axis at $x^{(2)}$, and so on. The formula can be readily derived from the figure. The ordinate of P_0 is $f(x_0)$, and the slope of the tangent line is $f'(x_0)$. From the triangle $x_0 P_0 x^{(1)}$, we see that

$$f'(x_0) = -\frac{f(x_0)}{x^{(1)} - x_0}$$

or

$$x^{(1)} = x_0 - \frac{f(x_0)}{f'(x_0)}$$

In like manner the succeeding steps can be derived.

From the graph it is clear that each step of the process is performed by following the line tangent to the x axis rather than the curve itself. If the derivative does not change too rapidly, or if the curve does not become nearly horizontal near the crossing point, convergence would be expected to be good.

8.53   Convergence of the Newton-Raphson Method

If we let

$$\phi(x) = x - \frac{f(x)}{f'(x)} \tag{8-18}$$

then expression (8-17), the nth iteration of the Newton-Raphson method, can be written

$$x^{(n)} = \phi(x^{(n-1)})$$

which is the nth step in the general iteration method discussed in Section 8.4. We found that that method converged when

$$|\phi'(x)| < 1$$

near the root.
    From (8.18),

$$\phi'(x) = 1 - \frac{[f'(x)]^2 - f(x)f''(x)}{[f'(x)]^2} = \frac{f(x)f''(x)}{f'(x)^2}$$

Hence the method converges when

$$\left| \frac{f(x)f''(x)}{[f'(x)]^2} \right| < 1$$

or

$$|f(x)f''(x)| < [f'(x)]^2$$

near the root.

8.6 SPEED OF CONVERGENCE

It was remarked in the discussion of convergence for the basic iteration process that each iteration tends to reduce the error by a fixed factor, that factor being the value of the derivative, $\phi'(x)$, at the root. A process which converges in this fashion is referred to as a first-order process. For the Newton-Raphson method, if the convergence criterion of Section 8.53 is satisfied, the speed of convergence tends to increase markedly as the error becomes small. This can be seen as follows.

The basic convergence scheme is

$$x^{(n+1)} = x^{(n)} - \frac{f(x^{(n)})}{f'(x^{(n)})} \tag{8-19}$$

Let the error at step n be denoted by e_n. Then

$$e_n = x - x^{(n)}$$

and

$$e_{n+1} = x - x^{(n+1)}$$

Subtracting both sides of (8-19) from x, we have

$$x - x^{(n+1)} = x - x^{(n)} + \frac{f(x^{(n)})}{f'(x^{(n)})}$$

or, in terms of the errors,

$$e_{n+1} = e_n + \frac{f(x - e_n)}{f'(x - e_n)}$$

Expanding in Taylor series,

$$e_{n+1} = e_n + \frac{f(x) - e_n f'(x) + (e_n^2/2)f''(x) + \cdots}{f'(x) - e_n f''(x) + (e_n^2/2)f'''(x) + \cdots}$$

Making use of the fact that $f(x) = 0$, and dividing the denominator into the numerator, we obtain

$$e_{n+1} = e_n - e_n - \frac{e_n^2}{2}\frac{f''(x)}{f'(x)} + \text{higher-order terms}$$

or

$$e_{n+1} \approx - \frac{e_n^2}{2} \frac{f''(x)}{f'(x)}$$

This says that the absolute error at step $n + 1$ is proportional to the *square* of the absolute error at step n. Hence if we have an answer correct to one decimal place at one step, it should be accurate to two places at the next step, four at the next, eight at the next, and so forth. This behavior is seen in Example 1, Section 8.51, where the tenths place was correct at step 3, the thousandths at step 4, and the millionths at step 5. This rapid convergence, where the error at one step is proportional to the square of the previous error, is called "second-order" convergence. It is to be contrasted with the first-order convergence exhibited in Example 1, Section 8.4, where the error was reduced at each step by the constant factor .57, and also the first-order convergence of the bisection method, where the error at each step is reduced by the factor .5. In computer programs which are to be subject to repeated use, it is usually well to use a second-order process. In programs which will receive relatively little use, it is frequently wise to use a first-order scheme if convergence can be more fully assured by doing so, accepting an assured but slow convergence rather than fast but uncertain convergence.

8.7   IMPLICIT FUNCTIONS

There are occasions when a dependent variable, y, is given in terms of an independent variable, x, but only in an implicit fashion. For example, the expression

$$ye^y = x$$

defines y in terms of x, but only in an implicit way. Given a value of x, we can find y by means of the techniques discussed in this chapter. By a few simple changes, the program of Example 1, Section 8.4, can be converted into a subroutine to do this. The subroutine given below will suffice:

```
      SUBROUTINE XINV(X,Y)
      Y=1.
    1 Z=X*EXP(−Y)
      IF(ABS(Z−Y)−.0001)3,3,2
    2 Y=Z
      GO TO 1
    3 Y=Z
      RETURN
      END
```

The concerns about speed of convergence and assurance of convergence expressed in the preceding pages must be considered in constructing and utilizing a subroutine such as this one. We noted, for example, that for $x = 1$ the method given in Example 2, Section 8-43, produced more rapid convergence than the above method. Also, we noted in Example 3, Section 8.4, how the program can be modified to prevent the computer from remaining in a loop if convergence does not occur in a reasonable number of steps. It is usually advisable to include such a feature in a subroutine designed to evaluate an implicit function.

8.71   Experimental Approach

Methods have been described that allow one to assess the convergence and speed of convergence of a proposed scheme for evaluating an implicit function. However, these methods are for the most part rather awkward and cumbersome. On the other hand, it is quite easy to write the program itself and is ordinarily a matter of seconds for the computer to attempt the program. In this circumstance it is frequently expedient, and even economical, to guess at a calculation method and try it out, and if it does not converge, try another.

Example 1.   Write a subroutine to calculate y for x between zero and one, where

$$y^2 + ye^{xy} = e^x$$

The following iteration schemes suggest themselves:

(1) $y = e^x/(y + e^{xy})$

(2) $y = \sqrt{e^x - ye^{xy}}$

(3) $y = \dfrac{1}{x} \ln \dfrac{e^x - y^2}{y}$

The following program can be used to test each of these, in order to select the best or eliminate all as possibilities:

```
      DO 1 J=1,10
      FJ=J
      X=.1*FJ
      CALL SQIZ(X,Y,I)
    1 PRINT 101,X,Y,I
      STOP
```

```
101 FORMAT(2E12.4,I5)
    END
    SUBROUTINE SQIZ(X,Y,I)
  1 QUNF(R,S)= (insert correct function)
    Y=1.
    I=1
  2 Z=QUNF(X,Y)
    IF(ABSF(Z−Y)−.0001)4,4,3
  3 Y=Z
    I=I+1
    IF(I−100)2,5,5
  4 Y=Z
    RETURN
  5 PRINT 201
    RETURN
201 FORMAT(17H DID NOT CONVERGE)
    END
```

We would run this program three times, once with

$$QUNF(R,S)=EXP(R)/(S+EXP(R*S))$$

once with

$$QUNF(R,S)=SQRT(EXP(R)-S*EXP(R*S))$$

and once with

$$QUNF(R,S)=ALOG((EXP(R)-S*S)/S)/R$$

corresponding to expressions (1), (2), and (3). Upon completion we would have values of y for $x = .1, .2$, etc., up to 1, as computed by each method, and also the number of iterations required to obtain each value, as well as a printout indicating any cases that did not converge. These results should give a good idea of the relative value of the three iteration schemes.

8.8 FUNCTIONS OF SEVERAL VARIABLES

The methods discussed in this chapter can be applied with only minor modification to functions of more than one variable. For example, if we wish to find a root of the system of equations

$$f(x, y) = 0$$

$$g(x, y) = 0$$

(8-20)

We may rewrite the equations as

$$x = \phi_1(x, y)$$

$$y = \phi_2(x, y)$$

and use the iteration scheme

$$x^{(n+1)} = \phi_1(x^{(n)}, y^{(n)})$$
$$y^{(n+1)} = \phi_2(x^{(n)}, y^{(n)})$$

A procedure analogous to the Newton-Raphson method for equations (8-20) can be obtained by writing

$$x = x^{(n)} + \Delta x$$
$$y = y^{(n)} + \Delta y$$
$$\tag{8.21}$$

where (x, y) are the coordinates of the true root and $x^{(n)}$, $y^{(n)}$ are the appropriate coordinates at the nth iteration. Then, using Taylor's theorem,

$$f(x^{(n)} + \Delta x, y^{(n)} + \Delta y) = 0 = f(x^{(n)}, y^{(n)})$$
$$+ \Delta x f_x(x^{(n)}, y^{(n)}) + \Delta y f_y(x^{(n)}, y^{(n)}) + \cdots$$

$$g(x^{(n)} + \Delta x, y^{(n)} + \Delta y) = 0 = g(x^{(n)}, y^{(n)})$$
$$+ \Delta x g_x(x^{(n)}, y^{(n)}) + \Delta y g_y(x^{(n)}, y^{(n)}) + \cdots$$

where f_x, f_y, etc., denote partial derivatives. Neglecting all terms involving terms higher than the first power in Δx and Δy, we have

$$\Delta x f_x + \Delta y f_y \simeq -f$$
$$\Delta x g_x + \Delta y g_y \simeq -g$$

Solving for Δx and Δy we have

$$\Delta x \simeq (g f_y - f g_y)/(f_x g_y - g_x f_y)$$
$$\Delta y \simeq (f g_x - g f_x)/(f_x g_y - g_x f_y)$$

If these approximate values for Δx and Δy are used in (8-21) we will (hopefully) obtain a better estimate for x and y. Denoting these new estimates by

$x^{(n+1)}$ and $y^{(n+1)}$, we have

$$x^{(n+1)} = x^{(n)} + [g(x^{(n)}, y^{(n)})f_y(x^{(n)}, y^{(n)}) - f(x^{(n)}, y^{(n)})g_y(x^{(n)}, y^{(n)})]/d$$

$$y^{(n+1)} = y^{(n)} + [f(x^{(n)}, y^{(n)})g_x(x^{(n)}, y^{(n)}) - g(x^{(n)}, y^{(n)})f_x(x^{(n)}, y^{(n)})]/d$$

(8-22)

where

$$d = f_x(x^{(n)}, y^{(n)})g_y(x^{(n)}, y^{(n)}) - g_x(x^{(n)}, y^{(n)})f_y(x^{(n)}, y^{(n)})$$

This method is easily extended to three or more variables. An example of its use is given in Section 11.42.

The methods given above have the same convergence problems, as do their single-variable counterparts. Convergence criteria can be derived but will not be given here.

EXERCISE 23

1. In the following problems (1) draw a graph and obtain a rough estimate of the roots, (2) draw a flow chart and write a FORTRAN program for the determination of the roots by one of the processes discussed in the text, (3) calculate by hand rough values for the first three steps of the process, and (4) estimate the computer time required for the solution to four significant figures.

   a. $x^3 = e^{-x}$ b. $x^3 + 3x - 1 = 0$
   c. $e^{\sin x} = e^x + e^{-3x}$ d. $x \ln x = e^x - 4$

2. In the following problems, y is defined implicitly as a function of x. In each case, write a FORTRAN subroutine that will compute y, given x. Include a feature that will print out "NON-CONVERGE" and continue if the iteration scheme does not converge in 100 steps. Estimate the maximum time for performing the subroutine.

   a. $y \ln y = x$ b. $y \sin y = x$
   c. $\sin (x + y) = y + \cos x$ d. $e^x \cos y + 2xy = 3$

3. Write a program that will employ the bisection method to find a root of $f(x) = 0$, given that $f(A)$ and $f(B)$ have opposite signs but not given the information as to which one is positive.

4. Write a FORTRAN program to find the root of

$$(x - 1)^{1/3} + 1 = 0$$

by the Newton-Raphson method. Test your program with the starting value $x_0 = -2.375$. Explain the result.

5. Rewrite the remote-terminal program of Example 1, Section 8.23, to provide a recommended next guess for the root at each step, based on interpolation between the last two values used. Construct the program so that this recommended guess will be used if $x = 0$ is entered at the input request but that the value of x typed in by the user will be used if that is other than zero.

6. In the program for the bisection method, Section 8.31, what would you expect to happen if the value entered for D were 10^{-9}, and the computer were calculating in seven-place floating decimal arithmetic?

7. Modify the program for the bisection method, Section 8.31, to use D as acceptable relative error in x rather than acceptable absolute error.

8. In the program of Example 1, Section 8.4, what would you expect to happen if the constant .0001 in the statement at line 3 were changed to 10^{-9} and the computer were calculating in seven-place floating decimal arithmetic?

9. Modify the program of Example 1, Section 8.4, to exit on a test of relative error rather than absolute error.

10. In each of the following, the constants in the function $f(x)$ are accurate only to the number of digits shown. An approximate root is also shown. Determine the number of correct decimal places to which the root can be determined.

    a. $f(x) = 1.3 - 1.3x,$ $x = 1.0$
    b. $f(x) = 1.25 - 1.25 \sin x,$ $x = 1.57$
    c. $f(x) = 1.1 - 1.1e^x$ $x = 0$
    d. $f(x) = 1.41x - 2.0 \sin x,$ $x = .785$

11. Using three-place floating decimal arithmetic, find a root in each of the following equations by the iteration method, using a starting value of $x - 1.00$ (that is, $.100 \times 10^1$). Keep performing iterations until the values of x form a repetitive pattern. To what accuracy can the root be determined with three-place arithmetic?

    a. $x = .500 + .300x$ b. $x = .400 + .400 \sin x$ c. $x = .200 - .300e^x$

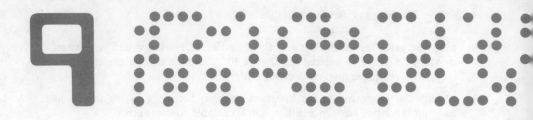

Solution of Polynomial Equations

9.1 INTRODUCTION

Chapter 8 treated the problem of finding roots of an equation $f(x) = 0$. Of frequent interest is the case in which $f(x)$ is a polynomial, $f(x) = a_1 x^n + a_2 x^{n-1} + \cdots + a_{n+1}$. This is a special case of the general problem studied in Chapter 8, and superficially it appears to be a more simple one than those in which $f(x)$ involves trigonometric, exponential, or other transcendental functions. The case deserves special consideration for two reasons, however: first, the problem of obtaining roots to a polynomial equation appears frequently in many fields of science and engineering; second, the methods of Chapter 8, while of some use for this particular problem, do not ordinarily give complete solutions. Other methods are frequently of greater use, although, unfortunately, there is no single method which is completely satisfactory for solving polynomial equations.

9.2 NATURE OF THE ROOTS OF A POLYNOMIAL EQUATION

A large amount of general information is available concerning roots of a polynomial equation. The first salient fact is the fundamental theorem of algebra:

Theorem I. *Every polynomial equation with arbitrarily given real or complex coefficients has at least one real or imaginary root.*

298

The second is a special case of the remainder theorem:

Theorem II.   *If c is a root of the polynomial equation*

$$f(x) = 0$$

then $f(x)$ is divisible by $x - c$, so that

$$f(x) = (x - c)f_1(x)$$

where $f_1(x)$ is a polynomial of degree $n - 1$.

We say that c is a root of multiplicity n if $f(x)$ is divisible by $(x - c)^n$ but not by $(x - c)^{n+1}$. With this definition it follows from the above two theorems that a polynomial equation of degree n has precisely n roots, each root being counted according to its multiplicity. If the coefficients $a_1, a_2, \ldots,$ a_{n+1} are real numbers (the case with which we shall concern ourselves), then these roots are either real or in complex conjugate pairs. The problem of solving a polynomial equation of degree n, then, involves isolating n different roots, some of which may be complex, and then utilizing some process for finding the value of each root to the required accuracy.

9.21   Real Roots

Frequently it is not necessary to find all the roots of a polynomial equation, but instead to find only a root or roots satisfying a special property, such as the largest real root, or all real roots. A few facts concerning real roots are worth noting.

Theorem III.   *Every polynomial equation of odd degree with real coefficients has at least one real root.*

This fact follows directly from the fact that the complex roots occur in conjugate pairs, so that there must be an even number of these. Another useful theorem is the famous Descartes rule of signs:

Theorem IV.   *The number of positive real roots of an equation*

$$f(x) = a_1 x^n + a_2 x^{n-1} + a_3 x^{n-2} + \cdots + a_{n+1} = 0$$

with real coefficients is never greater than the number of variations of sign in the sequence of its coefficients:

$$a_1, a_2, \ldots, a_{n+1}$$

and, if less, always by an even number.

Example 1. Determine the possible numbers of positive real roots for the equation

$$x^4 - x^2 + 2x - 1 = 0$$

The equation has three variations of sign in the sequence of coefficients 1, -1, 2, -1 (terms with zero coefficients can be ignored). Hence it has at most three positive roots, and at least one. The theorem can be used to obtain information concerning negative roots. If we make the substitution $x_1 = -x$, we obtain

$$x_1{}^4 - x_1{}^2 - 2x_1 - 1 = 0$$

There is only one variation, hence one positive real root for this equation. Thus the original equation had one negative real root.

The following theorem gives a quick, although somewhat crude, estimate of the maximum size of the roots.

Theorem V. *All roots are less than*

$$(|a_1| + |a_2| + \cdots + |a_{n+1}|)/|a_1|$$

in absolute value.

9.3 QUADRATIC EQUATIONS

The general quadratic equation is usually written in the form

$$ax^2 + bx + c = 0$$

The roots are given by the formula

$$x = \frac{-b \pm \sqrt{b^2 - 4ac}}{2a}$$

If the quantity $b^2 - 4ac$ is zero or positive, there are two real roots. If it is negative, the roots are complex conjugates. Hence the program to perform this calculation must distinguish between these two cases. In the first case it computes the square root of $b^2 - 4ac$ and combines with the term $-b$ to obtain the roots. In the second, it must take the square root of $4ac - b^2$

and print out both a real and imaginary part for the roots. The flow chart, Figure 9-1, describes how this might be done. In the case in which the roots are complex, some printout with the numbers to indicate this fact would be advisable. Since the roots are conjugates, only two numbers are required to describe them, a real part $x(rl)$ and an imaginary part $x(im)$. The roots are then $x(rl) + ix(im)$ and $x(rl) - ix(im)$.

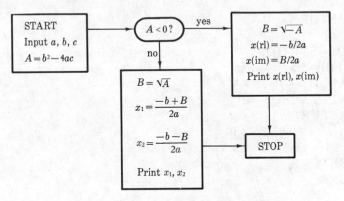

Figure 9-1: Roots of quadratic equation

A problem of accuracy can arise in using the quadratic formula if one of the roots is nearly equal to zero. For example, suppose b is positive. Then if the root

$$\frac{-b + \sqrt{b^2 - 4ac}}{2a}$$

is nearly zero, the subtraction in the numerator may cause the loss of leading significant digits. This can be avoided by multiplying both numerator and denominator of the above expression by $b + \sqrt{b^2 - 4ac}$, giving

$$\frac{-b^2 + b^2 - 4ac}{2a(b + \sqrt{b^2 - 4ac})} = \frac{-2c}{b + \sqrt{b^2 - 4ac}}$$

If b is negative, then the other root is the one which may be inaccurate. A similar protective measure can be used in this case. The flow chart, Figure 9-2, includes protection against this type of accuracy loss. It might be noted that this method is equivalent to computing the larger root (in absolute value) by the quadratic formula, and then computing the smaller one from the fact that the product of the roots is c/a.

The following FORTRAN subroutine will find the roots of a quadratic equation roughly following the flow chart, Figure 9-2:

```
SUBROUTINE QUADRO(A,B,C,I,X1,X2)
D=B*B-4.*A*C
IF(D)2,3,3
2 DS=SQRT(-D)
X1=-B/(2.*A)
X2=DS/(2.*A)
I=2
RETURN
3 I=1
DS=SQRT(D)
IF(B)4,5,5
4 X1=(DS-B)/(2.*A)
X2=2.*C/(DS-B)
RETURN
5 X1=-2.*C/(DS+B)
X2=-(DS+B)/(2.*A)
RETURN
END
```

In this subroutine, the output I serves to indicate whether the roots are real or imaginary. If $I = 1$, X1 and X2 are two real roots. If $I = 2$, they are real and imaginary parts of complex roots.

9.4   COMPUTER ARITHMETIC WITH POLYNOMIALS

Before proceeding to equations of higher degree, we require some background on the way polynomials can be handled on a computer. For use in a com-

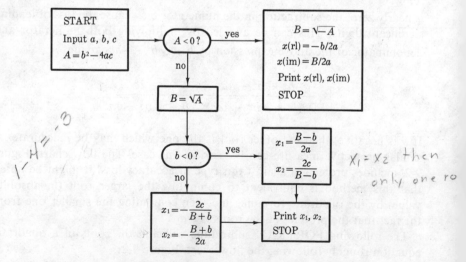

Figure 9-2: Roots of quadratic equation—improved method

puter, the polynomial $y = a_1 x^n + a_2 x^{n-1} + \cdots + a_{n+1}$ can be represented by the coefficients $a_1, a_2, \ldots, a_{n+1}$ and the degree, n. Thus a total of $n + 2$ numbers must be stored to represent the polynomial. If FORTRAN is being used, the coefficients can be stored as a subscripted variable A(I), where I will take on values from 1 to N + 1. The degree N can be stored as a separate, fixed-point number.

At this point it is important to make the distinction between the polynomials about which theorems were stated in Section 9.2 and the polynomials as treated within a computer. The pure theory in Section 9.2 concerned itself with polynomials in which the coefficients $a_1, a_2, \ldots, a_n$ were exact numbers, and which are to be handled using exact arithmetic. As has been emphasized in earlier chapters, on a computer one customarily does approximate arithmetic with approximate numbers. Hence the theorems of Section 9.2 are not directly applicable to polynomials as they are handled on a computer but require some interpretation. One should not infer that the computer method of handling polynomials is at fault. Actually, in practical problems, the coefficients of a polynomial of interest are usually obtained from measured quantities and hence are subject to error. Hence many polynomials of interest are only approximate. Where it is absolutely necessary to handle polynomials exactly on a computer, this can be done (as long as the coefficients are rational) by scaling the sizes of numbers and using integer arithmetic. Our interests in this book lie in the direction of applied problems, so we shall concern ourselves with approximate polynomials.

As a first point of interest note that the relationship

$$y = a_1 x^n + a_2 x^{n-1} + \cdots + a_n x + a_{n+1}$$

does not define a specific value of y if the a_i's are approximate numbers and if the arithmetic is approximate. For any x there is a range of values of y which must be considered acceptable.

Example 1. Draw a graph of the function

$$y = 1.0x^3 - 2.1x^2 + 1.1x + .0$$

where the coefficients are accurate to the number of places shown.

Figure 9-3 depicts the above equation. The solid line is the graph obtained under the assumption that the coefficients given above are exact. It shows roots at $x = 0$, 1, and 1.1. The shaded region depicts the possible values of y, considering the allowed region of uncertainty of the coefficients. For x greater than zero, y could be as large as

$$y = 1.05x^3 - 2.05x^2 + 1.15x + .05$$

or as small as

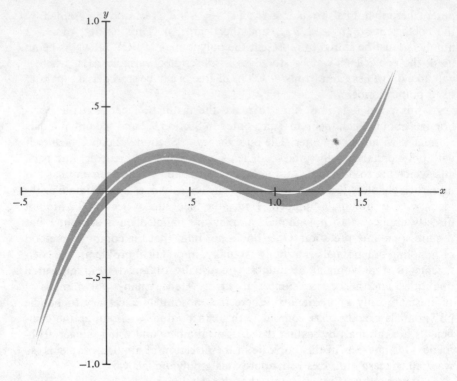

Figure 9-3: Graph of $y = 1.0x^3 - 2.1x^2 + 1.1x - .0$

$$y = .95x^3 - 2.15x^2 + 1.05x - .05$$

For x negative, y could be as large as

$$y = .95x^3 - 2.05x^2 + 1.05x + .05$$

or as small as

$$y = 1.05x^3 - 2.15x^2 + 1.15x - .05$$

These equations determine the bounds on the shaded region in Figure 9-3. Note that the root which was nominally at zero may actually be anywhere between about $-.05$ and $.05$. The roots which were nominally at 1 and 1.1 may range from about .9 to 1.2. They could be equal, or it could be that the true curve, corresponding to the unknown exact equation represented by the given equation does not touch the x axis at all in this region, so that there are no real roots in that area at all, and the true equation has one real root and two complex roots.

The example just given has demonstrated some of the difficulties introduced by working with polynomials having poorly known coefficients. In a computer the roundoff problem in storing the coefficients will introduce these sorts of difficulties—on a smaller scale, generally, since several significant digits are carried. The machine roundoff error can be troublesome, however, particularly so when there are two roots very close together. As was seen in Figure 9-3, roots which are very close together can sometimes be interpreted as two distinct real roots, or as multiple real roots, or a set of imaginary roots, all within the range of uncertainty of the coefficients. A polynomial with roots so close together that this situation may occur is termed ill-conditioned. In particular, a polynomial with multiple roots will always be ill-conditioned. Approximate arithmetic, even in double precision, can well confuse the nature of these roots.

9.41   Evaluation of a Polynomial

In earlier chapters it has been seen that the value of a polynomial for a particular value of the independent variable, x, can be found rather efficiently by grouping the terms:

$$y = ((\cdots (a_1 x + a_2)x + a_3) \cdots)x + a_{n+1}$$

A set of FORTRAN statements that will perform the evaluation is

```
    Y=A(1)
    DO 1 I=1,N
  1 Y=Y*X+A(I+1)
```

The number of multiplications required actually to find the value is equal to N, the degree of the polynomial.

9.42   Normal Form for a Polynomial

For a particular polynomial, some of the coefficients a_i may be zero. This is no problem unless the leading coefficient a_1 is zero. In this case, if n, the degree, is also zero, the polynomial is just the number zero. If n is different from zero, then the polynomial is incorrectly written. It is not really of degree n at all, but of some lower degree. Steps must be taken to assure that the computer always has polynomials correctly represented, with leading coefficient a_1 different from zero. If we are providing the coefficients as inputs, this is easy to do, but, if the computer obtains the polynomial as the result of some internal calculation, the machine itself must assure that the leading coefficient is not zero. Suppose the computer has performed a calculation which has resulted in the numbers $a_1, a_2, \ldots, a_{n+1}, n$, which are supposedly the coefficients and degree of a polynomial. The segment of flow chart shown in Figure 9-4 will assure that the representation is put in a form with leading coefficient different from zero, a form we might refer to as the " normal " form.

A program following this flow chart will redefine the coefficients so that a_1 is indeed the first nonzero coefficient, and will determine a new n which is the correct degree. A set of FORTRAN statements that will perform the same function is

```
1 IF(A(1))5,2,5
2 DO 3 I=1,N
3 A(I)=A(I+1)
  N=N-1
4 IF(N)5,5,1
5 .........continuation of program
```

9.43   Division of One Polynomial by Another

When one polynomial is divided by another of equal or lower degree, the quotient is again a polynomial, and ordinarily there is a remainder which is also a polynomial.

Example 1.   Divide $x^4 + 3x^3 + 6x + 5$ by $x^2 + 2x - 2$.

Performing the division in the normal manner we have

$$
\begin{array}{r}
x^2 + x \\
x^2 + 2x - 2 \overline{)x^4 + 3x^3 + 0x^2 + 6x + 5} \\
\underline{x^4 + 2x^3 - 2x^2 } \\
x^3 + 2x^2 + 6x \\
\underline{x^3 + 2x^2 - 2x } \\
8x + 5
\end{array}
$$

In this case the quotient is $x^2 + x$ and the remainder is $8x + 5$. Each step of the operation is a straightforward arithmetic operation, and clearly should be reducible to computer application. The powers of x that appear actually serve no useful purpose except to keep straight which terms are to be combined with which. The calculation can as well be done as follows:

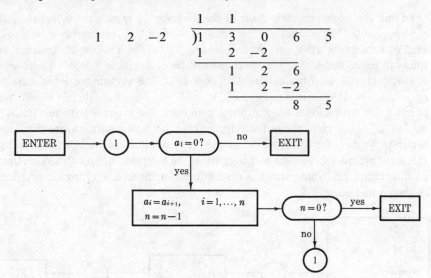

Figure 9-4: Rewriting a polynomial in normal form

All the numerical values are given by this procedure, and these can then be combined with the correct powers of x to give the quotient and remainder.

Example 2. Divide

$$a_1 x^n + a_2 x^{n-1} + \cdots + a_n x + a_{n+1} \quad \text{by} \quad b_1 x^m + b_2 x^{m-1} + \cdots + b_{m+1}$$

We can, of course, only do this symbolically since we do not have numerical values for the coefficients and degrees of the terms, but the procedure used in Example 1 can be followed.

$$
\begin{array}{r}
\begin{array}{cccccccc}
 & & & q_1 & q_2 \cdots\cdots\cdots & q_{n-m+1} \\
\hline
b_1 \quad b_2 \cdots b_{m+1} \,) a_1 & a_2 & a_3 \cdots\cdots\cdots\cdots\cdots\cdots\cdots\cdots & a_{n+1} \\
b_1 q_1 & b_2 q_1 & b_3 q_1 \cdots b_{m+1} q_1 \\
\hline
 & a_2' & a_3' & \cdots a_{m+1}' & a_{m+2} \\
 & b_1 q_2 & b_2 q_2 \cdots b_m q_2 & b_{m+1} q_2 \\
\hline
 & & a_3'' & \cdots a_{m+1}'' & a_{m+2}' \\
\end{array}
\end{array}
$$

$$a_{n-m+1}^* \quad\cdots\cdots\quad a_n^* \qquad a_{n+1}$$
$$b_1 q_{n-m+1} \quad\cdots\cdots\quad b_m q_{n-m+1} \quad b_{m+1} q_{n-m+1}$$
$$r_1 \cdots\cdots r_{m-1} \qquad r_m$$

In this symbolic representation of the division, q_1 represents a_1/b_1, so that $a_1 - b_1q_1 = 0$. The symbol a_2' represents $a_2 - b_2q_1$, a_3' represents $a_3 - b_3q_1$, etc., q_2 represents a_2'/b_1, so that $a_2' - b_2q_2 = 0$. The process is repeated as many times as necessary. The quotient will be of degree $n - m$, so $n - m + 1$ coefficients will be obtained in the quotient. The remainder is of degree $m - 1$, so m coefficients $r_1, r_2, \ldots, r_m$ will be obtained there. When we convert this operation to a computer program, the intermediate results a_2', a_3', a_3'', a_{n-m+1}^*, etc., are not needed; so in the flow chart, rather than use separate symbols for these, we can re-use the symbols, a_2, a_3, etc., thus re-using the same memory cells over and over for these intermediate results. A section of flow chart describing the process outlined in the above symbolic division is shown in Figure 9-5.

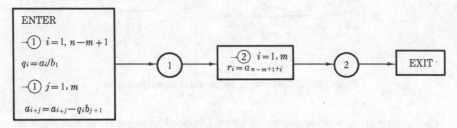

Figure 9-5: Polynomial division

We enter this section of a flow chart with $n + 2$ numbers $a_1, \ldots, a_{n+1}, n$, representing a polynomial of degree n, and $m + 2$ numbers $b_1, \ldots, b_{m+1}$, m, representing a polynomial of degree m. We leave with $n - m + 1$ coefficients $q_1, q_2, \ldots, q_{n-m+1}$ representing a quotient of degree $n - m$, and m coefficients $r_1, r_2, \ldots, r_m$, representing a remainder of degree $m - 1$.

In Section 6.3 there was a discussion of accuracy loss associated with the subtraction of two nearly equal numbers. Such a situation is likely at one point in Figure 9-5 where the operation $a_{i+j} = a_{i+j} - q_ib_{j+1}$ is to be performed. The major problem occurs when this result is supposed to be zero. Ordinarily, roundoff error in the computer will keep it from being exactly zero but will make it have some small value. This means, for example, that if the polynomial $a_1x^n + \cdots + a_{n+1}$ is exactly divisible by $b_1x^m + \cdots + b_{m+1}$, this result would not ordinarily appear on the computer following the above flow chart, since the remainder would be some polynomial with small but nonzero coefficients. The difficulty can be circumvented by checking for loss of significant figures as described in Section 6.31. For example, we might choose to set $a_{i+j} = 0$ every time the subtraction above introduces four leading zeros. This can be done by replacing that part of the flow chart in Figure 9-5 by Figure 9-6.

The FORTRAN subroutine given below divides the polynomial

$$A(1)x^N + A(2)x^{N-1} + \cdots + A(N + 1)$$

by

$$B(1)x^M + B(2)x^{M-1} + \cdots + B(M+1)$$

giving quotient

$$Q(1)x^K + Q(2)x^{K-1} + \cdots + Q(K+1)$$

and remainder

$$R(1)x^L + R(2)x^{L-1} + \cdots + R(L+1)$$

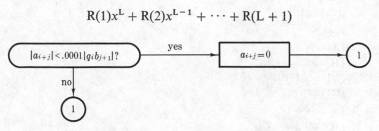

Figure 9-6: Error protection for polynomial division

The dimension statement allows for 20 values of $A(I)$, $B(I)$, $Q(I)$, and $R(I)$, so that polynomials of degree 0 to degree 19 could be handled. The statements up to statement 6 put the divisor into normal form, so that $B(I)$ will not be zero. If the divisor is identically zero, statement 3 prints that fact and the program is stopped. Statement 6 computes the degree of the quotient. If the degree of the dividend is less than that of the divisor, statements 7 through 8 set the quotient equal to zero and the remainder equal to the dividend. Statement 30 checks to see if the divisor is of degree zero, that is, merely a constant. If it is, statements 31 through 32 set the remainder equal to zero and obtain the quotient by dividing the coefficients of the dividend by the constant divisor. Statements 9 through 12 do the computation as outlined by the flow charts, Figures 9-5 and 9-6. Statements 13 through 18 put the remainder in normal form:

```
      SUBROUTINE POLDIV(A,N,B,M,Q,K,R,L)
      DIMENSION A(20),B(20),Q(20),R(20),F(20)
  1   IF(B(1))6,2,6
  2   IF(M)3,3,4
  3   PRINT 1001
      STOP
  4   DO 5 I=1,M
  5   B(I)=B(I+1)
      M=M-1
      GO TO 1
  6   K=N-M
```

```
         IF(K)7,30,30
    7  K=0
       L=N
       LL=N+1
       DO 8 I=1,LL
       Q(I)=0.
    8  R(I)=A(I)
       RETURN
   30  IF(M)31,31,9
   31  L=0
       R(1)=0.
       K=N
       LL=N+1
       DO 32 I=1,LL
   32  Q(I)=A(I)/B(1)
       RETURN
    9  LL=N+1
       DO 40 I=1,LL
   40  F(I)=A(I)
       LL=N-M+1
       DO 12 I=1,LL
       Q(I)=F(I)/B(1)
       DO 12 J=1,M
       W=Q(I)*B(J+1)
   10  F(I+J)=F(I+J)-W
       IF(ABS(F(I+J))-.0001*ABS(W))11,11,12
   11  F(I+J)=0.
   12  CONTINUE
       DO 13 I=1,M
   13  R(I)=F(N-M+I+1)
   14  L=M-1
   15  IF(R(1))19,16,19
   16  IF(L)19,19,17
   17  DO 18 I=1,L
   18  R(I)=R(I+1)
       L=L-1
       GO TO 15
   19  RETURN
 1001  FORMAT(16H DIVISOR IS ZERO)
       END
```

The operation of dividing one polynomial by another as outlined in this section involves $n - m + 1$ divisions and $(m + 1)(n - m + 1)$ multiplications, or about $(m + 2)(n - m + 1)$ multiplication times. For division of a twentieth-degree polynomial by a tenth-degree polynomial, this is about $12 \times 11 = 132$ multiplication times.

9.44   Highest Common Factor

The highest common factor of two polynomials is a polynomial of highest degree which will divide both the polynomials exactly, leaving zero remainder. For example, the highest common factor of $x^2 - 2x + 1$ and $x^2 - 3x + 2$ is $x - 1$. The highest common factor of $f(x)$ and $g(x)$, where the degree of $g(x)$ is equal to or less than that of $f(x)$, is obtained as follows.

Divide $f(x)$ by $g(x)$, obtaining $q_1(x)$ and remainder $f_1(x)$. Then

$$f(x) = g(x)q_1(x) + f_1(x)$$

Now divide $g(x)$ by $f_1(x)$, obtaining quotient $q_2(x)$ and remainder $f_2(x)$. Then

$$g(x) = f_1(x)q_2(x) + f_2(x)$$

Now divide $f_1(x)$ by $f_2(x)$, obtaining quotient $q_3(x)$ and remainder $f_3(x)$. Then

$$f_1(x) = f_2(x)q_3(x) + f_3(x)$$

If this process is continued, at some point the remainder will finally become zero, and in the last step we have

$$f_{r-2} = f_{r-1}q_r + f_r$$

$$f_{r-1} = f_r q_{r+1}$$

The last nonzero remainder, $f_r(x)$, is the highest common factor of $f(x)$ and $g(x)$.

Example 1.   Find the highest common factor of $f(x) = x^3 - 4x^2 + 5x - 2$ and $g(x) = 3x^2 - 8x + 5$.

Dividing $f(x)$ by $g(x)$,

$$
\begin{array}{r}
(1/3)x - 4/9 \\
3x^2 - 8x + 5 \overline{\smash{)}\,x^3 -\quad 4x^2 +\quad 5x\ - 2} \\
\underline{x^3 - (8/3)x^2 + (5/3)x} \\
-(4/3)x^2 + (10/3)x - 2 \\
\underline{-(4/3)x^2 + (32/9)x - 20/9} \\
-(2/9)x + 2/9
\end{array}
$$

We have $q_1(x) = (1/3)x - 4/9$ and $f_1(x) = -(2/9)x + 2/9$. Dividing $g(x)$ by $f_1(x)$,

$$
\begin{array}{r}
-(27/2)x + 45/2 \\
-(2/9)x + 2/9 \overline{)3x^2 - 8x + 5} \\
3x^2 - 3x \\
\hline
-5x + 5 \\
- 5x + 5 \\
\hline
0
\end{array}
$$

Since the remainder is zero, the highest common factor is the last nonzero remainder, $f_1(x) = -(2/9)x + 2/9$. The highest common factor is not uniquely defined. Any constant multiple of $f_1(x)$ could be regarded as the highest common factor. For example,

$$-(9/2)f_1(x) = x - 1$$

is a neater expression for the highest common factor. In the above example,

$$f(x) = (x - 1)(x^2 - 3x + 2) = (x - 1)^2(x - 2)$$

and

$$g(x) = 3x^2 - 8x + 5 = (x - 1)(3x - 5)$$

Once we have a program for dividing one polynomial by another, the finding of the highest common factor merely requires repeated application of that program. The flow chart, Figure 9-7, illustrates the process. In this flow chart, the assertion box entered by connector 1 represents the entire division process described in Section 9.43.

The FORTRAN subroutine given below would perform the process by repeated application of the subroutine POLDIV given in Section 9.43:

```
      SUBROUTINE  HIFACT(A,N,B,M,F,K)
      DIMENSION  A(20),B(20),Q(20),F(20),R(20),D(20)
      NN=N
      LL=N+1
      DO 10 I=1,LL
   10 D(I)=A(I)
      K=M
      LL=M+1
      DO 1 I=1,LL
    1 F(I)=B(I)
```

```
2 CALL POLDIV(D,NN,F,K,Q,KK,R,L)
  IF(R(1))3,6,3
3 NN=K
  LL=K+1
  DO 4 I=1,LL
4 D(I)=F(I)
  K=L
  LL=L+1
  DO 5 I=1,LL
5 F(I)=R(I)
  GO TO 2
6 RETURN
  END
```

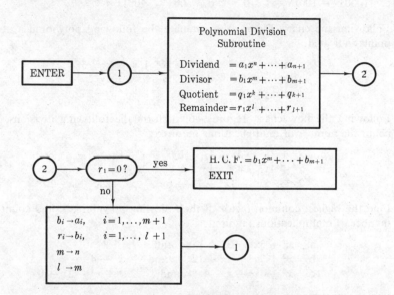

Figure 9-7: Finding highest common factor

The number of multiplications required to find the highest common factor depends on how many times the division of two polynomials must be done. The longest time is required when $m = n - 1$ and the highest common factor turns out to be of degree zero. In that case, we must divide an nth-degree polynomial by an $(n - 1)$st, then an $(n - 1)$st by an $(n - 2)$nd, etc. The total number of multiplication times in this case is

$$(n + 1)(2) + n(2) + (n - 1)(2) + \cdots + (1)(2) = (n + 1)(n + 2)$$

or roughly n^2 multiplications. For a tenth- and a ninth-degree polynomial, about 11×12, or 132 multiplication times may be required.

EXERCISE 24

1. By Descartes' rule find the maximum number of positive roots for each of the following equations. Find the maximum number of negative roots.

   a. $x^5 + 3x - 1 = 0$ b. $x^6 + x^4 + x^2 + 1 = 0$
   c. $x^7 - 2x^6 + x^4 - 3x^3 + 4 = 0$ d. $3x^4 - 4x^3 + 2 = 0$

2. By Theorem V of Section 9.2, find an upper bound on the absolute value of the roots for each of the equations in problem 1.

3. Following the second flow chart of Section 9.3, solve each of the following quadratic equations:

   a. $x^2 - 7x + 12 = 0$ b. $2x^2 + 6x + 5 = 0$
   c. $1000x^2 - 1001x + 1 = 0$ d. $500x^2 + 1001x + 2 = 0$

4. Following the chart, Figure 9–3, evaluate the following polynomials at the points indicated:

   $$\begin{aligned} &\text{a.} \quad y = 5x^3 + 3x^2 + 2x + 1, &&x = .5 \\ &\text{b.} \quad y = 3x^4 - 4x^3 - 2x + 6, &&x = 2 \\ &\text{c.} \quad y = x^6 - 1 &&x = 2 \end{aligned}$$

5. Following the flow chart, Figure 9–6, perform the following divisions, and count the number of multiplications required:

   $$\begin{aligned} &\text{a.} \quad x^3 + 3x^2 + 3x + 1 &&\text{by} \quad x + 1 \\ &\text{b.} \quad x^4 + 4x^2 + 4 &&\text{by} \quad x^3 + 3x + 1 \\ &\text{c.} \quad 2x^5 + 5x^4 - 3x^2 + 7 &&\text{by} \quad x^4 - 2x^3 + x - 1 \end{aligned}$$

6. Find the highest common factor of the following polynomials, and count the number of multiplications required:

   $$\begin{aligned} &\text{a.} \quad x^3 + 3x^2 + 3x + 1 &&\text{and} \quad x + 1 \\ &\text{b.} \quad x^3 + 3x^2 - 4x - 12 &&\text{and} \quad x^2 - 4 \\ &\text{c.} \quad x^3 + 4x^2 - x - 4 &&\text{and} \quad 2x^3 + x^2 - 2x - 1 \end{aligned}$$

7. Write a FORTRAN program that will input two polynomials, call subroutine POLDIV, and print the quotient and remainder.

8. Write a FORTRAN program that will input two polynomials, call subroutine HIFACT, and print the highest common factor.

9.5 SIMPLE REAL ROOTS

For equations of degree higher than four, there are no general formulas which give the roots directly in terms of the coefficients. Such methods as the halving of the interval, the iteration method, or the Newton-Raphson method discussed in Chapter 8 are frequently useful for finding real roots.

9.51   Real Root in a Known Interval

If $f(x)$ is of opposite signs for two values of x, $x = x_1$ and $x = x_2$, then by Theorem I of Chapter 8 there is at least one real root between x_1 and x_2. The method of bisection, described in Section 8.3, could be used to obtain a real root in this case. The FORTRAN subroutine given below would serve to find a real root X of

$$A(1)X^N + A(2)X^{N-1} + \cdots + A(N + 1)$$

known to lie between X1 and X2, to an accuracy specified by the number ACC:

```
     SUBROUTINE FINDRO(A,N,X1,X2,X,ACC)
     DIMENSION A(20)
     J=1
   1 X=X2
   2 Z=A(1)
     DO 3 I=1,N
   3 Z=Z*X+A(I+1)
     GO TO (4,5),J
   4 X=X1
     J=2
     Y=Z
     GO TO 2
   5 IF(ABS(X2-X)-ACC*ABS(X))10,10,6
   6 IF(Y*Z)7,10,8
   7 X1=X
     GO TO 9
   8 X2=X
   9 X=(X1+X2)/2.
     GO TO 2
  10 RETURN
     END
```

As mentioned in Chapter 8, the method of bisection is somewhat slow compared to the Newton-Raphson process, which is a second-order process, and converges extremely rapidly if the initial value is rather near the root. If one wishes to take advantage of this more rapid convergence (at some risk of having the process fail to converge at all), one could use a few steps of bisection to obtain a value fairly close to the root, and then switch to Newton-Raphson. Figure 9-8 illustrates the importance of obtaining a good initial

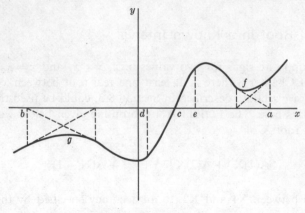

Figure 9-8

estimate of the root before switching to the Newton-Raphson method. Remembering the geometry of the Newton-Raphson process from Chapter 8, we see that an initial guess $x_0 = a$ will give a set of successive values that will oscillate about the minimum point at f but will not converge. An initial guess $x_0 = b$ will give values that will oscillate about the maximum point at g but will not converge. Convergence to the root c is assured only when the initial guess is in a rather narrow range about that value, say between d and e. It is clear from this illustration that, if there are other maxima or minima in the curve quite near the root, it may be hard to choose an initial value near enough to ensure convergence.

9.52   Reducing the Degree of an Equation

Once a real root to a polynomial equation has been found, Theorem II of Section 9.2 can be used to obtain a new equation for the remaining roots, having degree reduced by one.

If a root of the polynomial equation

$$a_1 x^n + a_2 x^{n-1} + \cdots + a_n x + a_{n+1} = 0 \qquad (9\text{-}1)$$

is c, then we must divide this polynomial by $x - c$, obtaining a new polynomial of degree $n - 1$.

This can be done by means of subroutine POLDIV developed in Section 9.43. However, there is an accuracy problem associated with this process which merits some further consideration. When we do this operation with approximate arithmetic, the reduced equation will be in error, and its roots will differ by some amount from those of the original equation. Let the reduced equation be

$$Q(x) = q_1 x^{n-1} + q_2 x^{n-2} + \cdots + q_{n-1} x + q_n = 0 \qquad (9\text{-}2)$$

When the q's are calculated from the a_i's and c according to subroutine POLDIV, we will have

$$q_1 = a_1$$
$$q_2 = a_2 + cq_1$$
$$q_3 = a_3 + cq_2 \qquad\qquad\qquad (9\text{-}3)$$
$$q_4 = a_4 + cq_3$$
$$\vdots$$
$$q_n = a_n + cq_{n-1}$$

It is quicker to perform the calculation with the simple loop

```
   Q(1)=A(1)
   DO 1 I=2, N+1
 1 Q(I)=A(I)+C*Q(I-1)
```

In this loop we have calculated $Q(N + 1)$ as well as the Q's used above. If this loop is compared with the one in Section 9.41 for evaluating a polynomial, it is seen that $Q(N + 1)$ is the value of the polynomial $P(x)$ at $x = c$, which should be zero. The process performed by the above loop is that usually known as synthetic division. Although it is possible to develop a general relationship for the errors in the roots of (9-2), using the techniques of Chapter 8, no especially useful result is obtained thereby. Therefore, we will restrict ourselves to a few examples and observations. Consider the following examples.

Example 1.   One root of the polynomial $1.0x^2 - 9.9x - 1.0$ has been found to be $x = 10$. Using two-place floating-decimal arithmetic, find the reduced equation.

We have

$$a_1 = .10 \times 10^1$$
$$a_2 = -.99 \times 10^1$$
$$a_3 = -.10 \times 10^1$$
$$c = .10 \times 10^2$$

so that

$$q_1 = .10 \times 10^1$$
$$q_2 = -.99 \times 10^1 + (.10 \times 10^2)(.10 \times 10^1)$$
$$= -.99 \times 10^1 + .10 \times 10^2$$

When the left-hand number is shifted to make its exponent become 2, we have

$$q_2 = -.10 \times 10^2 + .10 \times 10^2 = 0$$

Thus the reduced equation is

$$1.0x = 0$$

which will give the second root as $x = 0$. A plot of the original polynomial with error bounds is shown on Figure 9-9. It is seen that the nominal value of the second root is $x = -.1$, and that $x = 0$ is outside the range of possible

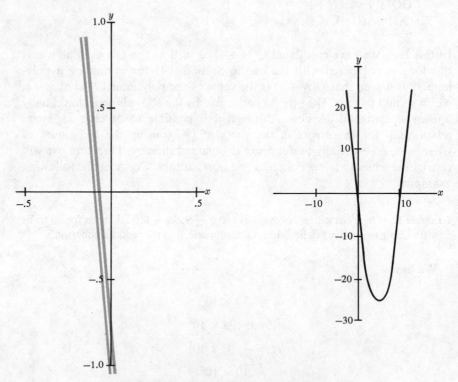

Figure 9-9: Graphs of $y = 1.0x^2 - 9.9x - 1.0$

values of this root. The process of reducing the equation has altered the remaining root by an appreciable amount.

Example 2.   One root of the equation $1.0x^2 - 9.9x - 1.0$ has been found to be $x = -.1$. Using two-place floating-decimal arithmetic, find the reduced equation.

This is the same problem as the one above, with the smaller root given instead of the larger one. In this case we have

$$c = -.10 \times 10^0$$

so that

$$q_1 = .10 \times 10^1$$
$$q_2 = -.99 \times 10^1 + (-.10 \times 10^0)(.10 \times 10^1)$$
$$= -.99 \times 10^1 - .01 \times 10^1$$
$$= -.10 \times 10^2$$

so the reduced equation is

$$1.0x - 10. = 0$$

and the nominal value of the second root is 10, with range from about 9.45 to 10.55.

In these two examples it was seen that removal of the larger root had a major impact on the location of the smaller root, whereas removal of the smaller root did not alter the location of the larger root. That this is generally the case can be inferred from formulas (9-3). If c has absolute error Δc and q_i has absolute error Δq_i and a_i has absolute error Δa_i, then from Table I of Section 3.6 the relative error in cq_i is

$$\frac{\Delta c}{|c|} + \frac{\Delta q_i}{|q_i|} + r$$

where r is the roundoff error in the multiplication. The relative error in $q_{i+1} = a_{i+1} + cq_i$ is then

$$\frac{\Delta q_{i+1}}{|q_{i+1}|} = \frac{|a_{i+1}|}{|q_{i+1}|}\frac{\Delta a_{i+2}}{|a_{i+2}|} + \frac{|cq_i|}{|q_{i+1}|}\left(\frac{\Delta c}{|c|} + \frac{\Delta q_i}{|q_i|} + r\right) + r$$

For given $\Delta c/|c|$, that is, given relative error in c, the relative error in q_{i+1} will increase as c increases, so that the q_i's will tend to be less accurate if c is large.

Many methods of root finding determine only the largest root of an equation, which seems to be in direct conflict with the desire we have just noted, to remove the smallest roots first. Fortunately there is a simple way around this difficulty. If we let $Z = 1/x$ in (9-1) we obtain

$$a_1 + a_2 Z + a_3 Z^2 + \cdots + a_n Z^{n-1} + a_{n+1} Z^n = 0 \qquad (9\text{-}4)$$

This is simply equation (9-1) with coefficients reversed.

If c is a root of (9-1), then $1/c$ is a root of (9-4). To reduce the equation for the root c, if c is large, we merely reduce the equation (9-4) for the root $1/c$. If $k = 1/c$, we simply replace (9-3) by the relations

$$q_n = a_{n+1}$$

$$q_{n-1} = a_n + kq_n$$

$$q_{n-2} = a_{n-1} + kq_{n-1}$$

$$\vdots$$

$$q_1 = a_2 + kq_2$$

This is easily done by the loop

```
CK=1./C
Q(N)=A(N+1)
DO 1 I=2,N
J=N+1-I
1 Q(J)=A(J+1)+CK*Q(J+1)
```

9.53   Root Squaring

One way of finding the largest root of an equation is that due to Graeff. It is based on replacing the given polynomial by one whose roots are the squares of those of the original polynomial. This can be done as follows. Suppose the polynomial (9-1) has roots $x_1, x_2, \ldots, x_n$. Then the polynomial obtained by replacing x by $-x$,

$$P(-x) = a_1(-1)^n x^n + a_2(-1)^{n-1} x^{n-1} + \cdots - a_n x + a_{n+1} \qquad (9\text{-}6)$$

has roots $-x_1, -x_2, \ldots, -x_n$.

We may write

$$P(x) = a_1(x - x_1)(x - x_2) \cdots (x - x_n) \qquad (9\text{-}7)$$

and

$$P(-x) = (-1)^n a_1 (x + x_1)(x + x_2) \cdots (x + x_n) \qquad (9\text{-}8)$$

Now consider a polynomial

$$S(x) = b_1 x^n + b_2 x^{n-1} + \cdots + b_n x + b_{n+1} \qquad (9\text{-}9)$$

whose roots are $x_1^2, x_2^2, \ldots, x_n^2$. Then

$$S(x) = b_1 (x - x_1^2)(x - x_2^2) \cdots (x - x_n^2)$$

Substituting x^2 for x in this expression we have

$$
\begin{aligned}
S(x^2) &= b_1 (x^2 - x_1^2)(x^2 - x_2^2) \cdots (x^2 - x_n^2) \\
&= b_1 (x - x_1)(x + x_1)(x - x_2)(x + x_2) \cdots (x - x_n)(x + x_n) \\
&= b_1 [(x - x_1)(x - x_2) \cdots (x - x_n)][(x + x_1)(x + x_2) \cdots (x + x_n)]
\end{aligned}
$$

Using (9-7) and (9-8), we may write this last relation as

$$S(x^2) = (b_1/a_1^2)(-1)^n P(x) P(-x) \qquad (9\text{-}10)$$

Thus given the coefficients of polynomial $P(x)$, it is quite simple to generate the coefficients of a polynomial whose roots are the squares of the roots of $P(x)$. The product $(-1)^n P(x) P(-x)$ is given by

$$a_1 x^n + a_2 x^{n-1} + a_3 x^{n-2} + a_4 x^{n-3} + a_5 x^{n-4} + \cdots$$
$$a_1 x^n - a_2 x^{n-1} + a_3 x^{n-2} - a_4 x^{n-3} + a_5 x^{n-4} - \cdots$$

$$\overline{}$$

$$a_1^2 x^{2n} + a_1 a_2 x^{2n-1} + a_1 a_3 x^{2n-2} + a_1 a_4 x^{2n-3} + a_1 a_5 x^{2n-4} + \cdots$$
$$\qquad - a_2 a_1 x^{2n-1} - a_2^2 x^{2n-2} \quad - a_2 a_3 x^{2n-3} - a_2 a_4 x^{2n-4} - \cdots$$
$$\qquad\qquad\qquad\qquad a_3 a_1 x^{2n-2} + a_3 a_2 x^{2n-3} + a_3^2 x^{2n-4} \quad + \cdots$$
$$\qquad\qquad\qquad\qquad\qquad\qquad - a_1 a_4 x^{2n-3} - a_2 a_4 x^{2n-4} - \cdots$$
$$\qquad\qquad\qquad\qquad\qquad\qquad\qquad\qquad a_1 a_5 x^{2n-4} + \cdots$$

$$\overline{}$$

$$a_1^2 x^{2n} + 0 + (2a_1 a_3 - a_2^2) x^{2n-2} + 0 + (2a_1 a_5 - 2a_2 a_4 + a_3^2) x^{2n-4} + \cdots$$

The polynomial $S(x^2)$ is

$$S(x^2) = b_1 x^{2n} + b_2 x^{2n-2} + \cdots + b_n x^2 + b_{n+1} \qquad (9\text{-}11)$$

Comparing coefficients of like powers of x, we have

$$b_1 = a_1{}^2$$
$$b_2 = -a_2{}^2 + 2a_1 a_3$$
$$b_3 = a_3{}^2 - 2a_2 a_4 + 2a_1 a_5$$
$$\vdots$$
$$b_i = (-1)^{i-1}[a_i{}^2 - 2a_{i-1}a_{i+1} + 2a_{i-2}a_{i+2} + \cdots]$$
$$\vdots$$
$$b_n = (-1)^{n-1}(a_n{}^2 - 2a_{n-1}a_{n+1})$$
$$b_{n+1} = (-1)^n a_{n+1}^2$$

(9-12)

In order to automate this calculation, note that b_i is made up of two kinds of terms: $a_i{}^2$, and products $a_{i-1}a_{i+1}$, $a_{i-2}a_{i+2}$, etc., extending until a_1 or a_{n+1} is reached. The sum of products can be formed by a set of statements of the form

```
  B(I)=0.
  K=I-1
  KK=N-I
  IF(K-KK)4,4,3
3 K=KK
4 DO 5 J=1,K
  L=K+1-J
5 B(I)=A(I+L)*A(I-L)-B(I)
```

This set of statements first determines K to be the lesser of $I-1$ and $N-I$. Then it forms, as successive values for B(I),

```
B(I)=A(I+K)*A(I-K)
B(I)=A(I+K-1)*A(I-K+1)-A(I+K)*A(I-K)      etc.
```

and finally

```
B(I)=A(I+1)*A(I-1)-A(I+2)*A(I-2)+A(I+3)*(I-3)- ···
```

In order to convert this last sum into the desired value it must be doubled, subtracted from A(I)**2, and the result provided with the correct sign. This can be accomplished by the statements

```
SIGN=2*(I-2*(I/2))-1
B(I)=SIGN*(A(I)**2-2.*B(I))
```

The rules of fixed-point arithmetic, discussed in Chapter 4, assure that if I is even, SIGN will be $-1.$, and if I is odd, it will be $+1$.

The above fragments can be used to construct a FORTRAN subroutine which, given coefficients A(1), A(2), ..., A(N + 1) of a polynomial of degree N, will return coefficients B(1), B(2), ..., B(N + 1) of a polynomial whose roots are the squares of those of the original polynomial. A suitable subroutine is given below:

```
SUBROUTINE ROOTSQ(A,B,N)
DIMENSION A(20),B(20)
B(1)=A(1)**2
DO 6 I=2,N
  B(I)=0
  K=I-1
  KK=N-I
  IF(K-KK)4,4,3
3 K=KK
4 DO 5 J=1,K
  L=K+1-J
5 B(I)=A(I+L)*A(I-L)-B(I)
  SIGN=2*(I-2*(I/2))-1
6 B(I)=SIGN*(A(I)**2-2.*B(I))
  SIGN=1+2*(2*(N/2)-N)
  B(N+1)=SIGN*A(N+1)**2
  RETURN
  END
```

If the above routine is applied to a polynomial $P(x)$ with roots x_1, x_2, ..., x_n, we obtain a polynomial with roots $x_1^2, x_2^2, \ldots, x_n^2$. If the root-squaring routine is applied m times, we obtain a polynomial with roots $x_1^{2^m}$, $x_2^{2^m}, \ldots, x_n^{2^m}$.

9.54   Root Estimation by Root Squaring

The use of root squaring to estimate the values of the roots is based on the relations that exist between the roots and coefficients of a polynomial equation. If $x_1, x_2, \ldots, x_n$ are the roots of equation (9-1), then

$$\frac{a_2}{a_1} = -(x_1 + x_2 + \cdots + x_n)$$

$$\frac{a_3}{a_1} = +(x_1 x_2 + x_1 x_3 + \cdots)$$

$$\frac{a_4}{a_1} = -(x_1 x_2 x_3 + x_1 x_2 x_4 + \cdots) \qquad\qquad \textbf{(9-13)}$$

$$\vdots$$

$$\frac{a_{n+1}}{a_1} = (-1)^n x_1 x_2 x_3 \cdots x_n$$

That is, the sum of the roots is $-a_2/a_1$, the sum of the products by pairs is a_3/a_1, the sum of the products by threes is $-a_4/a_1$, etc. Now suppose that root squaring has been applied several times and has resulted in the new equation

$$b_1(x^k)^n + b_2(x^k)^{n-1} + \cdots + b_n x^k + b_{n+1} = 0 \qquad (9\text{-}14)$$

Relations similar to (9-13) hold also for this equation, so that

$$\frac{b_2}{b_1} = -(x_1^k + x_2^k + \cdots + x_n^k)$$

$$= -x_1^k\left(1 + \frac{x_2^k}{x_1^k} + \cdots + \frac{x_n^k}{x_1^k}\right)$$

$$\frac{b_3}{b_1} = x_1^k x_2^k + x_1^k x_3^k + x_1^k x_4^k + \cdots$$

$$= x_1^k x_2^k\left(1 + \frac{x_3^k}{x_2^k} + \frac{x_4^k}{x_2^k} + \cdots\right) \qquad (9\text{-}15)$$

$$\frac{b_4}{b_1} = -(x_1^k x_2^k x_3^k + x_1^k x_2^k x_4^k + x_1^k x_3^k x_4^k + \cdots)$$

$$= -x_1^k x_2^k x_3^k\left(1 + \frac{x_4^k}{x_3^k} + \frac{x_4^k}{x_2^k} + \cdots\right)$$

$$\vdots$$

$$\frac{b_{n+1}}{b_1} = (-1)^n x_1^k x_2^k \ldots x_n^k$$

Consider first the case in which all roots are real and unequal, and suppose that

$$|x_1| > |x_2| > |x_3| > \ldots > |x_n|$$

Then if root squaring has been applied enough times that k is a large power (note that $k = 2^m$, where m is the number of times root squaring has been applied, so that k grows quite rapidly), then the ratios x_2^k/x_1^k, x_3^k/x_2^k, etc., are negligibly small compared to one, and the relations (9-15) are approximately

$$b_2/b_1 \approx -x_1^k,\; b_3/b_1 \approx x_1^k x_2^k,\; b_4/b_1 \approx -x_1^k x_2^k x_3^k, \ldots,$$

$$b_{n+1}/b_1 \approx (-1)^n x_1^k x_2^k \cdots x_n^k \qquad (9\text{-}16)$$

Dividing each of these equations after the first by the preceding one, we have

$$b_2/b_1 \approx -x_1{}^k, \; b_3/b_2 \approx -x_2{}^k, \; b_4/b_3 \approx -x_3{}^k, \ldots, \; b_{n+1}/b_n \approx -x_n{}^k \quad (9\text{-}17)$$

Hence if root squaring has been applied enough times, the roots may be found simply by taking the ratios of the coefficients, as indicated in (9-17) and taking kth roots.

The obvious question is: How large must k be in order for this approximation to be a valid one? The answer to this is rather clearly indicated by the coefficients themselves. Suppose we apply root squaring once more to (9-14), obtaining a new equation

$$c_1(x^{2k})^n + c_2(x^{2k})^{n-1} + \cdots + c_n x^{2k} + c_{n+1} = 0 \qquad (9\text{-}18)$$

For this equation, we have relations similar to (9-16),

$$\frac{c_2}{c_1} \approx -x_1^{2k}, \; \frac{c_3}{c_1} \approx x_1^{2k}x_2^{2k}, \; \frac{c_4}{c_1} \approx -x_1^{2k}x_2^{2k}x_3^{2k}, \text{ etc.} \qquad (9\text{-}19)$$

Now from (9-12), $c_1 = b_1{}^2$, so comparison of these relations shows that $c_2 \approx b_2{}^2, \; c_3 \approx b_3{}^2$, etc. Thus when root squaring has been applied enough times to make the approximations (9-17) good ones, the coefficients themselves will be approximately squared at each step.

Example 1.   Assuming that SUBROUTINE ROOTSQ(A,B,N) is stored under the file name RTSQ, write a remote-terminal program to find the absolute values of the roots of $x^3 - 6x^2 + 11x - 6 = 0$.

We can separate the problem into two parts. The first part is to perform root squaring until the coefficients very nearly become squared at each step. The second is to use (9-17) to compute the roots. A suitable program is the following:

```
1      DIMENSION A(20),B(20),ER(20)
2      PRINT, "ENTER N, LESS THAN 20"
3      INPUT, N
4      NN = N + 1
5      PRINT "ENTER A(1) TO A(N+1)"
6      DO 1 I=1,NN
7    1 INPUT, A(I)
8      K=2
9    2 CALL ROOTSQ(A,B,N)
10     DO 3 I=1,NN
11   3 ER(I)=A(I)**2/B(I)−1.
12     PRINT, (B(I),I=1,NN)
13     PRINT, (ER(I),I=1,NN)
14     PRINT, "ENTER NEG NR TO ROOTSQ, OTHERWISE
          PLUS"
```

```
15    INPUT, Z
16    IF(Z)5,5,7
17  5 DO 6 I=1,NN
18  6 B(I)=A(I)
19    K=2*K
20    GO TO 2
21  7 FK=K
22    PWR=1./FK
23    DO 8 I=1,N
24    X=B(I+1)/B(I)
25    X=X**PWR
26  8 PRINT, X
27    STOP
28    END
29    $USE RTSQ
```

In this program, the statements at lines 2 through 7 are arranged so that the coefficients of any polynomial up to degree 20 can be entered. Then the statements at lines 9 through 13 perform a root squaring and print the new coefficients as well as the difference between the ratio of the square of the old coefficients to new coefficients and one. These ratios should become very nearly one when root squaring has been performed enough times. Statements at lines 14 and 15 allow the user to select whether to root square again or to compute absolute values of the roots based on the coefficients as they now stand. If he enters a negative number, the routine will go to lines 16 through 20, where it will double K, the quantity which indicates the power being used, rename the B(I)'s as A(I)'s, and return to statement 2 at line 9 to perform root squaring again. If the user types in a positive number, the routine will go to statement 7 at line 21, where it will compute and print the absolute values of the roots and stop. For the problem at hand the inputs and outputs from an actual run might appear as follows.

```
    ENTER N, LESS THAN 20
?  3
    ENTER A(I) TO A(N+1)
?  1.
?  -6.
?  11.
?  -6.
    1.000000      -14.00000       49.00000       -36.00000
     .0.          1.571429        1.469388       0.
    ENTER NEG NR TO ROOTSQ, OTHERWISE PLUS
?  -1.
    1.000000      -98.00000       1393.000       -1296.000
    0.            1.000000        .7236181       0.
```

ENTER NEG NR TO ROOTSQ, OTHERWISE PLUS
? −1.
1.000000   6818.000   1.686433E7   −1.679616E7
0.   .4086242   .1506232   0.
ENTER NEG NR TO ROOTSQ, OTHERWISE PLUS
? −1.
1.000000   −4.311226E7   2.821153E12   −2.821110E12
0.   .07823450   .008118393   0.
ENTER NEG NR TO ROOTSQ, OTHERWISE PLUS
? −1.
1.000000   −1.853024E15   7.958661E24   −7.958661E24
0.   3.044918E−3   3.056601E−5 0.
ENTER NEG NR TO ROOTSQ, OTHERWISE PLUS
? −1.
1.000000   −3.433684E30   6.334029E49   −6.334029E49
0.   4.636124E−6   1.862645E−9 0.
ENTER NEG NR TO ROOTSQ, OTHERWISE PLUS
? 1.
3.000000
2.000000
1.000000

In the computer run, a minus one was entered five times. Note that by the last time through, the coefficients were becoming quite large, and an exponent overflow might have occurred from another attempt at root squaring. The allowed exponent range of the machine can place a limit on how many times root squaring can be performed for a particular equation. It might seem that the situation could be alleviated by dividing all the roots by some number at each stage, say by 2^k. This can help some, but only to a limited extent, because if it is done, some of the roots will eventually run into the opposite problem, that of exponent underflow. Consider the case where the exponent range of the machine will allow numbers from 10^{-65} to 10^{65}, and consider the sizes of coefficients for the problem of Example 1 after several root-squaring operations. From equations (9-16), after m root-squarings, operation we have

$$b_2 \approx -3^{2^m}, \; b_3 \approx 3^{2^m} \cdot 2^{2^m}, \; b_4 \approx -3^{2^m} \cdot 2^{2^m} \cdot 1^{2^m}$$

or, making use of the fact that $3 = 10^{\log_{10} 3} \approx 10^{.5}$ and $2 = 10^{\log_{10} 2} \approx 10^{.3}$,

$$b_2 \approx -10^{.5 \times 2^m}, \; b_3 \approx 10^{.8 \times 2^m}, \; b_4 \approx -10^{.8 \times 2^m}$$

When $m = 6$, b_3 and b_4 are about 10^{50}, as was seen in the printouts of Example 1. When $m = 7$, they are about 10^{100}, or have exceeded the overflow condition on the above machine. Suppose we attempt to remedy this situation by dividing all the roots by two, so that they will not grow so rapidly. This is equivalent to starting with an equation with roots half as large, and after m root-squaring operations we will have

$$b_2 \approx -(3/2)^{2^m}, \; b_3 \approx (3/2)^{2^m} \cdot 1^{2^m}, \; b_4 \approx -(3/2)^{2^m} \cdot 1^{2^m} \cdot (1/2)^{2^m}$$

or, more approximately,

$$b_2 \approx -10^{.2 \times 2^m}, \; b_3 \approx 10^{.2 \times 2^m}, \; b_4 \approx -10^{-.1 \times 2^m}$$

When $m = 7$, b_2 and b_3 are about 10^{25} in absolute value and b_4 is about 10^{-13}, so all numbers are in range and this squaring operation can be done successfully. If $m = 8$, b_2 and b_3 are about 10^{50} and b_4 is about 10^{-25}, so all numbers are still in range. If $m = 9$, b_2 and b_3 are about 10^{100}, and we again have exponent overflow. Dividing the roots by two allowed root squaring to be done eight times instead of six.

Suppose we try to obtain the capability to perform still more root-squaring operations by dividing all roots by three, so that they will grow still less rapidly. In this case the roots would be 1, 2/3, and 1/3, and after m root-squaring operations we will have

$$b_2 \approx -(1)^{2^m}, \; b_3 \approx 1^{2^m} \cdot (2/3)^{2^m}, \; b_4 \approx -(1)^{2^m} \cdot (2/3)^{2^m} \cdot (1/3)^{2^m}$$

or, more approximately,

$$b_2 \approx -1, \; b_3 \approx 10^{-.2 \cdot 2^m}, \; b_4 \approx -10^{-.7 \cdot 2^m}$$

When $m = 6$, b_3 is about 10^{-16} and b_4 is about -10^{-45}. All numbers are within computer range. When $m = 7$, b_3 is about 10^{-32} and b_4 is about -10^{-90}, a number too small to be stored with the exponent sizes allowed. Two things might occur at this point, depending on the details of the computer software. The machine might print out a warning of exponent underflow, or it might simply set the value to zero and proceed. If it does the latter, then b_4 is taken to be zero, so when the corresponding root x_3 is computed, its value will be zero instead of the true value of 1/3.

The problem of rapid growth of exponents, as just indicated, is one of considerable importance in practical application of root squaring. One way of alleviating the problem to some extent is to use double-precision arithmetic, which ordinarily allows for twice the normal number of bit places for exponent storage and hence twice the number of root-squaring operations. Another is to reduce the size of the larger exponents by dividing the roots by some number as in the above discussion, allowing the coefficients representing the smaller coefficients to underflow if necessary. Use the resulting equation to obtain those roots which have not been set to zero by exponent underflow.

Then deflate the equation as described in Section 9.52, and repeat the entire process on the new equation to obtain the remaining roots.

When root squaring is performed as in the above example, the roots themselves are not found, only their absolute values. There still remains the task of determining the signs. Sometimes these can be obtained directly by application of Descartes' rule. Frequently, however, it is necessary to substitute both the number and its negative into the equation and accept the value which most nearly satisfies the equation.

9.6 MULTIPLE AND COMPLEX ROOTS

In the root-squaring method as described in Section 9.54, it was seen that after enough applications of the root-squaring process, an additional application causes all coefficients to merely become the squares of their previous values. This is the case only when all roots are real and simple. When there are multiple roots or complex roots, all coefficients will not display this behavior. Suppose, for example, that the j roots $x_1, x_2, \ldots, x_j$ have the same absolute value and that all other roots have lesser absolute value. These largest roots may represent one real value of multiplicity j, or two real values, positive and negative, of some multiplicity, and/or pairs of complex roots. Consider the sum of the roots of equation (9-14), obtained by several root-squaring operations. We have

$$x_1{}^k + x_2{}^k + x_3{}^k + \cdots + x_j{}^k + x_{j+1}^k + \cdots + x_n{}^k = -b_2/b_1 \qquad \text{(9-20)}$$

All the terms from x_{j+1}^k on should be negligible. Of the first j terms, each real root will contribute the same amount to the sum. If all j of these roots are real, then we have

$$jx_1{}^k \approx -b_2/b_1 \qquad \text{(9-21)}$$

an expression identical to the single-root case, equation (9-16), except for the factor j. A simple modification of the method described in Section 9.54 would handle this case. Suppose, however, that x_2 and x_3 are complex conjugates,

$$x_2 = pe^{i\theta}, \quad x_3 = pe^{-i\theta} \qquad \text{(9-22)}$$

Then

$$x_2{}^k + x_3{}^k = p^k(e^{ik\theta} + e^{-ik\theta}) = 2p^k \cos k\theta \qquad \text{(9-23)}$$

a quantity which can change enormously in value for different values of k.

Hence the contribution of these two terms to the sum (9-20) is essentially unpredictable, and (9-20) cannot be used to estimate the absolute value of the roots. However, the estimate can still be obtained, using other coefficients in the equation. Consider the relation among the set (9-15) in which products of exactly j roots at a time are taken,

$$x_1{}^k x_2{}^k \cdots x_j{}^k + x_1{}^k x_2{}^k \cdots x_{j-1}^k x_{j+1}^k + \cdots = (-1)^j b_{j+1}/b_1 \qquad \text{(9-24)}$$

the first term contains all roots having the largest absolute value, including pairs of complex conjugates, if any are present. If, for example, x_2 and x_3 are as given by (9-22), then

$$x_2{}^k x_3{}^k = p_1^{2k} \qquad \text{(9-25)}$$

and these two contribute this factor in the first term of (9-24). Hence

$$x_1{}^k x_2{}^k \cdots x_j{}^k = p_1^{jk} \qquad \text{(9-26)}$$

where p_1 is the common absolute value of these j roots. Now each term after the first contains at least one of the small roots x_{j+1}^k through $x_n{}^k$ in place of one of the large roots, and so will be negligible in comparison with the first term. Hence, using (9-26), we may write (9-24) as

$$p_1^{jk} \approx (-1)^j b_{j+1}/b_1 \qquad \text{(9-27)}$$

and the value p_1, which is the absolute value of the first j roots, can be obtained from this equation. Further, if root squaring were applied again, as in relation (9-18), the absolute value of all roots would be squared, and (9-27) would be replaced by

$$p_1^{2jk} \approx (-1)^j c_{j+1}/c_1 \qquad \text{(9-28)}$$

Comparing these two relations, we conclude that $c_{j+1} \approx b_{j+1}^2$. Thus even though some coefficients may not have become squared, this particular one did.

Continuing the above analysis further, it can be shown that if the next h roots have the same absolute value p_2,

$$x_{j+1} = x_{j+2} = \cdots = x_{j+h} = p_2$$

and if all remaining roots are still smaller in absolute value, then $c_{j+h+1} \approx b_{j+h+1}^2$ and

$$p_1^{jk} p_2^{hk} = (-1)^{j+h} b_{j+h+1}/b_1 \qquad \text{(9-29)}$$

or, combining this with (9-27),

$$p_2^{hk} = (-1)^h b_{j+h+1}/b_{j+1} \tag{9-30}$$

We are now in a position to make general statements about behavior of the coefficients in root squaring and about the absolute values of the roots. *If root squaring is applied a sufficient number of times to a polynomial, certain of the coefficients eventually become nearly squared at each application of root squaring. These are termed pivotal coefficients. If b_r and b_{r+s} are two adjacent pivotal coefficients, then there are exactly s roots having absolute value p given by*

$$p^{sk} = (-1)^s b_{r+s}/b_r \tag{9-31}$$

Example 1. The coefficients obtained by performing root squaring four times on the equation $x^3 - 4x^2 + 5x - 2 = 0$ are

1	−6	9	−4
1	−18	33	−16
1	−258	513	−256
1	−65538	131073	−65536

Locate the pivotal coefficients and estimate the absolute values of the roots.

The squares of the next-to-last set of coefficients are

1	66564	263169	65536

The first and fourth coefficients are clearly pivotal, and the second appears to be also. Thus for this last set of coefficients there should be a root with absolute value $|b_2/b_1|$ or 65538. There should be two roots with absolute value $|b_4/b_2|^{1/2}$ or $\sqrt{65536/65538} \approx 1$. Since these coefficients are the result of four root-squaring operations, the original roots have been raised to the 2^4, or 16th, power. Then the original equation should have a root with absolute value $(65538)^{1/16} \approx 2$ and two with absolute value $1^{1/16} = 1$.

9.61 Real Roots

Once the absolute values of the roots have been obtained, there still remains some work to find the actual values. If there are j roots having a given absolute value, we do not know in advance how many may be real and how many complex. A straightforward approach is to substitute a positive, real value and if it satisfies the equation, deflate the equation as described in Section 9.52. Then try the same root again, and if it works, deflate the equation again.

When this no longer works, try the negative value in a similar fashion. When this no longer works, if we have extracted less than j roots, the remaining ones must be complex (if an odd number remains, we have an accuracy problem or an error).

9.62 Complex Roots

The complex roots for a polynomial with real coefficients occur in conjugate pairs, each pair being the solution of a quadratic equation

$$x^2 + px + q = 0 \qquad \text{(9-32)}$$

where p and q are real numbers. Hence the process of finding a pair of complex roots for a polynomial

$$f(x) = a_1 x^n + a_2 x^{n-1} + \cdots + a_{n+1}$$

is equivalent to finding a quadratic factor of $f(x)$. If we already know the absolute value of this pair of roots from the root squaring, then we already know the value of q in (9-32) and need only to determine p.

If the quadratic expression (9-32) is divided into $f(x)$, we have, in symbolic form,

$$x^2 + px + q / a_1 x^n + a_2 x^{n-1} + a_3 x^{n-2} + a_4 x^{n-3} + \cdots + a_{n+1}$$
$$a_1 x^n + a_1 p x^{n-1} + a_1 q x^{n-2}$$

$$(a_2 - a_1 p) x^{n-1} + (a_3 - a_1 q) x^{n-2}$$
$$(a_2 - a_1 p) x^{n-1} + (a_2 - a_1 p) p x^{n-2} + (a_2 - a_1 p) q x^{n-3}$$

$$\cdot$$

$$\cdot$$

$$\cdot$$

$$r_1 x + r_2$$

The remainder is $r_1 x + r_2$. In the above symbolic division, q and a_1 through a_{n+1} are known. If we use the correct value of p, both r_1 and r_2 will be zero. Thus we wish to choose p so that this will happen. We can obtain a clearer picture of the dependence of r_1 and r_2 on p by noting that in each step of the above division, the remainder consists of just two terms of different powers of x. Let $R_1{}^i(p)$ be the coefficient of the higher power of x in the ith-step remainder and $R_2{}^i(p)$ be the coefficient of the lower power of x. Then by inspection of the above tableau we see that

$$R_1{}^1(p) = a_2 - a_1 p, \quad R_2{}^1(p) = a_3 - a_1 q \qquad \text{(9-33)}$$

Further, if we reconstruct the tableau with these symbols, we have

$$\frac{a_1 x^{n-2} + R_1^1(p)x^{n-1}}{x^2 + px + q/a_1 x^n + a_2 x^{n-1} + a_3 x^{n-2} + a_4 x^{n-3} + \cdots + a_{n+1}}$$

$$\frac{a_1 x^n + a_1 p x^{n-1} + a_1 q x^{n-2}}{R_1^1(p)x^{n-1} + R_2^1(p)x^{n-2}}$$

$$\frac{R_1^1(p)x^{n-1} + pR_1^1(p)x^{n-2} + qR_1^1(p)x^{n-3}}{R_1^2(p)x^{n-2} + R_2^2(p)x^{n-3}}$$

$$\cdot$$
$$\cdot$$
$$\cdot$$

$$R_1^{n-1}(p)x + R_2^{n-1}(p)$$

We note that

$$R_1^2(p) = R_2^1(p) - pR_1^1(p)$$

$$R_2^2(p) = a_4 - qR_1^1(p)$$

and further, that at the ith step of division we will have

$$R_1^{i+1}(p) = R_2^i(p) - pR_1^i(p) \tag{9-34}$$

$$R_2^{i+1}(p) = a_{i+3} - qR_1^i(p) \tag{9-35}$$

These relations, along with (9-33), show us that $R_1^{n-1}(p)$ and $R_2^{n-1}(p)$ are polynomials in p. Hence we are looking for real value p which are roots of both of these two polynomials. Hence a way to find p is to generate the polynomials $R_1^{n-1}(p)$ and $R_2^{n-1}(p)$ and then find the value or values of p which satisfy both of these. Once p is known, SUBROUTINE QUADRO of Section 9.3 can be used to obtain the actual roots.

The roots which are common to two polynomials are roots also of their highest common factor. Thus SUBROUTINE HIFACT of Section 9.44 can be used to obtain a single polynomial whose real roots are the desired values of p. Then root squaring as described above can be used to find these roots.

The first task is to find the polynomials $R_1^{n-1}(p)$ and $R_2^{n-1}(p)$, and for computer use this means finding an algorithm for computing the coefficients. From (9-33) to (9-35) it is clear that $R_1^i(p)$ is a polynomial of degree i in p and $R_2^i(p)$ is a polynomial of degree $i - 1$. Let

$$R_1^i(p) = c_1^i p^i + c_2^i p^{i-1} + \cdots + c_{i+1}^i$$

$$R_2^i(p) = d_1^i p^{i-1} + d_2^i p^{i-2} + \cdots + d_i^i \tag{9-36}$$

These expressions can be written

$$R_1{}^i(p) = \sum_{j=0}^{i} c_{i+1-j}^i p^j, \; R_2{}^i(p) = \sum_{j=0}^{i-1} d_{i-j}^i p^j \qquad (9\text{-}37)$$

If equations (9-34) and (9-35) are to be satisfied when relations (9-36) are substituted into them, then coefficients of like powers of p must be equal. Equating coefficients of p^{i+1} in (9-37) we have

$$c_1^{i+1} = -c_1{}^i \qquad (9\text{-}38)$$

Equating coefficients of p^i, we have

$$c_2^{i+1} = -c_2{}^i \qquad (9\text{-}39)$$

Equating coefficients of p^j for $j = 0, 1, 2, \ldots, i - 1$, we have

$$c_{i+2-j}^{i+1} = d_{i-j}^i - c_{i+2-j}^i \qquad (9\text{-}40)$$

Equating constant terms in (9-35), we have

$$d_{i+1}^{i+1} = a_{i+3} - qc_{i+1}^i \qquad (9\text{-}41)$$

Equating coefficients of p^j for $j = 1, 2, \ldots, i$, we have

$$d_{i+1-j}^{i+1} = -qc_{i+1-j}^i \qquad (9\text{-}42)$$

The above five equations can be used repetitively, starting with the values for $i = 1$, to determine the coefficients in (9-37) for $i = n - 1$. For this purpose, they can be written in a more easily handled form. In the third relation, let $k = i - j$. Then the first three relations for taking the c's from one value of i to the next can be summarized by

$$-c_1 \rightarrow c_1, \; -c_2 \rightarrow c_2, \; d_k - c_{k+2} \rightarrow c_{k+2} \qquad \text{for } k = 1, 2, \ldots, i \quad (9\text{-}43)$$

In the last relation, let $k = i + 1 - j$. Then the final two relations for taking the a's from one value of i to the next can be summarized by

$$-qc_k \rightarrow d_k \qquad \text{for } k = 1, 2, \ldots, i, \qquad a_{i+3} - qc_{i+1} \rightarrow d_{i+1} \qquad (9\text{-}44)$$

Now from (9-33),

$$c_1{}^1 = -a_1, \quad c_2{}^1 = a_2, \quad d_1{}^1 = a_3 - a_1 q \qquad (9\text{-}45)$$

The entire process of finding the coefficients of $r_1^{n-1}(p)$ and $r_2^{n-1}(p)$ can be performed by the subroutine given below. The subroutine also calls SUB-ROUTINE HIFACT to find the highest common factor of these two polynomials and returns the coefficients P(1), P(2), ..., P(M + 1) and degree M of a single polynomial whose roots are the acceptable values of p in (9-32). As inputs it requires the coefficients A(1), A(2), ..., A(N + 1) and degree N of the polynomial having the complex roots and the absolute value ABX of the complex roots.

```
      SUBROUTINE  PCOEF(A,N,ABX,P,M)
      DIMENSION  A(20),C(20),D(20),P(20)
      Q=ABX**2
      NM2=N-2
      C(1)=-A(1)
      C(2)=A(2)
      DO 1 I=3,N
    1 C(I)=0.
      DO 10 I=1,NM2
      D(I+1)=A(I+3)-Q*C(I+1)
      DO 5 K=1,I
      DD=D(K)
      D(K)=-Q*C(K)
    5 C(K+2)=DD-C(K+2)
      C(1)=-C(1)
   10 C(2)=-C(2)
      NM2=N-1
      CALL HIFACT(C,NM1,D,NM2,P,M)
      RETURN
      END
```

In this subroutine, considerable care had to be used in ordering the computation steps so as to avoid storing over a C(I) or D(I) value while it is still needed in a later step. This is the price of discarding the i and $i + 1$ super-scripts in relations (9-38) through (9-42).

EXERCISE 25

1. In each of the following, find the only real root to an accuracy of .01 by the bisection method.

   a. $x^5 + x + 1 = 0$ b. $x^3 - 3x^2 - 2 = 0$ c. $x^7 + x^6 + x^2 + 1 = 0$

2. Write a FORTRAN program that will input a polynomial and two limits X_1 and X_2, call subroutine FINDRO given in Section 9.51, and print the value of the root to four places.

3. Write a modified version of subroutine FINDRO that will switch to the Newton-Raphson method when an accuracy of .1 has been reached.

4. Perform the root-squaring process four times on each of the following polynomials. Identify the pivotal coefficients and estimate the absolute values of the roots.

   a. $x^3 + 3x^2 - x - 3 = 0$
   b. $x^3 - x^2 - x - 2 = 0$
   c. $x^4 + 2x^2 - x + 2 = 0$

5. Two roots of a polynomial equation stand in the ratio $x_i/x_j = .99$. How many root-squaring operations will be required to separate these roots by a factor of 1000? If a third root x_k is about $.01x_j$, what range of floating-point exponent must the computer be using if both exponent overflow and exponent underflow are to be avoided?

6. A certain polynomial is known to have a pair of complex roots having absolute value equal to 10, and hence has a quadratic factor of the form $x^2 + px + 100$. Find an upper limit for the size of p.

7. In relations (9-14) and (9-18), show that if x_1 is the largest root and if

$$|b_2^2/c_2 - 1| < \epsilon \tag{1}$$

where

$$\epsilon < 1/n \tag{2}$$

then the value z, given by $z_1{}^k = b_2/b_1$, represents $|x_1|$ with relative error less than $2\epsilon/k$. (*Hint*:

$$x_1{}^k + x_2{}^k + \cdots + x_n{}^k = -b_2/b_1 \tag{3}$$

$$x_1^{2k} + x_2^{2k} + \cdots + x_n^{2k} = -c_2/c_1 \tag{4}$$

Let

$$E = \frac{x_2{}^k}{x_1{}^k} + \frac{x_3{}^k}{x_1{}^k} + \cdots + \frac{x_n{}^k}{x_1{}^k}$$

a. Show that

$$E^2 < \frac{x_2^{2k}}{x_1^{2k}} + \frac{x_3^{2k}}{x_1^{2k}} + \cdots + \frac{x_n^{2k}}{x_1^{2k}} \tag{6}$$

b. Show that (1) implies

$$\frac{x_1{}^k x_2{}^k + x_2{}^k x_3{}^k + \cdots}{x_1^{2k} + x_2^{2k} + \cdots} < \epsilon/2 \tag{7}$$

c. Show that (6) implies

$$E/(1 + E^2) < \epsilon/2 \tag{8}$$

d. Show that

$$E < 1/n \tag{9}$$

e. Show that (2), (8), and (9) imply $E < 2\epsilon$.

f. Show that (3) can be written $x_1{}^k(1 + E) = z_1{}^k$.

g. Hence show that

$$\left| \frac{z_1 - |x_1|}{|x_1|} \right| = (1 + E)^{1/k} - 1 < 2\epsilon/k$$

h. Show that $z_1 > |x_1|$ and hence that

$$\frac{z_1 - |x_1|}{x_1} < 2\epsilon/k.)$$

8. In relations (9-17) and (9-18), show that if x_1 through x_j are the j largest roots in absolute value and if

$$|b_{j+1}^2/c_{j+1} - 1| < \epsilon$$

where $\epsilon < 1/n^2$, that the value of z given by $z^j = (-1)^j b_{j+1}/b_1$ represents the absolute value of the product of the roots $x_1 x_2 \cdots x_j$ with relative error less than $2\epsilon/k$. (*Hint*: Show that the sum $x_1{}^k x_2{}^k \cdots x_j{}^k + x_1{}^k x_2{}^k \cdots x_{j-1}^k x_{j+1}^k + \cdots$ of all products of roots taken j at a time contains less than n^2 terms. Then proceed in a fashion analogous to problem 7.)

9. Using the result of problem 8, show that if for two adjacent pivotal coefficients b_j and b_{j+h},

$$|b_j{}^2/c_j - 1| < \epsilon \qquad \text{and} \qquad |b_{j+h}^2/c_{j+h} - 1| < \epsilon$$

where $\epsilon < 1/n^2$, then there are h roots whose absolute value is $|b_{j+h}/b_j|^{1/h}$ with relative error less than $2\epsilon/hk$.

10. Show that if the absolute values of two adjacent roots x_j and x_{j+1} have the relationship $|x_{j+1}| = |x_j|(1 - \delta)$, then the number of root-squaring operations to make

$$|b_{j+1}^2/c_{j+1} - 1| < \epsilon$$

is at least m, where $(1 - \delta)^{2^m} < \epsilon$. Find m for the following cases.

  a. $\delta = .001,$ $\epsilon = .001$
  b. $\delta = .1,$ $\epsilon = .001$
  c. $\delta = .9,$ $\epsilon = .001$

11. A given fifth-degree equation has roots $x_1 = 10.1$, $x_2 = 10$, $x_3 = 1$, $x_4 = x_5 = .9$. You are performing root-squaring operations, using the remote-terminal program of Example 1, Section 9.54, and plan to consider a coefficient b_i to be pivotal if $|b_i{}^2/c_i - 1| < .001$. In how many root-squaring steps should b_2 become pivotal? b_3? b_4? b_5?

12. Rewrite the program of Example 1, Section 9.54, to ask for the actual roots as inputs and construct the polynomial coefficients from the relations (9-13). Use this program to solve problem 11 experimentally.

13. The polynomial equation $x^4 + 4x^2 + 3x + 4 = 0$ has a pair of complex roots having absolute value equal to one, that is, has a quadratic factor $x^2 + px + 1$.

  a. Use the recursion relations (9-43), (9-44), and (9-45) to find the polynomials of which p must be a root.

b. Find the highest common factor of these polynomials.

c. Find the roots.

14. By direct substitution show that $y = ce^{\alpha t}$ is a solution of the differential equation

$$\frac{d^4 y}{dt^4} + 3\frac{d^3 y}{dt^3} - 2\frac{d^2 y}{dt^2} + 3\frac{dy}{dt} + y = 0$$

if α is a root of the polynomial equation

$$\alpha^4 + 3\alpha^3 - 2\alpha^2 + 3\alpha + 1 = 0$$

If $\alpha_1, \alpha_2, \alpha_3$, and α_4 are the four roots of this equation, show that

$$y = c_1 e^{\alpha_1 t} + c_2 e^{\alpha_2 t} + c_3 e^{\alpha_3 t} + c_4 e^{\alpha_4 t}$$

also satisfies the differential equation for any values of c_1, c_2, c_3, and c_4.

15. Show that for t sufficiently large and $\alpha_1, \alpha_2, \alpha_3$, and α_4 all real, the value of y will be determined by the largest positive α_i.

16. Show that if $y = ce^{\alpha t}$, then y is less than or equal to c in absolute value for all $t > 0$, if the real part of α is zero or negative.

17. In a mechanical system of springs and masses, the motion of any part after a sudden impulse acceleration is governed by a differential equation of the form

$$a_1 \frac{d^n y}{dt^n} + a_2 \frac{d^{n-1} y}{dt^{n-1}} + \cdots + a_n \frac{dy}{dt} + a_{n+1} y = 0$$

The system will be stable, that is, will not tend to shake itself apart, if none of the solutions $y = ce^{\alpha t}$ grow very large as t increases. Show that the system will be stable if all the roots of the polynomial equation

$$a_1 \alpha^n + a_2 \alpha^{n-1} + \cdots + a_n \alpha + a_{n+1} = 0$$

have zero or negative real parts.

18. In an electrical circuit of resistors, capacitors, and inductances, the current at any point after a sudden initial impulse current is governed by a differential equation of the form

$$a_1 \frac{d^n i}{dt^n} + a_2 \frac{d^{n-1} i}{dt^{n-1}} + \cdots + a_n \frac{di}{dt} + a_{n+1} = 0$$

The system will be stable, that is, will not tend to develop very large local currents and burn out components if none of the solutions of the form $i = ce^{\alpha t}$ grow very large as t increases. Show that the system will be stable if all the roots of the polynomial equation

$$a_1 \alpha^n + a_2 \alpha^{n-1} + \cdots + a_n \alpha + a_{n+1} = 0$$

have zero or negative real parts.

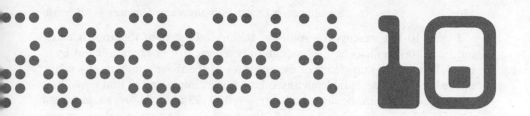

Simultaneous Linear Equations and Matrices

10.1 INTRODUCTION

In this chapter we turn to a problem of finding the values of unknowns, x_1, x_2, etc., which satisfy systems of equations of type

$$a_{11}x_1 + a_{12}x_2 + a_{13}x_3 + \cdots + a_{1n}x_n = b_1$$
$$a_{21}x_1 + a_{22}x_2 + a_{23}x_3 + \cdots + a_{2n}x_n = b_2$$
$$a_{31}x_1 + a_{32}x_2 + a_{33}x_3 + \cdots + a_{3n}x_n = b_3 \qquad \textbf{(10-1)}$$
$$\vdots$$
$$a_{n1}x_1 + a_{n2}x_2 + a_{n3}x_3 + \cdots + a_{nn}x_n = b_n$$

When the number of equations is equal to the number of unknowns, there will ordinarily be a unique solution; that is, one set of values of x_1, x_2, ..., x_n which satisfy all of the equations. At least such is the concept in the world of exact numbers and exact arithmetic. When the coefficients are approximate numbers, the concept of a solution becomes less clear, as the following example demonstrates.

Example 1. Find the solution of

$$1.0x - 2.0y = 1.0$$
$$.5x - 4.0y = 1.0$$

Figure 10-1 represents the solution, taking into account the approximate nature of the coefficients. Each equation is represented not by a line but by a band. Within our knowledge of the accuracy of the above numbers, any value in the band is as acceptable as any other. For example, in the first equation, when $x = 0$, y can be as small as $-1.05/1.95 \approx -.54$ or as large as $-.95/2.05 \approx -.46$. Thus at $x = 0$, the band for the first equation covers the region from $y = -.54$ to $y = -.46$. The two bands intersect not in a unique point but in a region, and any point in this region might be accepted as a solution. The nominal solution, for the above system of equations, obtained by accepting the coefficients as exact, is $x = 2/3$, $y = -1/6$, or approximately $x = .67$, $y = -.17$. However, the points $x = .86$, $y = -.12$

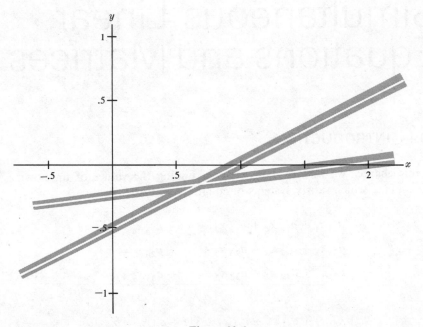

Figure 10-1

and $x = .5$, $y = -.21$ are also within the acceptable region. It is somewhat disconcerting to note that in this rather straightforward case, with the co-efficients known to 10% or better, the solution is uncertain by 30% or more. It can be seen that if the equations represent lines that are nearly parallel, the region of overlap of the two bands representing the equations can be quite extended, as illustrated in Figure 10-2(a). In this case, even if the coefficients were exact, a small change in one of them can make a sizable difference in the solution, as illustrated in Figure 10-2(b) and (c). Equations having this property are termed ill-conditioned. An accurate solution can be found only by performing the computation with great care, since even small roundoff errors may influence the answer greatly. Further, in practical problems, the answer itself must be viewed with some circumspection, since any inherent inaccuracies in the values of the coefficients may cause large changes in the answers.

The above example concerned itself with two equations and two unknowns, but analogous situations exist for higher numbers of equations and unknowns.

In this chapter, three general methods of solving a set of simultaneous linear equations are discussed: direct methods, in which the solution is found by a finite number of algebraic manipulations of the coefficients; iterative methods, which produce a set of successive approximations to the solution which hopefully become very close to the solution but never actually reach it; and matrix

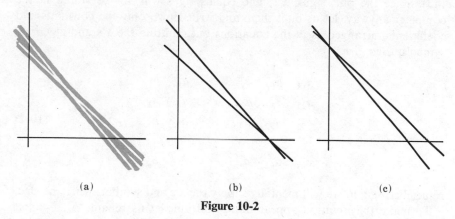

(a)   (b)   (c)

Figure 10-2

inversion methods, which are quite similar to the direct methods in numerical content but which provide conceptually more elegant bases for such methods. As was indicated in Chapter 5, no one of these methods is always best. The direct methods and matrix methods can have accuracy problems for some values of the coefficients and constant terms. The iterative methods can fail to converge to a solution. An attempt will be made to indicate the conditions under which the various methods can be expected to give satisfactory results.

10.2 THE ELIMINATION METHOD

The elimination method consists of multiplying various of the equations by appropriate constants and adding to other equations so as to obtain zero coefficients in some locations and eventually obtain equations that can be solved directly. The particular form of the elimination we shall use is that known as the Gauss-Jordan method. In this method, an appropriate multiple of the first equation is added to each of the other equations so that the result-ing $n - 1$ equations have zero coefficients for the x_1 term. (If the first equation does not have a term involving x_1, we must first interchange two equations to obtain one with an x_1 term as the first equation.) Then an appropriate multiple of the next equation is added to all equations to eliminate the x_2 term from all but one equation. The process is continued until each equation contains only one unknown, and the equations are solved. At each step, the coefficient being used to eliminate other coefficients is called the pivotal coefficient.

To demonstrate how a machine program can be organized to perform this process, we shall construct some diagrams. Equation (10-1) will be represented internally in a computer only by the stored value of the coefficients a_{11} through a_{nn} and b_1 through b_n, perhaps as subscripted variables A(I,J) and B(J). Since the plus signs, x's, and equals signs will not be stored in the computer anyway, let us omit them and write down only the constants and coefficients, arranged as in the equations but omitting the x's and algebraic symbols, thus:

$$
\begin{array}{cccccc}
a_{11} & a_{12} & a_{13} & \cdots & a_{1n} & b_1 \\
a_{21} & a_{22} & a_{23} & \cdots & a_{2n} & b_2 \\
a_{31} & a_{32} & a_{33} & \cdots & a_{3n} & b_3 \\
\vdots & & & & & \\
a_{n1} & a_{n2} & a_{n3} & \cdots & a_{n4} & b_n
\end{array}
\qquad \textbf{(10-2)}
$$

remembering that we will mentally supply the x's and symbols where needed.

To make the notation appear more uniform, let us rename b_1, b_2, $\ldots$, b_n as a_{1n+1}, a_{2n+1}, $\ldots$, a_{nn+1}. Then the array can be written

$$
\begin{array}{cccccc}
a_{11} & a_{12} & a_{13} & \cdots & a_{1n} & a_{1n+1} \\
a_{21} & a_{22} & a_{23} & \cdots & a_{2n} & a_{2n+1} \\
a_{31} & a_{32} & a_{33} & \cdots & a_{3n} & a_{3n+1} \\
\vdots & & & & & \\
a_{n1} & a_{n2} & a_{n3} & \cdots & a_{nn} & a_{nn+1}
\end{array}
\qquad \textbf{(10-3)}
$$

As a first step in the elimination process we can divide the first equation by a_{11} to make the coefficient of x_1 become 1, and obtain the equations represented by

$$
\begin{array}{cccccc}
1 & \dfrac{a_{12}}{a_{11}} & \dfrac{a_{13}}{a_{11}} & \cdots & \dfrac{a_{1n}}{a_{11}} & \dfrac{a_{1n+1}}{a_{11}} \\
a_{21} & a_{22} & a_{23} & \cdots & a_{2n} & a_{2n+1} \\
a_{31} & a_{32} & a_{33} & \cdots & a_{3n} & a_{3n+1} \\
\vdots & & & & & \\
a_{n1} & a_{n2} & a_{n3} & \cdots & a_{nn} & a_{nn+1}
\end{array}
$$

Now we can eliminate the x_1 term from each of the other equations by multiplying the first equation by a_{21} and subtracting from the second, by a_{31}, and subtracting from the third, etc., giving

$$1 \qquad \frac{a_{12}}{a_{11}} \qquad \frac{a_{13}}{a_{11}} \qquad \cdots \qquad \frac{a_{1n}}{a_{11}} \qquad \frac{a_{1n+1}}{a_{11}}$$

$$0 \quad a_{22} - a_{21}\left(\frac{a_{12}}{a_{11}}\right) \quad a_{23} - a_{21}\left(\frac{a_{13}}{a_{11}}\right) \quad \cdots \quad a_{2n} - a_{21}\left(\frac{a_{1n}}{a_{11}}\right) \quad a_{2n+1} - a_{21}\left(\frac{a_{1n+1}}{a_{11}}\right)$$

$$0 \quad a_{32} - a_{31}\left(\frac{a_{12}}{a_{11}}\right) \quad a_{33} - a_{31}\left(\frac{a_{13}}{a_{11}}\right) \quad \cdots \quad a_{3n} - a_{31}\left(\frac{a_{1n}}{a_{11}}\right) \quad a_{3n+1} - a_{31}\left(\frac{a_{1n+1}}{a_{11}}\right)$$

$$\vdots$$

$$0 \quad a_{n2} - a_{n1}\left(\frac{a_{12}}{a_{11}}\right) \quad a_{n3} - a_{n1}\left(\frac{a_{13}}{a_{11}}\right) \quad \cdots \quad a_{nn} - a_{n1}\left(\frac{a_{1n}}{a_{11}}\right) \quad a_{nn+1} - a_{n1}\left(\frac{a_{1n+1}}{a_{11}}\right)$$

At this point, we have eliminated the x_1 term from all but the first equation, using a_{11} as the pivotal coefficient. Note that in the computer, the new coefficients may as well be stored in the locations which held the old ones; that is, a_{12}/a_{11} simply replaces a_{12}, etc. If this is done, the above array becomes

$$
\begin{array}{cccccc}
1 & a_{12} & a_{13} & \cdots & a_{1n} & a_{1n+1} \\
0 & a_{22} & a_{23} & \cdots & a_{2n} & a_{2n+1} \\
0 & a_{32} & a_{33} & \cdots & a_{3n} & a_{3n+1} \\
\vdots & & & & & \\
0 & a_{n2} & a_{n3} & \cdots & a_{nn} & a_{nn+1}
\end{array}
$$

and the process which gives this array from the original one can be described by

$$a_{1j}/a_{11} \to a_{1j} \qquad \text{for } j = 2, \ldots, n+1$$
$$a_{ij} - a_{i1}a_{1j} \to a_{ij} \qquad \text{for } i = 2, \ldots, n$$
$$j = 2, \ldots, n+1$$

Note that these steps will not actually put $a_{11} = 1$ and $a_{i1} = 0$ for $i > 1$, that is, will not set the first column to one and zeros. Since we know they should be there, we can simply remember the fact, and not force the computer to take the extra steps to actually put them there.

Now we need to eliminate x_2 from equations 3 through n and from equation 1 by an analogous process. The steps are described by

$$a_{2j}/a_{22} \to a_{2j} \qquad \text{for } j = 3, \ldots, n+1$$
$$a_{ij} - a_{i2}a_{2j} \to a_{ij} \qquad \text{for } i = 3, \ldots, n, \text{ and } i = 1$$
$$j = 3, \ldots, n+1$$

and produce an array of the form

$$
\begin{array}{cccccc}
1 & 0 & a_{13} & \cdots & a_{1n} & a_{1n+1} \\
0 & 1 & a_{23} & \cdots & a_{2n} & a_{2n+1} \\
0 & 0 & a_{33} & \cdots & a_{3n} & a_{3n+1} \\
\vdots & & & & & \\
0 & 0 & a_{n3} & \cdots & a_{nn} & a_{nn+1}
\end{array}
$$

If the process is continued, we eventually obtain the array

$$
\begin{array}{cccccc}
1 & 0 & 0 & \cdots & 0 & a_{1n+1} \\
0 & 1 & 0 & \cdots & 0 & a_{2n+1} \\
0 & 0 & 1 & \cdots & 0 & a_{3n+1} \\
\vdots & & & & & \\
0 & 0 & 0 & \cdots & 1 & a_{nn+1}
\end{array}
\tag{10-4}
$$

and so $x_1 = a_{1n+1}$, $x_2 = a_{2n+1}$, etc. The process can be summarized in flow chart form as in Figure 10-3. A remote-terminal routine which would perform the process for systems up to 10 by 10 can be written as follows:

```
 1       DIMENSION  A(10,11)
 2     1 PRINT, "NUMBER  OF  EQUATIONS"
 3       INPUT, N
 4       NN=N+1
 5       PRINT, "A(1,1),A(1,2),,,A(1,N),B(1),A(2,1),ETC"
 6       INPUT, ((A(I,J),J=1,NN),I=1,N)
 7       DO 3 K=1,N
 8       KK=K+1
 9       DO 3 J=KK,NN
10       A(K,J)=A(K,J)/A(K,K)
11       DO 3 I=1,N
12       IF(K−I)2,3,2
13     2 A(I,J)=A(I,J)−A(I,K)*A(K,J)
14     3 CONTINUE
15       PRINT, "SOLUTION",(A(I,NN),I=1,N)
16       GO TO 1
17       END
```

Example 1. Show all inputs and machine responses for running the above program to solve the set of equations

$$
\begin{aligned}
2x_1 + 3x_2 + 5x_3 &= 5 \\
3x_1 + 4x_2 + 7x_3 &= 6 \\
x_1 + 3x_2 + 2x_3 &= 5
\end{aligned}
$$

The inputs and responses would appear as follows:

```
RUN
N
? 3
   A(1,1),A(1,2),,,A(1,N),B(1),A(2,1),ETC
? 2,3,5,5,3,4,7,6,1,3,2,5
   SOLUTION   -3.000000          2.000000          1.000000
```

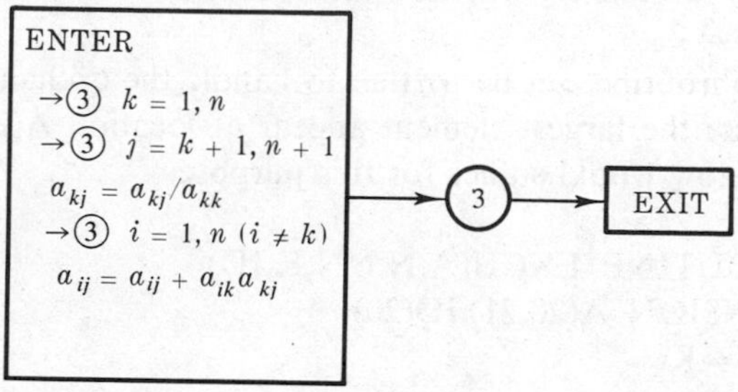

Figure 10-3: Flow chart for Gauss-Jordan method

The program given above will run into trouble if any of the coefficients $A(K,K)$ are zero, since it will attempt to divide by zero. One way to avoid this problem is to rearrange the equations any time a zero element on the diagonal is encountered.

Another way, not much more difficult to execute, is to rearrange the equations at each step so that the pivotal coefficient at each step is not only nonzero but is actually the largest coefficient. This approach not only avoids division by zero but also tends to enhance accuracy by minimizing round-off error. It has the disadvantage that the rearrangement will cause the unknowns to be scrambled at the end of the process. Suppose, for example, that initially the largest coefficient is a_{32}. Then we would like to arrange the equations as

$$a_{32}x_2 + a_{31}x_1 + a_{33}x_3 + \cdots + a_{3n}x_n = b_3$$
$$a_{22}x_2 + a_{21}x_1 + a_{23}x_3 + \cdots + a_{2n}x_n = b_2$$
$$a_{12}x_2 + a_{11}x_1 + a_{13}x_3 + \cdots + a_{1n}x_n = b_1$$
$$a_{42}x_2 + a_{41}x_1 + a_{43}x_3 + \cdots + a_{4n}x_n = b_4$$
$$\vdots$$
$$a_{n2}x_2 + a_{n1}x_1 + a_{n3}x_3 + \cdots + a_{nn}x_n = b_n$$

In terms of the original set of equations, (10-1), we have interchanged the first and third equations and have also interchanged the positions of x_1 and x_2 in all equations. In terms of the array of coefficients (10-3) we have interchanged the first and third rows and the first and second columns. If we continued the process to the end with no further rearrangement, the final value in a_{1n+1} when we reach the stage represented by (10-4) is not x_1 but x_2. Thus when we interchange rows or columns to obtain a large pivotal coefficient, we must also keep track of which unknown is represented by a particular column. This can be done by storing an identification number, ID, for each column which indicates the number of the unknown represented by that column. For example, in the rearrangement shown above, the information that the variable x_2 was now in the first column would be indicated by setting ID(1) = 2.

A separate subroutine can be written to handle the exchange of rows and columns to make the largest element appear at location A(K,K). The subroutine given below would suffice for this purpose.

```
      SUBROUTINE EXCH(A,N,NN,K,ID)
      DIMENSION A(20,21),ID(20)
      NROW = K
      NCOL = K
      B = ABS(A(K,K))
      DO 2 I = 1,N
      DO 2 J = 1,NN
      IF(ABS(A(I,J) − B))2,2,21
   21 NROW = I
      NCOL = J
      B = ABS(A(I,J))
    2 CONTINUE
      IF(NROW − K)3,3,31
   31 DO 32 J = K,NN
      C = A(NROW,J)
      A(NROW,J) = A(K,J)
   32 A(K,J) = C
    3 CONTINUE
      IF(NCOL − K)4,4,41
   41 DO 42 I = 1,N
      C = A(I,NCOL)
      A(I,NCOL) = A(I,K)
   42 A(I,K) = C
      I = ID(NCOL)
      ID(NCOL) = ID(K)
      ID(K) = I
    4 CONTINUE
      RETURN
      END
```

In this subroutine, the statements up to number 2 locate the element having the largest absolute value and identify its location as NROW,NCOL. The statements from 2 to 3 interchange rows K and NROW if they are not the same row. The statements from 3 to 4 interchange columns K and NCOL if they are not the same column, and also interchange the ID numbers to record this fact. Using this subroutine, one to solve the set of linear equations can be written as follows:

```
      SUBROUTINE ELIM(AA,N,BB,X)
      DIMENSION AA(20,20),BB(20),A(20,21),X(20),ID(20)
      NN=N+1
      DO 100 I=1,N
      A(I,NN)=BB(I)
      ID(I)=I
      DO 100 J=1,N
  100 A(I,J)=AA(I,J)
      K=1
    1 CALL EXCH(A,N,NN,K,ID)
    2 IF(A(K,K))3,999,3
    3 KK=K+1
      DO 4 J=KK,NN
      A(K,J)=A(K,J)/A(K,K)
      DO 4 I=1,N
      IF(K-I)41,4,41
   41 A(I,J)=A(I,J)-A(I,K)*A(K,J)
    4 CONTINUE
      K=KK
      IF(K-N)1,2,5
    5 DO 10 I=1,N
      DO 10 J=1,N
      IF(ID(J)-I)10,6,10
    6 X(I)=A(J,NN)
   10 CONTINUE
      RETURN
  999 PRINT 1000
      RETURN
 1000 FORMAT(19H NO UNIQUE SOLUTION)
      END
```

In this subroutine, the input coefficients are identified as AA(I,J) and the input constants as BB(I). The statements up to 100 reidentify these quantities as A(I,J), so the original values will not be destroyed by the subroutine. Statement 1 calls subroutine EXCH to make the largest coefficient the pivotal coefficient. If the largest coefficient is zero, the message "NO UNIQUE SOLUTION" is printed and an exit is taken. Otherwise, statements 3 through 4 solve the equations as in the remote-terminal program given earlier. Statements 5 through 10 use the identification numbers to unscramble the unknowns and return them in proper order.

10.3   GAUSS-SEIDEL METHOD

Another and quite different method of solving a system of linear equations is the so-called Gauss-Seidel method, in which equations (10-1) are rewritten in the following form:

$$a_{11}x_1 = b_1 - a_{12}x_2 - a_{13}x_3 \qquad - \cdots - a_{1n}x_n$$

$$a_{22}x_2 = b_2 - a_{21}x_1 - a_{23}x_3 \qquad - \cdots - a_{2n}x_n$$

$$a_{33}x_3 = b_3 - a_{31}x_1 - a_{32}x_2 - a_{34}x_4 - \cdots - a_{3n}x_n \qquad \text{(10-5)}$$

$$\vdots$$

$$a_{nn}x_n = b_n - a_{n1}x_1 - a_{n2}x_2 \qquad - \cdots - a_{nn-1}x_{n-1}$$

In words, in each of the equations all but one unknown is taken to the right-hand side of the equation. We then guess a set of values for $x_2, x_3, \ldots, x_n$ and substitute these in the right-hand side of the first equation and solve for x_1. Then we substitute this value and the original values of $x_3, \ldots, x_n$ in the right-hand side of the second equation and solve for x_2. We discard the old value of x_2 and keep this as a better one. We then substitute in the right-hand side of the third equation and obtain a new value for x_3. After we have proceeded through all the equations in this fashion, we have a new set of values $x_1, x_2, \ldots, x_n$. (We must first arrange the equations so that none of the $a_{ii} = 0$.) We then start again with the first equation and find a new x_1, then a new x_2, etc. Each time through this process gives us a new, and, we hope, better set of values for $x_1, x_2, \ldots, x_n$. When the new values obtained agree with the previous set to within the accuracy we desire, we have the solution. This is an iteration process similar in nature to those discussed in Chapter 8. It is not absolutely certain that this process will converge, that is, that the differences between succeeding sets of values will get smaller and smaller. We shall discuss the convergence problem more fully a little later. It is not certain, either, how many multiplications will be required to obtain the solution to a desired accuracy. Each trip through the set of equations, or iteration, requires n^2 multiplications. If $(1/3)n$ iterations happen to be required, then the method will take about as long as the elimination method. It may take more or less time, depending entirely on the speed of convergence and accuracy required.

Example 1.   Solve the system

$$x_1 - 2x_2 = 1$$
$$x_1 + 4x_2 = 4$$

by the Gauss-Seidel method.

We write the equations as

$$x_1 = 1 + 2x_2 \qquad\qquad \textbf{(10-6)}$$

$$x_2 = 1 - x_1/4 \qquad\qquad \textbf{(10-7)}$$

Let us take as starting values $x_1 = x_2 = 0$.
  Putting $x_2 = 0$ in equation (10-6), we obtain

$$x_1 = 1$$

Putting $x_1 = 1$ in equation (10-7), we obtain

$$x_2 = 3/4$$

At the end of the first iteration, then, we have

$$x_1 = 1, \qquad x_2 = 3/4$$

Putting $x_2 = 3/4$ in equation (10-6), we have

$$x_1 = 5/2$$

Putting $x_1 = 5/2$ in equation (10-7), we have

$$x_2 = 3/8$$

At the end of the second iteration, then, we have

$$x_1 = 5/2, \qquad x_2 = 3/8$$

We can continue this process. The results for the first several steps, starting from the beginning, are

x_1	x_2
0	0
1	.75
2.5	.375
1.75	.5625
2.125	.46875
1.9375	.515625
2.03125	.4921875
1.984375	.51390625

It is easily verified from the equation that the correct solution is $x_1 = 2$, $x_2 = 1/2$. This solution is slowly converging toward those values.

Example 2.   Solve the system

$$x_1 + 4x_2 = 4$$
$$x_1 - 2x_2 = 1$$

by the Gauss-Seidel method.

This is the same problem as Example 1, with the equations reversed. We write the equations as

$$x_1 = 4 - 4x_2$$
$$x_2 = -1/2 + x_1/2$$

Then the successive interations give the following values:

x_1	x_2
0	0
4	1.5
-2	-1
8	3.5
-10	-5.5
26	12.5
-46	-23.5

It is clear that the process is diverging, and the solution will not be obtained.

Example 3.   Apply the Gauss-Seidel method to Example 1, Section 10.2.

The equations are

$$2x_1 + 3x_2 + 5x_3 = 5$$
$$3x_1 + 4x_2 + 7x_3 = 6$$
$$x_1 + 3x_2 + 2x_3 = 5$$

We write them as

$$2x_1 = 5 - 3x_2 - 5x_3$$
$$4x_2 = 6 - 3x_1 - 7x_3$$
$$2x_3 = 5 - x_1 - 3x_2$$

Successive iterations give (to four decimal places)

x_1	x_2	x_3
0	0	0
2.5	−.375	1.8125
−1.4688	.5703	2.3789
−4.3027	.5640	3.8054
−7.8595	.7352	5.3270
−11.9203	1.1180	6.7831

In Section 10.2 we found that the solution to this system was

$$x_1 = -3, \qquad x_2 = 2, \qquad x_3 = 1$$

Our iteration scheme is not converging toward those values.

10.31   Convergence of the Gauss-Seidel Method

Some insight into the convergence problem can be obtained by following Examples 1 and 2, Section 10.3, in graphical form. Figure 10-4 illustrates the scheme followed in Example 1. Starting at the point P_0, we change x_1 (that is, move horizontally) to arrive on the line $x_1 - 2x_2 = 1$, and then change x_2 (that is, move vertically) to arrive on the line $x_1 + 4x_2 = 4$, bringing us to the point P_1. This is the point given by the first iteration. On the second iteration we move horizontally, then vertically to arrive at P_2. On the third we move horizontally, then vertically to arrive at P_3, etc. It is clear from the figure that this process is bringing us closer and closer to the true point of intersection.

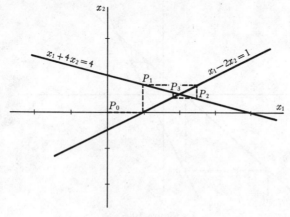

Figure 10-4

Figure 10-5 illustrates the scheme followed in Example 2. The same straight lines are involved, but this time we always move horizontally to reach the line $x_1 + 4x_2 = 4$ and vertically to reach the line $x_1 - 2x_2 = 1$. The points $P_0, P_1, P_2, \ldots$ are the results of the successive iterations in this case.

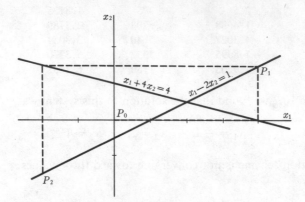

Figure 10-5

It appears that graphically the Gauss-Seidel method for two equations in two unknown consists of following the above boxlike pattern about the point of intersection of the two lines: If this pattern is followed in the correct direction the intersection will be approached, but if it is followed in the wrong direction the process will diverge from the intersection. This is the case if the slopes of the lines have opposite signs. If the signs of the slopes are the same, the situation is a little different, as depicted in Figure 10-6. The sequence of points P_0, P_1, P_2 is part of a convergent process, in which we proceed horizontally to line (b), then vertically to line (a). The points P_0,

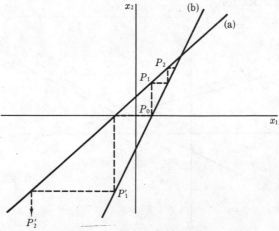

Figure 10-6

P_1', P_2' are part of a divergent process, in which we proceed horizontally to line (a) then vertically to line (b).

As indicated by the above figures, the situation regarding convergence for the Gauss-Seidel method for two equations in two unknowns is as follows: The process will converge for the equations arranged in one order and diverge for the equations arranged in the opposite order. The only exception occurs when the equations represent perpendicular lines, in which case the process will not converge for either arrangement. It is interesting to note that, contrary to our experience with iteration methods in the preceding chapters, the convergence or nonconvergence for these linear equations does *not* depend on choice of initial estimate.

For larger systems of equations the situation becomes much more complex. The necessary and sufficient conditions for convergence are known but are not easily expressed in a very usable form. Sometimes a rearrangement of the equations will produce convergence, but this is not at all guaranteed. The likelihood of convergence is usually increased if the equations are re-arranged so that the coefficients $a_{11}, a_{22}, a_{33}, \ldots, a_{nn}$ which appear on the left-hand side in the system as written in Section 10.3 are the largest coefficients in absolute value. In fact, convergence is assured in this case if in each equation the absolute value of the coefficient a_{ii} is larger than the sum of the absolute values of the remaining coefficients. This condition is not often met. In fact, as in Example 3, Section 10.3, it is often impossible even to write all the equations with largest terms on the left-hand side.

10.32   Flow Chart and Program for the Gauss-Seidel Method

The flow chart in Figure 10-7 describes the Gauss-Seidel method. This flow chart uses the equations arranged just as they are, with no attempt to rearrange the equations to increase the likelihood of convergence. If desired, it could be preceded by another section of flow chart which would rearrange the equations in attempt to enhance the likelihood of convergence. In order to cut down on the number of divisions required, each of the equations is first divided through by the coefficient a_{ii}, so that in the set of new coefficients, c_{ij}, the c_{ii}'s are all one. This flow chart computes at each iteration a quantity

$$E = \sum_{i=1}^{n} |x_i^{\text{new}} - x_i^{\text{old}}|$$

and when this quantity becomes smaller than the given number d, the iteration stops. Note that, in the way the expression for P is written, P is precisely $x_i^{\text{new}} - x_i^{\text{old}}$.

The FORTRAN subroutine below uses the Gauss-Seidel method to solve an N by N system of linear equations, following the flow chart. Again, if a rearrangement of the equations were desired, it could be accomplished by using a subroutine for that purpose just prior to using the one given below:

```
      SUBROUTINE GAUSID(A,N,B,X,ERR)
      DIMENSION A(20,20), B(20),C(20,21),X(20)
      K=0
      NN=N+1
      DO 11 I=1,N
      IF(A(I,I))12,6,12
   12 X(I)=1.
      C(I,NN)=B(I)/A(I,I)
      DO 11 J=1,N
   11 C(I,J)=A(I,J)/A(I,I)
    1 CONTINUE
      E=0.
      DO 3 I=1,N
      P=C(I,NN)
      DO 2 J=1,N
      P=P-C(I,J)*X(J)
    2 CONTINUE
      X(I)=X(I)+P
      E=E+ABSF(P)
    3 CONTINUE
      IF(E-ERR)4,4,5
    4 RETURN
    5 K=K+1
      IF(100-K)6,1,1
    6 PRINT 1000
      RETURN
 1000 FORMAT(25H GAUSID DOES NOT CONVERGE)
      END
```

EXERCISE 26

1. Following the remote-terminal program of Section 10.2, solve the following systems of equations,

   a. $x - y = 2$ b. $x + 2y = 7$
      $x - y = 4$ $4x + y = 5$

   c. $2x + 3y + z = 2$ d. $x + y + z = 4$
      $x + 2y - 4z = 3$ $3x - y - z = 1$
      $4x - 2y + z = -2$ $x + 2y - z = 5$

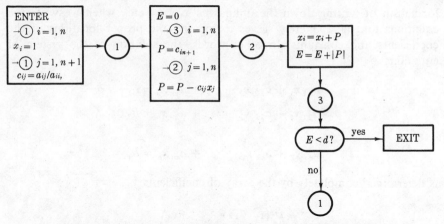

Figure 10-7: Solution of linear equations—Gauss-Seidel method

2. Following the flow chart, Figure 10-7, perform the first four iterations for the following systems of equations.

   a. $2x + y = 3$
       $x + 2y = 3$

   b. $4x - y = 6$
       $x + 3y = -5$

   c. $3x + 2y + z - 5$
       $2x + 5y + 4z = 8$
       $x + 4y + 6z = 4$

   d. $x + 2y + 4z = 6$
       $3x + y + 2z = 5$
       $2x + 4y + z = 4$

3. Write a FORTRAN program that will input a system of linear equations up to size 20 by 20, call subroutine ELIM of Section 10.2 to solve them, and print the result.

4. a. Write a FORTRAN subroutine which will rearrange a set of linear equations, for use of subroutine GAUSID of Section 10.3, so that after rearrangement

   $$a_{ii} \geqslant a_{ki} \qquad \text{for } k > i$$

   b. Show that your subroutine will correctly arrange the equations

   $$x_1 + 8x_2 + x_3 = 10$$
   $$x_1 + x_2 + 7x_3 = 9$$
   $$9x_1 + x_2 + x_3 = 11$$

   so that the Gauss-Seidel method will converge.

   c. Explain what may go wrong with this method of arrangement if some of the coefficients are zero.

10.4   MATRICES

In all the methods of solving linear equations by computer, we have seen that only the coefficients and the constants appear within the machine. The formalism of writing down the unknowns x_1, x_2, etc., when we write the equations longhand, merely serves to identify the proper locations of the coefficients and constants. In other words, the solution of the system of equations

$$a_{11}x_1 + a_{12}x_2 + \cdots + a_{1n}x_n = b_1$$

$$a_{21}x_1 + a_{22}x_2 + \cdots + a_{2n}x_n = b_2$$

$$\vdots$$

$$a_{n1}x_1 + a_{n2}x_2 + \cdots + a_{nn}x_n = b_n$$

is determined completely by the array of coefficients

$$\begin{pmatrix} a_{11} & a_{12} & \cdots & a_{1n} \\ a_{21} & a_{22} & \cdots & a_{2n} \\ \vdots & & & \\ a_{n1} & a_{n2} & \cdots & a_{nn} \end{pmatrix}$$

and the array of constants

$$\begin{pmatrix} b_1 \\ b_2 \\ \vdots \\ b_n \end{pmatrix}$$

If we are given any two such arrays, we can write the set of equations they represent. If we were to change the numerical value of any number in one of these arrays, a different set of equations would be represented. Further, if we were to interchange the position of any two of the numbers, a still different set of equations would be represented. All this suggests that it may be useful to consider these arrays of numbers as separate entities, establish rules for manipulating them, and perhaps free ourselves somewhat of the repetitious writing of the basically nonessential symbols $x_1 +$, $x_2 +$, $x_3 +$, etc. Considerations such as these have led to the definition of a matrix as an array of numbers, and to the development of an "algebra" of matrices, a set of rules for combining matrices to form other matrices. Once developed, matrix algebra has come to have far-reaching applications, completely apart from systems of linear equations.

10.5   DEFINITIONS AND ELEMENTARY OPERATIONS

A matrix is a rectangular array of quantities or numbers, such as

$$\begin{matrix} a_{11} & a_{12} & a_{13} \\ a_{21} & a_{22} & a_{23} \\ a_{31} & a_{32} & a_{33} \end{matrix}$$

In order to distinguish a matrix from a determinant, which also frequently *looks like* an array of numbers, it is customary to enclose a matrix in brackets, or large parentheses, or double bars, as

$$
\begin{bmatrix} a_{11} & a_{12} & a_{13} \\ a_{21} & a_{22} & a_{23} \\ a_{31} & a_{32} & a_{33} \end{bmatrix}, \quad
\begin{pmatrix} a_{11} & a_{12} & a_{13} \\ a_{21} & a_{22} & a_{23} \\ a_{31} & a_{32} & a_{33} \end{pmatrix}, \quad \text{or} \quad
\left\| \begin{matrix} a_{11} & a_{12} & a_{13} \\ a_{21} & a_{22} & a_{23} \\ a_{31} & a_{32} & a_{33} \end{matrix} \right\|
$$

A determinant is usually written between single bars, as

$$
\begin{vmatrix} a_{11} & a_{12} & a_{13} \\ a_{21} & a_{22} & a_{23} \\ a_{31} & a_{32} & a_{33} \end{vmatrix}
$$

This determinant only looks like an array. Really the symbol only stands for a single quantity, which is obtained by multiplying and adding the individual a_{ij}'s in the manner described in Section 10.54. The matrix, on the other hand, has no single numerical value but is instead the entire array. We shall be using a single letter or symbol to stand for a matrix, such as

$$
A = \begin{pmatrix} a_{11} & a_{12} & \cdots & a_{1n} \\ a_{21} & a_{22} & \cdots & a_{2n} \\ \vdots & & & \\ a_{m1} & a_{m2} & \cdots & a_{mn} \end{pmatrix}
$$

When we do this, it is important to remember that A is not a number, and so does not act like a number; that is, it does not obey the ordinary laws of algebra.

Occasionally, we will be interested in the value of a determinant made up of exactly the same elements as some square matrix A. When we do we shall refer to it as the determinant of the matrix A.

A matrix of m rows and n columns is an m by n matrix. If $m = n$, the matrix is a square matrix of order m.

The sum of the diagonal elements of a square matrix is called the "trace" of the matrix, tr $A = a_{11} + a_{22} + \cdots + a_{nn}$.

If a matrix consists of a single column it is called a column matrix, or sometimes a column vector.

If the elements in the main diagonal of a square matrix are ones, and all the other elements are zeros, the matrix is called a unit matrix, or identity matrix. Thus

$$
\begin{pmatrix} 1 & 0 & 0 \\ 0 & 1 & 0 \\ 0 & 0 & 1 \end{pmatrix}
$$

is a unit matrix of order 3. Unit matrices of any order are usually denoted by the symbol I.

If all the elements are zero, the matrix is called a zero matrix.

Two matrices A and B are said to be equal if:
(*1*)   They have the same number of rows.
(*2*)   They have the same number of columns.
(*3*)   Each pair of corresponding elements are equal.

10.51   Addition and Subtraction of Matrices

The operations of addition and subtraction are defined for two matrices A and B if:
(*1*)   They have the same number of rows.
(*2*)   They have the same number of columns.

The sum of two matrices is the matrix obtained by adding corresponding pairs of elements. Thus, if

$$A = \begin{pmatrix} a_{11} & a_{12} & a_{13} \\ a_{21} & a_{22} & a_{23} \\ a_{31} & a_{32} & a_{33} \end{pmatrix} \quad \text{and} \quad B = \begin{pmatrix} b_{11} & b_{12} & b_{13} \\ b_{21} & b_{22} & b_{23} \\ b_{31} & b_{32} & b_{33} \end{pmatrix}$$

then

$$A + B = \begin{pmatrix} a_{11} + b_{11} & a_{12} + b_{12} & a_{13} + b_{13} \\ a_{21} + b_{21} & a_{22} + b_{22} & a_{23} + b_{23} \\ a_{31} + b_{31} & a_{32} + b_{32} & a_{33} + b_{33} \end{pmatrix}$$

The difference $A - B$ is the matrix obtained by subtracting the elements of B from the corresponding elements of A.

$$A - B = \begin{pmatrix} a_{11} - b_{11} & a_{12} - b_{12} & a_{13} - b_{13} \\ a_{21} - b_{21} & a_{22} - b_{22} & a_{23} - b_{23} \\ a_{31} - b_{31} & a_{32} - b_{32} & a_{33} - b_{33} \end{pmatrix}$$

Example 1.   Find $A + B$ and $A - B$, where

$$A = \begin{pmatrix} 3 & 0 & -2 \\ 1 & 3 & 1 \end{pmatrix}, \qquad B = \begin{pmatrix} 2 & 1 & 2 \\ -1 & 3 & -2 \end{pmatrix}$$

SOLUTION:

$$A + B = \begin{pmatrix} 5 & 1 & 0 \\ 0 & 6 & -1 \end{pmatrix}, \qquad A - B = \begin{pmatrix} 1 & -1 & -4 \\ 2 & 0 & 3 \end{pmatrix}$$

Example 2.   Find $A + B$ and $A - B$, where

$$A = \begin{pmatrix} 3 & 0 & -2 \\ 1 & 3 & 2 \end{pmatrix}, \qquad B = \begin{pmatrix} 2 & -1 \\ 1 & 3 \\ 2 & -2 \end{pmatrix}$$

Since there are not the same number of rows or columns in A and B, they cannot be added or subtracted. The symbols $A + B$ and $A - B$ are meaningless in this case.

Example 3.   Given two matrices A and B, each with N columns and M rows, write FORTRAN statements which would form the sum, $C = A + B$.

The matrix A can be represented by a single subscripted variable A(I,J), where I runs from 1 to M and J runs from 1 to N. The same is true for B and C. Then the required FORTRAN statements are

```
      DO  20  I = 1M,
      DO  20  J = 1,N
   20 C(I,J) = A(I,J) + B(I,J)
```

A total of $N \times M$ additions are required to obtain C.

As a direct extension of addition, it would be natural to be able to say

$$A + A = 2A$$

This leads to the definition of multiplication of a matrix by a constant as follows: A constant times a matrix is the matrix obtained by multiplying *all* elements of the original matrix by the constant.

10.52   Multiplication of Matrices

At first acquaintance, the operation of multiplication of two matrices seems to be defined in a most peculiar way. There are very good reasons for choosing to call this seemingly awkward process "multiplication," and these will appear shortly.

The product AB of two matrices, A and B, is defined only if the number of *columns* in A is equal to the number of *rows* in B. In all other cases the product is undefined. If the number of columns in A is equal to the number of rows in B, then A and B are said to be "conformable" in the order AB.

The product AB of two conformable matrices is itself a matrix, whose elements are found according to the following rule: The element in the ith row and the jth column of the product is the sum of the products by pairs of the elements of the ith row of A and jth column of B.

Example 1. If

$$A = \begin{pmatrix} 1 & 2 \\ 3 & -1 \end{pmatrix}, \qquad B = \begin{pmatrix} 3 & -2 \\ 2 & 1 \end{pmatrix}$$

find AB.

Since A has 2 columns and B has 2 rows, A and B are conformable in the order AB, so the product is indeed defined. To find the element in the first row, first column of the product matrix, we take the first row of A, which is

$$1 \quad 2$$

and the first column of B, which is

$$\begin{matrix} 3 \\ 2 \end{matrix}$$

and form the sum of the products by pairs:

$$1 \times 3 + 2 \times 2 = 7$$

Hence 7 is the element in the first row, first column of the product.

In like manner, the element in the first row and second column of the product is obtained from combining the first row of A with the second column of B, thus:

$$1 \times (-2) + 2 \times 1 = 0$$

and for the second row, first column,

$$3 \times 3 + (-1) \times 2 = 7$$

and the second row, second column,

$$3 \times (-2) + (-1) \times 1 = -7$$

Hence the product is

$$\begin{pmatrix} 1 & 2 \\ 3 & -1 \end{pmatrix} \begin{pmatrix} 3 & -2 \\ 2 & 1 \end{pmatrix} = \begin{pmatrix} 7 & 0 \\ 7 & -7 \end{pmatrix}$$

Example 2. If

$$A = \begin{pmatrix} 1 & 3 & 1 \\ -2 & 1 & -1 \end{pmatrix}, \qquad B = \begin{pmatrix} 1 \\ 2 \\ 3 \end{pmatrix}$$

find AB.

Since A has 3 columns and B has 3 rows, they are conformable in the order AB. We can expedite the process of finding the product by writing the two matrices side by side, and then going across a row of A and down a column of B forming products by pairs, thus:

$$\begin{pmatrix} 1 & 3 & 1 \\ -2 & 1 & -1 \end{pmatrix}\begin{pmatrix} 1 \\ 2 \\ 3 \end{pmatrix} = \begin{pmatrix} 1 \times 1 + 3 \times 2 + 1 \times 3 \\ -2 \times 1 + 1 \times 2 - 1 \times 3 \end{pmatrix} = \begin{pmatrix} 10 \\ -3 \end{pmatrix}$$

Example 3. For the matrices A and B of Example 2, find BA.

Since B has 1 column and A has 2 rows, they are not conformable in the order BA. The product BA is not defined!

Example 4. If

$$A = \begin{pmatrix} a_{11} & a_{12} & \cdots & a_{1n} \\ a_{21} & a_{22} & \cdots & a_{2n} \\ \vdots & & & \\ a_{m1} & a_{m2} & \cdots & a_{mn} \end{pmatrix}, \qquad B = \begin{pmatrix} b_{11} & b_{12} & \cdots & b_{1l} \\ b_{21} & b_{22} & \cdots & b_{2l} \\ \vdots & & & \\ b_{n1} & b_{n2} & \cdots & b_{nl} \end{pmatrix}$$

and

$$AB = C$$

write a formula for finding c_{ij}, the element in the ith row and jth column of C.

The ith row of A is

$$a_{i1} \quad a_{i2} \quad \cdots \quad a_{in}$$

and the jth column of B is

$$b_{1j}$$
$$b_{2j}$$
$$\vdots$$
$$b_{nj}$$

and the sum of the products by pairs gives

$$c_{ij} = a_{i1}b_{1j} + a_{i2}b_{2j} + \cdots + a_{in}b_{nj}$$

or, in more abbreviated form,

$$c_{ij} = \sum_{k=1}^{n} a_{ik}b_{kj}$$

Example 5. Given matrix A with M rows and N columns and matrix B with N rows and L columns, write a set of FORTRAN statements which will form the product $C = AB$.

A suitable set of statements is

```
      DO 10 I=1,M
      DO 10 J=1,L
      C(I,J)=0.
      DO 10 K=1,N
   10 C(I,J)=C(I,J)+A(I,K)*B(K,J)
```

We note that statement 10 is in three DO loops, and will be performed $N \times M \times L$ times, or $N \times M \times L$ multiplications are required to find the product matrix C.

Example 6. If

$$A = \begin{pmatrix} a_{11} & a_{12} & a_{13} \\ a_{21} & a_{22} & a_{23} \\ a_{31} & a_{32} & a_{33} \end{pmatrix}, \qquad x = \begin{pmatrix} x_1 \\ x_2 \\ x_3 \end{pmatrix}$$

write the product Ax.

SOLUTION:

$$Ax = \begin{pmatrix} a_{11}x_1 + a_{12}x_2 + a_{13}x_3 \\ a_{21}x_1 + a_{22}x_2 + a_{23}x_3 \\ a_{31}x_1 + a_{32}x_2 + a_{33}x_3 \end{pmatrix}$$

Note that this product Ax is actually a column vector, having three elements.

Example 7. Write the system of linear equations

$$a_{11}x_1 + a_{12}x_2 + a_{13}x_3 = b_1$$
$$a_{21}x_1 + a_{22}x_2 + a_{23}x_3 = b_2$$
$$a_{31}x_1 + a_{32}x_2 + a_{33}x_3 = b_3$$

in matrix form.

From Example 6, if we define

$$A = \begin{pmatrix} a_{11} & a_{12} & a_{13} \\ a_{21} & a_{22} & a_{23} \\ a_{31} & a_{32} & a_{33} \end{pmatrix}, \qquad x = \begin{pmatrix} x_1 \\ x_2 \\ x_3 \end{pmatrix}$$

then the left-hand sides of the equations above are just the three elements of the column vector Ax. Now let us define the column vector

$$b = \begin{pmatrix} b_1 \\ b_2 \\ b_3 \end{pmatrix}$$

We recall that two matrices are equal if and only if every pair of corresponding elements are equal. Thus, the statement

$$Ax = b$$

is a matrix equation. The expressions on each side of the equals sign are matrices. The equation means that

(1)   The first element of Ax, that is $a_{11}x_1 + a_{12}x_2 + a_{13}x_3$, is equal to b_1.

(2)   The second element of Ax, that is, $a_{21}x_1 + a_{22}x_2 + a_{23}x_3$, is equal to b_2.

(3)   The third element of Ax, that is, $a_{31}x_1 + a_{32}x_2 + a_{33}x_3$, is equal to b_3.

Hence the matrix equation

$$Ax = b$$

says exactly the same thing as the system of linear equations above.

We see from Examples 6 and 7 that any system of linear equations, with any number of unknowns, can be represented by a matrix equation

$$Ax = b$$

where A is a matrix and x and b are column vectors of the correct order. This simple expression is one of the several happy results of the seemingly odd definition of multiplication.

10.53   Laws of Matrix Algebra

We have defined three operations with matrices and have given them the names " addition," " subtraction," and " multiplication "—names we use in the ordinary algebra of numbers. Actually this is a little dangerous, since it suggests that these new matrix operations will obey the same rules as the ordinary arithmetic operations, and we really have no right to expect that they will do so.

The fundamental laws of ordinary algebra are the following:

(1).   Addition is *commutative*. $a + b = b + a$; that is, if we add b to a, or a to b, we will get the same result.

(2).   Addition is *associative*. $(a + b) + c = a + (b + c)$; that is, if we add $a + b$, and then add c to this sum, we get the same result as if we add b and c first, and then add a to the sum.

(3).   Multiplication is *distributive* with respect to addition. $a(b + c) = ab + ac$; that is, if we add b to c and then multiply by a, we get the same result as if we multiply a by b, multiply a by c, and then add the result.

(4).   Multiplication is *commutative*. $ab = ba$; that is, if we multiply a by b or b by a, we get the same result.

(5).   Multiplication is *associative*. $(ab)c = a(bc)$; that is, if we take the product ab and multiply by c we get the same answer as if we take the product bc and multiply by a.

When these laws for the algebra of numbers are investigated for matrices, it is found that they all hold *except* law 4, the commutative law for multiplication. As was seen in Examples 2 and 3, Section 10.52, it is possible to have two matrices whose product AB could be found but whose product BA was not even defined.

In summary, then, we can say that, in expressions involving sums, differences, and products of matrices, we can use the same laws for combining these operations as for ordinary numbers except that the order of any two matrices in a product cannot be reversed. In a matrix equation, we may add the same matrix to both sides or subtract the same matrix from both sides without changing the equality. We also may *multiply* both sides by the same matrix, provided that:

(*1*)   The matrix is conformable with those by which it is to be multiplied.

(*2*)   The order of multiplication is made the same on both sides of the equation.

Example 1.   If A, B, and C are square matrices of order n, and if

$$A + B = C$$

solve for A.

Subtracting B from both sides, we have

$$A = C - B$$

Example 2.   If A, B, C, and D are square matrices of order n, and if $A = B + C$, find AD and DA.

Multiplying the equation

$$A = B + C$$

on the right by D, we have

$$AD = (B + C)D = BD + CD$$

Multiplying the above equation on the left by D, we have

$$DA = D(B + C) = DB + DC$$

10.54   Determinants

The determinant of a square matrix A is defined to be the number obtained in the following manner: From the elements of A, we form all possible products containing exactly one element from each row and column in A. To each such term we assign a plus or minus sign in accordance with a rule to be stated shortly. The sum of these terms is the value of the determinant. The sign to be assigned to a term is determined by the following procedure. The factors in the term are arranged in order according to the row from which each factor was chosen:

$$a_{1k_1} a_{2k_2} a_{3k_3} \cdots a_{nk_n}$$

We then rearrange these factors so that they are in order according to the column from which each was chosen, that is, so that the subscripts $k_1, k_2, \ldots,$ k_n are in their natural order, and count the number of interchanges required to do this. We assign the term a plus sign if the number of interchanges was even and a minus sign if it was odd. For a 2 by 2 determinant, then,

$$\begin{vmatrix} a_{11} & a_{12} \\ a_{21} & a_{22} \end{vmatrix} = a_{11}a_{22} - a_{21}a_{12}$$

For a 3 by 3 system,

$$\begin{vmatrix} a_{11} & a_{12} & a_{13} \\ a_{21} & a_{22} & a_{23} \\ a_{31} & a_{32} & a_{33} \end{vmatrix} = \begin{aligned} & a_{11}a_{22}a_{33} - a_{11}a_{23}a_{32} - a_{12}a_{21}a_{33} \\ & + a_{12}a_{23}a_{31} + a_{13}a_{21}a_{32} - a_{13}a_{22}a_{31} \end{aligned}$$

It is clear that, by utilizing the programming methods of the earlier chapters, we can cause a computer to perform such calculations and provide the solution to a system of equations. It is not so obvious, but it can be shown that such a procedure is quite inefficient in machine time, particularly for systems involving a very large number of unknowns. According to the rule just stated for evaluating a determinant, an n by n determinant is the sum of $n!$ terms, each of which is the product of n numbers. If we were to calculate the value of a determinant by the most direct method, then, about $n \times n!$ multiplications would be required. For even a 10 by 10 determinant, several million multiplications would be required, and for a 20 by 20 determinant, over 10^{18} multiplications would be needed. This would require over 100,000 years even on the fastest computers.

There is another method of evaluation of a determinant that is very much faster than the brute-force approach. If all elements on one row of a determinant are changed by adding or subtracting a constant multiple of the corresponding elements of another row, the value of the determinant is unchanged. By repeated application of this rule, we can reduce a determinant to a "triangular" form, in which all elements below the main diagonal are zero. For example,

$$
\begin{vmatrix}
b_{11} & b_{12} & b_{13} & \cdots & & b_{1n} \\
0 & b_{22} & b_{23} & \cdots & & b_{2n} \\
0 & 0 & b_{33} & \cdots & & b_{3n} \\
0 & 0 & 0 & b_{44} & \cdots & b_{4n} \\
& & \vdots & & & \\
0 & 0 & 0 & \cdots & & b_{nn}
\end{vmatrix}
$$

The value of a determinant when written in this form turns out to be just the product of the diagonal elements, $b_{11}b_{22}b_{33} \cdots b_{nn}$, since all other terms formed in accordance with the definition of a determinant's value contain at least one factor whose value is zero. Hence, after a determinant is written in triangular form, only $n - 1$ multiplications are required to find its value.

The process is quite similar to that used in Section 10.2 to solve a system of linear equations by the elimination method. We start out with the array

$$
\begin{matrix}
a_{11} & a_{12} & a_{13} & \cdots & a_{1n} \\
a_{21} & a_{22} & a_{23} & \cdots & a_{2n} \\
a_{31} & a_{32} & a_{33} & \cdots & a_{3n} \\
\vdots & & & & \\
a_{n1} & a_{n2} & a_{n3} & \cdots & a_{nn}
\end{matrix}
$$

and perform the operations

$$
a_{ij} - \frac{a_{ik}a_{kj}}{a_{kk}} \rightarrow a_{ij} \qquad \text{for } i \text{ and } j = k + 1, k + 2, \ldots, n
$$

$$
k = 1, \ldots, n
$$

Figure 10-8 is a flow chart of the process.

A calculation based on the flow chart, Figure 10-8, could run into trouble if a_{kk} ever becomes zero, since there is a division by this quantity. This problem can be avoided by taking the additional precaution of checking to see if a_{kk} is zero and if so interchanging two rows to obtain a nonzero value for a_{kk}. Since interchanging two rows in a determinant changes the sign of the determinant's value, we must also change the signs of the elements in one of the rows to correct this.

There can also be accuracy problems associated with evaluating a determinant using the above flow chart, particularly for determinants of large order. These problems tend to be alleviated if the rows and columns are rearranged at each step so that a_{kk} is not only nonzero but is actually the largest element in absolute value. SUBROUTINE EXCH of Section 10.2

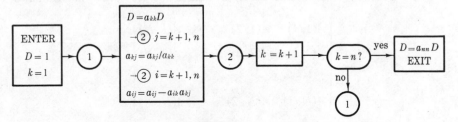

Figure 10-8: Evaluation of a determinant

provided this service for the elimination method, and with minor modifications it can be made to work in the present case. The main difference is that interchanging rows or columns in a determinant changes the sign of the determinant. We can correct for this by changing statement 32 to read

  32 $A(K,J) = -C$

and statement 42 to read

  42 $A(I,K) = -C$

We also need the dimension statement to read A(20,20) instead of A(20,21). We do not need the quantity ID as output, so we can eliminate the three statements following statement 42 and change the first statement to read

  SUBROUTINE EXCH2(A,N,NN,K)

We changed the name as well, to ensure that the old routine of Section 10.2 is not used by mistake.

Another step which is useful to avoid undetected accuracy loss is in connection with the computation

$$a_{ij} = a_{ij} - a_{ik} a_{kj}$$

If the result of this subtraction is supposed to be zero, then this subtraction will be subject to the trouble mentioned many times earlier in the text, loss of accuracy caused by introduction of leading zeros. The method of protection against this trouble is the same one used in division of polynomials in Section 9.43. We check the result of the subtraction, and if the difference is much smaller than the numbers being subtracted, we set the difference equal to zero. The operation can be described by a section of flow chart (as in Figure 10-9) in which a_{ij} is set equal to zero if more than four significant figures have been lost in the subtraction.

The FORTRAN subroutine below evaluates the Nth-order determinant

$$\begin{vmatrix} A(1,1) & A(1,2) & \cdots & A(1,N) \\ A(2,1) & A(2,2) & \cdots & A(2,N) \\ \vdots & & & \\ A(N,1) & A(N,2) & \cdots & A(N,N) \end{vmatrix}$$

for values of N to 20. In the first statement, the determinant is given the name AA, and the statements up to 100 redefine the elements so that the original determinant will not be destroyed during the calculation. State-

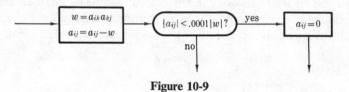

Figure 10-9

ment 1 calls EXCH2 to interchange rows and columns if necessary to move the largest element to location A(K,K). Statements 3 through 4 perform the actual calculation required in the main part of the flow chart, Figure 10-8.

```
      SUBROUTINE DETERM(AA,N,D)
      DIMENSION AA(20,20),A(20,20)
      DO 100 I=1,N
      DO 100 J=1,N
  100 A(I,J)=AA(I,J)
      D=1.
      K=1
    1 CALL EXCH2(A,N,N,K)
      D=A(K,K)*D
      IF(A(K,K))3,10,3
    3 KK=K+1
      DO 4 J=KK,N
      A(K,J)=A(K,J)/A(K,K)
```

```
      DO 4 I=KK,N
      W=A(I,K)*A(K,J)
      A(I,J)=A(I,J)-W
      IF(ABS(A(I,J))-.0001*ABS(W))42,4,4
   42 A(I,J)=0.
    4 CONTINUE
      K=KK
      IF(K-N)1,9,10
    9 D=A(N,N)*D
   10 RETURN
      END
```

10.55   Matrix Inversion

We have given definitions and rules for the addition, subtraction, and multiplication of matrices which parallel to some extent the rules of ordinary algebra. As yet we have not mentioned division, for the very good reason that division as such is not defined for matrices. There is another operation which serves a somewhat analogous purpose, however. That operation is the "inversion" of a matrix.

In ordinary algebra, b/a stands for the number which, when multiplied by a, gives b. Thus, if $ax = b$, we can say that $x = b/a$. Instead of treating division in this manner, we could define an "inverse" of a number as follows: For any number a, the inverse, a^{-1}, is that number which, when multiplied by a gives 1. Every nonzero number has a unique inverse; for example, the inverse of 2 is .5, and .5 is the *only* inverse of 2. Then if we have $ax = b$, we do not even have to have a process of division in order to find x, for we can multiply both sides of the equation by a^{-1}, giving

$$a^{-1}ax = a^{-1}b \qquad \text{or} \qquad x = a^{-1}b$$

For square matrices, we define the inverse in a manner analogous to that above. For a square matrix A of order n, the inverse matrix, A^{-1} is that matrix which when multiplied by A gives the identity matrix of order n; that is,

$$AA^{-1} = I$$

Example 1.   Show that the inverse of

$$\begin{pmatrix} 2 & -1 \\ -1 & 1 \end{pmatrix}$$

is

$$\begin{pmatrix} 1 & 1 \\ 1 & 2 \end{pmatrix}$$

We form the product

$$\begin{pmatrix} 2 & -1 \\ -1 & 1 \end{pmatrix}\begin{pmatrix} 1 & 1 \\ 1 & 2 \end{pmatrix} = \begin{pmatrix} 1 & 0 \\ 0 & 1 \end{pmatrix}$$

Since the product is the identity matrix, the second matrix is indeed the inverse of the first.

It can be shown that any square matrix A has a unique inverse if and only if its determinant is different from zero. It can also be shown that A commutes with its inverse; that is,

$$AA^{-1} = A^{-1}A = I$$

The inverse is not defined for nonsquare matrices.

A formula for the inverse of a matrix A can be found as follows. Consider the set of linear equations

$$y_1 = a_{11}x_1 + a_{12}x_2 + \cdots + a_{1n}x_n$$
$$y_2 = a_{21}x_1 + a_{22}x_2 + \cdots + a_{2n}x_n$$
$$\vdots$$
$$y_n = a_{n1}x_1 + a_{n2}x_2 + \cdots + a_{nn}x_n$$

(10-8)

connecting one set of variables $x_1, x_2, \ldots, x_n$ with another set $y_1, y_2, \ldots, y_n$. In matrix form we can write this set of equations as

$$y = Ax$$

where

$$y = \begin{pmatrix} y_1 \\ y_2 \\ \vdots \\ y_n \end{pmatrix}, \qquad x = \begin{pmatrix} x_1 \\ x_2 \\ \vdots \\ x_n \end{pmatrix}$$

If we multiply this set of equations by A^{-1}, we obtain

$$A^{-1}y = A^{-1}Ax$$

or

$$A^{-1}y = Ix = x$$

Hence the elements of A^{-1} are just the coefficients of the y's if we solve the set of equations (10-8) for the x's in terms of the y's.

A rather efficient way of solving for the x's in terms of the y's is to proceed as in the elimination method, Section 10.2. Instead of the n by $n + 1$ array of constants shown at (10-3) we start with the n by $2n$ array of the form

$$
\begin{array}{cccccccccc}
a_{11} & a_{12} & a_{13} & \cdots & a_{1n} & 1 & 0 & 0 & \cdots & 0 \\
a_{21} & a_{22} & a_{23} & \cdots & a_{2n} & 0 & 1 & 0 & \cdots & 0 \\
a_{31} & a_{32} & a_{33} & \cdots & a_{3n} & 0 & 0 & 1 & \cdots & 0 \\
\vdots & & & & & & & & & \\
a_{n1} & a_{n2} & a_{n3} & \cdots & a_{nn} & 0 & 0 & 0 & \cdots & 1
\end{array}
\qquad \textbf{(10-9)}
$$

Then we proceed exactly as in Section 10.2, and end up with an array of the form

$$
\begin{array}{cccccccccc}
1 & 0 & 0 & \cdots & 0 & a_{1n+1} & a_{1n+2} & a_{1n+3} & \cdots & a_{12n} \\
0 & 1 & 0 & \cdots & 0 & a_{2n+1} & a_{2n+2} & a_{2n+3} & \cdots & a_{22n} \\
0 & 0 & 1 & \cdots & 0 & a_{3n+1} & a_{3n+2} & a_{3n+3} & \cdots & a_{32n} \\
\vdots & & & & & & & & & \\
0 & 0 & 0 & \cdots & 1 & a_{nn+1} & a_{nn+2} & a_{nn+3} & \cdots & a_{n2n}
\end{array}
$$

The solution, instead of being the single column a_{1n+1} to a_{nn+1}, is the entire right-hand side of the above array. If we have interchanged any rows in the process, the unscrambling is just as indicated in subroutine ELIM, except in the present case whole rows must be unscrambled. The subroutine can be written immediately as an almost direct paraphrase of SUBROUTINE ELIM

```
      SUBROUTINE MATINV(AA,N,AINV)
      DIMENSION AA(20,20),AINV(20,20),A(20,40),ID(20)
      NN=N+1
      N2=2*N
      DO 100 I=1,N
      ID(I)=I
      DO 100 J=1,N
  100 A(I,J)=AA(I,J)
      DO 200 I=1,N
      DO 200 J=NN,N2
```

```
200  A(I,J)=0.
     DO 300 I=1,N
300  A(I,N+I)=1.
     K=1
  1  CALL EXCH3(A,N,N2,K,ID)
  2  IF(A(K,K))3,999,3
  3  KK=K+1
     DO 4 J=KK,N2
     A(K,J)=A(K,J)/A(K,K)
     DO 4 I=1,N
     IF(K-I)41,4,41
 41  W=A(I,K)*A(K,J)
     A(I,J)=A(I,J)-W
     IF(ABS(A(I,J))-.0001*ABS(W))42,4,4
 42  A(I,J)=0.
  4  CONTINUE
     K=KK
     IF(K-N)1,2,5
  5  DO 10 J=1,N
     DO 10 J=1,N
     IF(ID(J)-I)10,8,10
  8  DO 10 K=1,N
     AINV(I,K)=A(J,N+K)
 10  CONTINUE
     RETURN
999  PRINT 1000
     RETURN
1000 FORMAT(19H MATRIX IS SINGULAR)
     END
```

In this subroutine the statements through 300 move the quantities to working storage to form the array depicted by (10-9). Statement 1 calls a version of SUBROUTINE EXCH given in Section 10.2. It is called EXCH3, to indicate that it must be a modified version of that subroutine with dimension statement changed to read

DIMENSION A(20,40)

Statements down to statement 4 parallel SUBROUTINE ELIM, except that the accuracy flag shown in Figure 10-9 has been inserted at statement 42. In the loops terminating on statement 10, instead of setting individual values of the X(I)'s, the subroutine sets entire rows of the inverse matrix AINV(I,K).

Once an inverse matrix has been obtained, an improved accuracy version can be obtained in a relatively straightforward manner. Let A be the matrix to be inverted, and let D_1 be the approximate inverse produced by the above routine. Then, because of inaccuracies,

$$AD_1 \neq I$$

but instead

$$I - AD_1 = F_1$$

where F_1 is a matrix which, if D_1 was a reasonably good estimate, has small elements. If all the elements of F_1 are less than one in absolute value, then the matrix D_2 defined by

$$D_2 = D_1(I + F_1)$$

is an improved estimate of A^{-1}. If the error matrix $F_2 = I - AD_2$ still has elements which are too large, then the matrix D_3 defined by $D_3 = D_2(I + F_2)$ is a still better estimate, and so on. Thus repetition of a process involving some matrix multiplications can be used to improve the accuracy of the inverse to the extent desired, within the limits imposed by the usual problems of approximate arithmetic on computers.

EXERCISE 27

1. Given the following matrices

$$A = \begin{pmatrix} 2 & 2 \\ 1 & -1 \\ -2 & 1 \end{pmatrix}, \quad B = \begin{pmatrix} 1 & -1 \\ 2 & 3 \end{pmatrix}, \quad C = \begin{pmatrix} 2 \\ 1 \\ -2 \end{pmatrix}$$

$$D = \begin{pmatrix} 4 & 1 & 3 \\ 2 & -1 & 1 \\ -3 & 2 & 1 \end{pmatrix}, \quad E = \begin{pmatrix} 1 \\ 2 \end{pmatrix}, \quad F = \begin{pmatrix} 1 & -1 & 2 \\ 2 & -3 & 1 \end{pmatrix}$$

evaluate the following expressions, or, if the expression is meaningless, so state.

a. AB b. DC c. BE
d. BA e. ABE f. $DA + A$
g. $FA + B$ h. $FC + BE$ i. $FDABE$
j. $AF + D$

2. Using the method of Section 10.55, invert the following matrices.

a. $\begin{pmatrix} 3 & 2 \\ 4 & 3 \end{pmatrix}$ b. $\begin{pmatrix} 1 & -3 \\ 2 & 4 \end{pmatrix}$

c. $\begin{pmatrix} 2 & 3 & 1 \\ 1 & -1 & 2 \\ -3 & 1 & -1 \end{pmatrix}$ d. $\begin{pmatrix} 1 & 0 & 0 \\ 0 & 3 & 2 \\ 0 & 4 & 3 \end{pmatrix}$

3. Find A^{-1}, then solve $Ax = b$ by multiplying both sides by A^{-1}, if

   a. $A = \begin{pmatrix} 2 & 3 \\ 3 & 4 \end{pmatrix}, \quad b = \begin{pmatrix} 2 \\ 1 \end{pmatrix}$

   b. $A = \begin{pmatrix} 2 & 4 & -1 \\ -1 & -3 & 1 \\ 3 & -1 & 2 \end{pmatrix}, \quad b = \begin{pmatrix} 4 \\ -2 \\ 6 \end{pmatrix}$

4. Write a FORTRAN subroutine INVIMP that will take the trial inverse obtained from MATINV and use the method described at the end of Section 10.55 to improve the inverse until all the elements of the error matrix F_n are less than .001 in absolute value.

10.6   OVERDETERMINED AND UNDERDETERMINED SYSTEMS OF LINEAR EQUATIONS

In several of the preceding sections, methods were discussed for solving systems of linear equations. In all these discussions it was assumed that there was a unique solution and that there were just as many equations as unknowns. Further, it was tacitly assumed that the equations were nonhomogeneous, that is, not all the constant terms were zero, and also that the determinant of the coefficients was not zero. With these conditions satisfied there *is* a unique solution. In many important cases, however, these conditions are not all satisfied—yet there may still be a unique solution, or there may be no solution or an infinite number of solutions. In this section we will discuss a method for finding which situation prevails and for completely describing the solutions when there is an infinite number of them.

10.61   Rank of a Matrix

As a tool for further study of systems of equations we will need the concept of rank of a matrix.

Definition. The rank of a matrix is the order of the highest-order nonvanishing determinant within the matrix.

By a "determinant within the matrix" we mean any determinant that can be made by crossing out rows or columns in the matrix.

Example 1. Find the rank of the matrix

$$\begin{pmatrix} -1 & 1 & 2 \\ -3 & 3 & 1 \end{pmatrix}$$

The largest-order determinant we can construct is second order, so the rank is 2 or less. To see if it is 2, we must check all second-order determinants. If we cross out the third column, we can construct the determinant

$$\begin{vmatrix} -1 & 1 \\ -3 & 3 \end{vmatrix}$$

which has the value zero. Since this one vanishes, we must check other second-order determinants. Crossing out the second column in the matrix, we obtain the determinant

$$\begin{vmatrix} -1 & 2 \\ -3 & 1 \end{vmatrix}$$

which has the value 5. Since there is a nonvanishing second-order determinant, the rank is 2.

Example 2.   Find the rank of the matrix

$$\begin{pmatrix} 1 & 2 & 3 \\ -1 & -2 & -3 \\ 2 & 4 & 6 \end{pmatrix}$$

The largest-order determinant we can construct is third order, so the rank is 3 or less. The only third-order determinant is

$$\begin{vmatrix} 1 & 2 & 3 \\ -1 & -2 & -3 \\ 2 & 4 & 6 \end{vmatrix} = 0$$

so the rank is not 3. If we cross out the third row and third column, we have the determinant

$$\begin{vmatrix} 1 & 2 \\ -1 & -2 \end{vmatrix} = 0$$

Similarly, if we check all other second-order determinants, we find that they all vanish.

Hence the rank is less than 2. If we cross out the second and third rows, and the second and third columns, we can form the determinant $|1| = 1$. Since the highest-order nonvanishing determinant is first order, the rank of the matrix is 1.

It is seen from the above examples that finding the rank of a matrix is a straightforward process. For matrices of higher order, however, the process as just demonstrated is extremely laborious, sometimes involving the evaluation of many determinants. Fortunately, however, a less laborious method is available, based on the following theorem:

Theorem 1.   *The rank of a matrix is unchanged if any multiple of the elements of one row (or columns) is added to the corresponding elements of another row (or column).*

This theorem means that we can proceed, just as in evaluating a determinant, to combine rows or columns to obtain zeros where we choose.

Example 3.   Find the rank of

$$\begin{pmatrix} 1 & -1 & -1 & -2 \\ 2 & 1 & -2 & 2 \\ 4 & 3 & -4 & 6 \end{pmatrix}$$

Using Theorem 1, we may proceed as follows:

$$\text{rank}\begin{pmatrix} 1 & -1 & -1 & -2 \\ 2 & 1 & -2 & 2 \\ 4 & 3 & -4 & 6 \end{pmatrix} = \text{rank}\begin{pmatrix} 1 & -1 & -1 & -2 \\ 0 & 3 & 0 & 6 \\ 4 & 3 & -4 & 6 \end{pmatrix} \begin{matrix} \text{(twice first} \\ \text{row subtracted} \\ \text{from second)} \end{matrix}$$

$$= \text{rank}\begin{pmatrix} 1 & -1 & -1 & -2 \\ 0 & 3 & 0 & 6 \\ 0 & 7 & 0 & 14 \end{pmatrix} \begin{matrix} \text{(four times first} \\ \text{row subtracted} \\ \text{from third} \end{matrix}$$

$$= \text{rank}\begin{pmatrix} 1 & -1 & -1 & -2 \\ 0 & 3 & 0 & 6 \\ 0 & 0 & 0 & 0 \end{pmatrix} \begin{matrix} \text{(7/3 times second} \\ \text{row subtracted} \\ \text{from third)} \end{matrix}$$

It is obvious in this last matrix all third-order determinants are zero, but at least one second-order determinant,

$$\begin{vmatrix} 1 & -1 \\ 0 & 3 \end{vmatrix}$$

is not zero. Hence the rank of the original matrix is 2.

Note that in the above example, we have *not* said the *matrices* obtained at each step are equal, but only that the *ranks* are equal. Each step has created a new matrix, one differing from the preceding in many respects, but having the rank in common.

It is seen that the method of determining rank as demonstrated in Example 3 is closely akin to the method of evaluation of a determinant given in Section

10.54. Minor modifications to the program given there will give a program for finding the rank of a matrix with no more effort than that involved in evaluating the largest determinant in the matrix.

The FORTRAN subroutine below finds the rank, K, of a matrix having N rows and M columns, where neither N nor M exceed 20.

```
        SUBROUTINE MARANK(AA,N,M,K)
        DIMENSION AA(20,20),A(20,20)
        DO 100 I=1,N
        DO 100 J=1,M
100 A(I,J)=AA(I,J)
        K=1
  1 CALL EXCH2(A,N,M,K)
        IF(A(K,K))2,10,2
  2 IF(K-N)3,11,11
  3 IF(K-M)40,11,11
 40 KK=K+1
        DO 4 J=KK,M
        A(K,J)=A(K,J)/A(K,K)
        DO 4 I=KK,N
        W=A(I,K)*A(K,J)
        A(I,J)=A(I,J)-W
        IF(ABS(A(I,J))-.0001*ABS(W))42,4,4
 42 A(I,J)=0.
  4 CONTINUE
        K=KK
        GO TO 1
 10 K=K-1
 11 RETURN
        END
```

10.62   Consistent and Inconsistent Equations

A set of linear equations

$$a_{11}x_1 + \cdots + a_{1m}x_m = b_1$$

$$a_{21}x_1 + \cdots + a_{2m}x_m = b_2$$

$$\vdots$$

$$a_{n1}x_1 + \cdots + a_{nm}x_m = b_n$$

is said to be consistent if there exists at least one solution and inconsistent

if there is no solution. We are now in a position to give a criterion for determining whether a set of equations is consistent or inconsistent. We will
refer to the matrix

$$\begin{pmatrix} a_{11} & a_{12} & \cdots & a_{1m} \\ a_{21} & a_{22} & \cdots & a_{2m} \\ \vdots & & & \\ a_{n1} & a_{n2} & \cdots & a_{nm} \end{pmatrix}$$

as the *coefficient* matrix, and to the matrix

$$\begin{pmatrix} a_{11} & a_{12} & \cdots & a_{1m} & b_1 \\ a_{21} & a_{22} & \cdots & a_{2m} & b_2 \\ \vdots & & & & \\ a_{n1} & a_{n2} & \cdots & a_{nm} & b_n \end{pmatrix}$$

as the *augmented* matrix. Then the following theorem applies:

Theorem 2.   *A set of linear equations is consistent if and only if the coefficient
matrix and augmented matrix have the same rank.*

Example 1.   Determine if the following equations are consistent:

$$x + 3y = 4$$
$$2x + 6y = 2$$

The coefficient matrix is

$$\begin{pmatrix} 1 & 3 \\ 2 & 6 \end{pmatrix}$$

which has rank 1.
   The augmented matrix is

$$\begin{pmatrix} 1 & 3 & 4 \\ 2 & 6 & 2 \end{pmatrix}$$

which has rank 2.
   Hence the system is inconsistent.

Example 2.   Determine if the following equations are consistent:

$$x + 2y = 3$$
$$2x - y = 2$$
$$3x + y = 5$$

The coefficient matrix is

$$\begin{pmatrix} 1 & 2 \\ 2 & -1 \\ 3 & 1 \end{pmatrix}$$

which has rank 2.

The augmented matrix is

$$\begin{pmatrix} 1 & 2 & 3 \\ 2 & -1 & 2 \\ 3 & 1 & 5 \end{pmatrix}$$

which has rank 2.

Hence the equations are consistent, and there is a solution, despite the fact that there are more equations than unknowns! Upon closer scrutiny, it will be observed that the third equation is merely the sum of the first two.

The last example illustrates an important principle, that consistency or inconsistency cannot be ascertained merely from the numbers of equations and unknowns. A system with more equations than unknowns can be consistent, and a system with more unknowns than equations can be inconsistent. The subroutine for finding rank given in Section 10.62 is the tool needed to investigate consistency in the larger systems.

10.63   Linear Independence of Vectors

Consistent systems of linear equations may have infinitely many solutions. It is possible, however, to investigate these solutions systematically and to characterize them completely. To do so we need first the concept of linear dependence and independence. Consider the set of column vectors

$$\mathbf{u}_1 = \begin{pmatrix} u_{11} \\ u_{21} \\ \vdots \\ u_{n1} \end{pmatrix}, \qquad \mathbf{u}_2 = \begin{pmatrix} u_{12} \\ u_{22} \\ \vdots \\ u_{n2} \end{pmatrix}, \dots, \qquad \mathbf{u}_r = \begin{pmatrix} u_{1r} \\ u_{2r} \\ \vdots \\ u_{nr} \end{pmatrix}$$

If $c_1, c_2, \dots, c_r$ are any constants, the expression

$$c_1 \mathbf{u}_1 + c_2 \mathbf{u}_2 + \cdots + c_r \mathbf{u}_r$$

is called a "linear combination" of the vectors $\mathbf{u}_1, \dots, \mathbf{u}_r$. If there is some set of constants $c_1, \dots, c_r$, not all zero, such that

$$c_1 u_1 + c_2 u_2 + \cdots + c_r u_r = 0$$

then the vectors are said to be "linearly dependent." If, on the other hand, every linear combination of the vectors $u_1, \ldots, u_r$ is nonzero except for the case $c_1 = c_2 = \cdots = c_r = 0$, then the vectors are said to be "linearly independent."

Example 1. Are the vectors

$$\begin{pmatrix} 0 \\ 1 \end{pmatrix} \quad \text{and} \quad \begin{pmatrix} 1 \\ 0 \end{pmatrix}$$

linearly independent?

The sum

$$c_1 \begin{pmatrix} 0 \\ 1 \end{pmatrix} + c_2 \begin{pmatrix} 1 \\ 0 \end{pmatrix} = \begin{pmatrix} c_2 \\ c_1 \end{pmatrix}$$

is zero only if both c_1 and c_2 are zero. Hence they are linearly independent.

Example 2. Are the vectors

$$\begin{pmatrix} 1 \\ -1 \\ 2 \end{pmatrix}, \quad \begin{pmatrix} 2 \\ 1 \\ 1 \end{pmatrix}, \quad \text{and} \quad \begin{pmatrix} 3 \\ 0 \\ 3 \end{pmatrix}$$

linearly independent?

The sum

$$c_1 \begin{pmatrix} 1 \\ -1 \\ 2 \end{pmatrix} + c_2 \begin{pmatrix} 2 \\ 1 \\ 1 \end{pmatrix} + c_3 \begin{pmatrix} 3 \\ 0 \\ 3 \end{pmatrix} = \begin{pmatrix} c_1 + 2c_2 + 3c_3 \\ -c_1 + c_2 \\ 2c_1 + c_2 + 3c_3 \end{pmatrix}$$

is zero if $c_1 = 1$, $c_2 = 1$, $c_3 = -1$. Hence the vectors are not linearly independent.

10.64   Complete Solution of Systems of Linear Equations

The following theorem gives a complete picture of the situation regarding solutions for systems of linear equations.

Theorem 3. *Let $Ax = b$ be a consistent system having m unknowns, and let the rank of A be r. Then:*

(1) If $r = m$, there is a unique solution vector x.

(2) If $r < m$, then there is at least one solution vector x. In addition, $m - r$ linearly independent vectors $u_1, u_2, \ldots, u_{m-r}$ can be found which are solutions to the set of homogeneous equations $Ax = 0$. The vector x plus any linear combination of these is also a solution of the given equation, and there are no other solutions. If $b = 0$, the vector x can be taken as $x = 0$.

Hereafter we will refer to the vector x described in this theorem as the *particular* solution.

A method of obtaining all these solutions in a systematic fashion is illustrated by the example below.

Example 1. Solve the system

$$\begin{pmatrix} 4 & 2 & -1 & 1 \\ 1 & -1 & 2 & -1 \\ 3 & 3 & -3 & 2 \\ 2 & -2 & 4 & -2 \end{pmatrix} \begin{pmatrix} x_1 \\ x_2 \\ x_3 \\ x_4 \end{pmatrix} = \begin{pmatrix} 6 \\ 1 \\ 5 \\ 2 \end{pmatrix}$$

We will proceed as in the elimination method as illustrated in Section 10.2. Dividing the first equation by 4 and using it to eliminate x_1 from the remaining equations,

$$\begin{pmatrix} 1 & .5 & -.25 & .25 \\ 0 & -1.5 & 2.25 & -1.25 \\ 0 & 1.5 & -2.25 & 1.25 \\ 0 & -3 & 4.5 & -2.5 \end{pmatrix} \begin{pmatrix} x_1 \\ x_2 \\ x_3 \\ x_4 \end{pmatrix} = \begin{pmatrix} 1.5 \\ -.5 \\ .5 \\ -1 \end{pmatrix}$$

Rearranging to make the largest element to be in the proper position,

$$\begin{pmatrix} 1 & -.25 & .5 & .25 \\ 0 & 4.5 & -3 & -2.5 \\ 0 & -2.25 & 1.5 & 1.25 \\ 0 & 2.25 & -1.5 & -1.25 \end{pmatrix} \begin{pmatrix} x_1 \\ x_3 \\ x_2 \\ x_4 \end{pmatrix} = \begin{pmatrix} 1.5 \\ -1 \\ .5 \\ -.5 \end{pmatrix}$$

Dividing the second equation by 4.5 and using it to eliminate x_3 from the other equations,

$$\begin{pmatrix} 1 & 0 & 1/3 & 1/9 \\ 0 & 1 & -2/3 & -5/9 \\ 0 & 0 & 0 & 0 \\ 0 & 0 & 0 & 0 \end{pmatrix} \begin{pmatrix} x_1 \\ x_3 \\ x_2 \\ x_4 \end{pmatrix} = \begin{pmatrix} 13/9 \\ -2/9 \\ 0 \\ 0 \end{pmatrix}$$

At this point we see that the rank of A is 2 and that the system now has two equations. If the system had been inconsistent, there would be more than two nonzero elements remaining on the right-hand side of the equation at this point.

Since there are four unknowns and the rank of A is 2, Theorem 2 tells us that the complete solution is made up of a particular solution and any linear combination of two linearly independent solution vectors.

We can find the particular solution by setting $x_2 = x_4 = 0$. Then the system becomes

$$x_1 = 13/9$$
$$x_3 = -2/9$$

Hence the particular solution is

$$\begin{pmatrix} x_1 \\ x_2 \\ x_3 \\ x_4 \end{pmatrix} = \begin{pmatrix} 13/9 \\ 0 \\ -2/9 \\ 0 \end{pmatrix}$$

To find two linearly independent solution vectors, we take the homogeneous equation

$$\begin{pmatrix} 1 & 0 & 1/3 & 1/9 \\ 0 & 1 & -2/3 & -5/9 \\ 0 & 0 & 0 & 0 \\ 0 & 0 & 0 & 0 \end{pmatrix} \begin{pmatrix} x_1 \\ x_3 \\ x_2 \\ x_4 \end{pmatrix} = \begin{pmatrix} 0 \\ 0 \\ 0 \\ 0 \end{pmatrix}$$

and choose arbitrary values for x_2 and x_4.

Taking $x_2 = 1$, $x_4 = 0$, we have

$$x_1 + 1/3 = 0$$
$$x_3 - 2/3 = 0$$

which has the solution

$$x_1 = -1/3, \qquad x_3 = 2/3$$

so one of the linearly independent vectors is

$$u_1 = \begin{pmatrix} -1/3 \\ 1 \\ 2/3 \\ 0 \end{pmatrix}$$

Taking $x_2 = 0$, $x_4 = 1$, we have

$$x_1 + 1/9 = 0$$
$$x_3 - 5/9 = 0$$

which has the solution

$$x_1 = -1/9, \qquad x_3 = 5/9$$

and so the other solution is

$$u_2 = \begin{pmatrix} -1/9 \\ 0 \\ 5/9 \\ 1 \end{pmatrix}$$

and the general solution is

$$x = \begin{pmatrix} 13/9 \\ 0 \\ -2/9 \\ 0 \end{pmatrix} + c_1 \begin{pmatrix} -1/3 \\ 1 \\ 2/3 \\ 0 \end{pmatrix} + c_2 \begin{pmatrix} -1/9 \\ 0 \\ 5/9 \\ 1 \end{pmatrix}$$

where c_1 and c_2 are arbitrary constants.

For convenience in organizing a computer solution, we note that these vectors (apart from a constant multiple of -1 in some cases) can be obtained from the last set of equations by the following somewhat artificial steps:

(1)   Add -1's down the last two columns of the diagonal of the coefficient matrix so that it becomes

$$\begin{pmatrix} 1 & 0 & 1/3 & 1/9 \\ 0 & 1 & -2/3 & -5/9 \\ 0 & 0 & -1 & 0 \\ 0 & 0 & 0 & -1 \end{pmatrix}$$

(2)   Rearrange these last two columns and the column of constants as if they were ordered just as the x's are:

$$\begin{pmatrix} x_1 \\ x_3 \\ x_2 \\ x_4 \end{pmatrix}$$

and needed to be correctly ordered. They become

$$\begin{pmatrix} 1/3 & 1/9 & 13/9 \\ -1 & 0 & 0 \\ -2/3 & -5/9 & -2/9 \\ 0 & -1 & 0 \end{pmatrix}$$

The column of constants has become the particular solution and the other two columns two linearly independent vectors that can be used to form the complete solution.

The method just demonstrated is a general one, and can be used for computer solution of larger systems. It requires only a few modifications and extensions of the elimination method given in Section 10.2.

The FORTRAN subroutine given below solves a system of N equations in M unknowns, where N and M are both 20 or less. Inputs are AA, the coefficient matrix; BB, the constant vector; and NI and M, the dimensions of the system. Outputs are: X, a particular solution vector, K, the number of linearly independent solution vectors for the homogeneous system, and U, a set of linearly independent solution vectors.

```
      SUBROUTINE  LINEQ(AA,NI,M,BB,X,K,U)
      DIMENSION  AA(20,20),BB(20),A(20,21),X(20),ID(20),U(20,20)
      N=NI
      MM=M+1
      DO  100  I=1,N
      A(I,MM)=BB(I)
      DO  100  J=1,M
  100 A(I,J)=AA(I,J)
      K=1
      IF(N-M)200,1,1
  200 NP=N+1
      N=M
      DO  300  I=NP,M
      DO  300  J=1,MM
  300 A(I,J)=0.
      K=1
    1 CALL  EXCH(A,M,MM,K,ID)
      IF(A(K,K))2,5,2
    2 KK=K+1
      DO  3  J=KK,MM
      A(K,J)=A(K,J)/A(K,K)
      DO  3  I=1,N
      IF(K-I)31,3,31
   31 W=A(I,K)*A(K,J)
      A(I,J)=A(I,J)-W
      IF(ABS(A(I,J))- .0001*ABS(W))32,3,3
```

```
   32 A(I,J)=0.
    3 CONTINUE
      K=KK
      IF(K-M)1,2,7
    5 DO 6 J=K,M
      A(J,J)=-1.
      DO 7 I=K,N
      IF(A(I,MM))999,7,999
    7 CONTINUE
      DO 10 I=1,M
      DO 10 J=1,M
      IF(ID(J)-I)10,8,10
    8 X(I)=A(J,MM)
      IF(K-MM)9,10,10
    9 KM=K-1
      DO 10 IP=K,M
      U(I,IP-KM)=A(J,IP)
   10 CONTINUE
      K=M-K
      RETURN
  999 PRINT 1000
      RETURN
 1000 FORMAT(27H EQUATIONS ARE INCONSISTENT)
      END
```

10.7   EIGENVALUES AND EIGENVECTORS

A surprisingly large number of problems in physics and engineering can be reduced to the following mathematical problem: Given a square nth-order matrix A, find a nonzero vector x and a constant λ such that

$$Ax = \lambda x$$

That is, find a vector x such that Ax is simply a multiple of the vector x itself. We can rewrite this equation as

$$Ax - \lambda x = 0$$

or

$$(A - \lambda I)x = 0 \tag{10-10}$$

In this form, the equation appears as a set of homogeneous, linear equations for $x_1, x_2, \ldots, x_n$. The matrix of coefficients is $(A - \lambda I)$, and the augmented matrix is the same with a column of zeros added, so by Theorem 2 of Section 10.62 the equations are consistent. By Theorem 3 of Section 10.63 there is a unique solution if the rank of the coefficient matrix is n. We already know that solution; it is $x_1 = x_2 = \cdots = x_n = 0$. Hence there is a nonzero vector x only if the rank of $(A - \lambda I)$ is less than n. This will be true if

$$\det(A - \lambda I) = 0 \tag{10-11}$$

If this determinant is zero, then by Theorem 3 of Section 10.64 there are one or more linearly independent solution vectors that can be used to describe the complete solution. Thus we are interested in the values of λ for which

$$\begin{vmatrix} a_{11} - \lambda & a_{12} & \cdots & a_{1n} \\ a_{21} & a_{22} - \lambda & & a_{2n} \\ \vdots & & & \vdots \\ a_{n1} & a_{n2} & \cdots & a_{nn} - \lambda \end{vmatrix} = 0$$

In Section 10.54 it was stated that the value of a determinant could be obtained by forming all possible terms containing as factors exactly one element from each row and each column. If we were to attempt to do this with the determinant above, we would find that the various terms would contain different powers of λ. If we were to collect the terms having like powers, we would obtain an expression of the form

$$(-1)^n[\lambda^n - p_1\lambda^{n-1} - p_2\lambda^{n-2} - \cdots - p_n] \tag{10-12}$$

where the constants $p_1, p_2, \ldots, p_n$ are numbers resulting from some very complicated manipulations of the numbers a_{ij} in the determinant.

From Chapter 9, there are exactly n values of λ (not necessarily distinct) which will make (10-12) be equal to zero. These values are called the "eigenvalues" (or "characteristic roots," or "latent roots," or "proper values") of the matrix A. For any eigenvalue λ_i, the vector x which satisfies equation (10-10) is called the "eigenvector" (or "characteristic vector," or "latent vector," or "proper vector") corresponding to λ_i. The polynomial (10-12) is called the "characteristic polynomial" of the matrix A, and the equation

$$\lambda^n - p_1\lambda^{n-1} - p_2\lambda^{n-2} - \cdots - p_n = 0 \tag{10-13}$$

is called the "characteristic equation."

Example 1. Find the eigenvalues and eigenvectors for the matrix

$$\begin{pmatrix} 1 & 3 \\ 2 & 2 \end{pmatrix}$$

To find the eigenvalues, we set

$$\begin{vmatrix} 1 - \lambda & 3 \\ 2 & 2 - \lambda \end{vmatrix} = 0$$

Expanding, we obtain the characteristic equation

$$(1 - \lambda)(2 - \lambda) - 6 = \lambda^2 - 3\lambda - 4 = 0$$

This factors into

$$(\lambda - 4)(\lambda + 1) = 0$$

so the eigenvalues are

$$\lambda_1 = 4, \qquad \lambda_2 = -1$$

To find the eigenvector corresponding to λ_1, we set

$$\begin{pmatrix} 1 - \lambda_1 & 3 \\ 2 & 2 - \lambda_1 \end{pmatrix} \begin{pmatrix} x_1 \\ x_2 \end{pmatrix} = 0$$

or

$$\begin{pmatrix} -3 & 3 \\ 2 & -2 \end{pmatrix} \begin{pmatrix} x_1 \\ x_2 \end{pmatrix} = 0$$

Since there are two unknowns and the coefficient matrix has rank 1, Theorem 3 of Section 10.63 tells us that these equations have one linearly independent vector solution U_1, and that all other solutions are multiples of this one. We see by inspection that the vector

$$\begin{pmatrix} 1 \\ 1 \end{pmatrix}$$

is a solution, and hence is an eigenvector corresponding to λ_1. All other solutions are of the form

$$c_1 \begin{pmatrix} 1 \\ 1 \end{pmatrix}$$

where c_1 is an arbitrary constant. Hence the eigenvector is really determined only up to an arbitrary constant multiple.

To find the eigenvector corresponding to λ_2, we set

$$\begin{pmatrix} 1 - \lambda_2 & 3 \\ 2 & 2 - \lambda_2 \end{pmatrix}\begin{pmatrix} x_1 \\ x_2 \end{pmatrix} = 0$$

or

$$\begin{pmatrix} 2 & 3 \\ 2 & 3 \end{pmatrix}\begin{pmatrix} x_1 \\ x_2 \end{pmatrix} = 0$$

Again there is one linearly independent solution vector. We see by inspection that

$$\begin{pmatrix} 3 \\ -2 \end{pmatrix}$$

is a solution. All solutions are of the form

$$c_2\begin{pmatrix} 3 \\ -2 \end{pmatrix}$$

where c_2 is an arbitrary constant.

Hence the eigenvalues are

$$4, \quad -1$$

and the corresponding eigenvectors are

$$\begin{pmatrix} 1 \\ 1 \end{pmatrix} \quad \text{and} \quad \begin{pmatrix} 3 \\ -2 \end{pmatrix}$$

(We ordinarily ignore the arbitrary constant multiple when writing an eigenvector.)

Example 2. Find the eigenvalues and eigenvectors for the matrix

$$\begin{pmatrix} 3 & 2 & 4 \\ 1 & 4 & 4 \\ -1 & -2 & -2 \end{pmatrix}$$

To determine the eigenvalues, we set

$$\begin{vmatrix} 3 - \lambda & 2 & 4 \\ 1 & 4 - \lambda & 4 \\ -1 & -2 & -2 - \lambda \end{vmatrix} = 0$$

Expanding, we obtain the characteristic equation

$$-\lambda^3 + 5\lambda^2 - 8\lambda + 4 = 0$$

which has the roots

$$\lambda_1 = 1, \qquad \lambda_2 = \lambda_3 = 2$$

To find the eigenvector corresponding to λ_1, we set

$$\begin{pmatrix} 3 - \lambda_1 & 2 & 4 \\ 1 & 4 - \lambda_1 & 4 \\ -1 & -2 & -2 - \lambda_1 \end{pmatrix} \begin{pmatrix} x_1 \\ x_2 \\ x_3 \end{pmatrix} = 0$$

or

$$\begin{pmatrix} 2 & 2 & 4 \\ 1 & 3 & 4 \\ -1 & -2 & -3 \end{pmatrix} \begin{pmatrix} x_1 \\ x_2 \\ x_3 \end{pmatrix} = 0$$

The coefficient matrix has rank 2, so this system has one linearly independent vector solution. If we solve by the method of Section 10.64, we find that the eigenvector is

$$\begin{pmatrix} 1 \\ 1 \\ -1 \end{pmatrix}$$

To find the eigenvector corresponding to λ_2, we set

$$\begin{pmatrix} 3 - \lambda_2 & 2 & 4 \\ 1 & 4 - \lambda_2 & 4 \\ -1 & -2 & -2 - \lambda_2 \end{pmatrix} \begin{pmatrix} x_1 \\ x_2 \\ x_3 \end{pmatrix} = 0$$

or

$$\begin{pmatrix} 1 & 2 & 4 \\ 1 & 2 & 4 \\ -1 & -2 & -4 \end{pmatrix} \begin{pmatrix} x_1 \\ x_2 \\ x_3 \end{pmatrix} = 0$$

The coefficient matrix has rank 1, so this system has two linearly independent vector solutions. Solving by the method of Section 10.64, we find two linearly independent eigenvectors,

$$\begin{pmatrix} -4 \\ 0 \\ 1 \end{pmatrix} \quad \text{and} \quad \begin{pmatrix} -2 \\ 1 \\ 0 \end{pmatrix}$$

The root λ_3, being the same as λ_2, has the same eigenvectors. Hence we have a single root, 1, with its eigenvector

$$\begin{pmatrix} 1 \\ 1 \\ -1 \end{pmatrix}$$

and a double root, 2, with two eigenvectors:

$$\begin{pmatrix} -4 \\ 0 \\ 1 \end{pmatrix} \quad \text{and} \quad \begin{pmatrix} -2 \\ 1 \\ 0 \end{pmatrix}$$

Example 3. Find the eigenvalues and eigenvectors for the matrix

$$\begin{pmatrix} 0 & 1 & 0 \\ 0 & 0 & 1 \\ -8 & -12 & -6 \end{pmatrix}$$

We set

$$\begin{vmatrix} -\lambda & 1 & 0 \\ 0 & -\lambda & 1 \\ -8 & -12 & -6-\lambda \end{vmatrix} = 0$$

and obtain the characteristic equation

$$-\lambda^3 - 6\lambda^2 - 12\lambda - 8 = 0$$

which has the roots

$$\lambda_1 = -2, \qquad \lambda_2 = -2, \qquad \lambda_3 = -2$$

To find the eigenvectors, we set

$$\begin{pmatrix} 2 & 1 & 0 \\ 0 & 2 & 1 \\ -8 & -12 & -4 \end{pmatrix} \begin{pmatrix} x_1 \\ x_2 \\ x_3 \end{pmatrix} = 0$$

Since the coefficient matrix has rank 2, there is only one linearly independent eigenvector. It turns out to be

$$\begin{pmatrix} 1 \\ -2 \\ 4 \end{pmatrix}$$

Since all roots are the same, we can obtain no more eigenvectors. Hence in this case we have a triple eigenvalue, -2, and only one eigenvector (which might be considered an eigenvector of multiplicity 3):

$$\begin{pmatrix} 1 \\ -2 \\ 4 \end{pmatrix}$$

The above examples have illustrated all the possibilities concerning real eigenvalues and their corresponding eigenvectors. These possibilities can be summarized in the following theorem.

Theorem 4. *An nth-order square matrix has n eigenvalues. If these are discrete, there is one eigenvector corresponding to each eigenvalue. If an eigenvalue is of multiplicity r, it may have from one to r linearly independent eigenvectors associated with it.*

10.71   Program for Largest Eigenvalue and Eigenvector

Suppose that the matrix A has one eigenvalue λ, which is larger than all others in absolute value, and y is any nonzero column vector conformable with A. Let the vectors y_1, y_2, etc., be defined by

$$
\begin{aligned}
y_1 &= Ay_1 \\
y_2 &= Ay_1 \\
&\;\;\vdots \\
y_n &= Ay_{n-1}
\end{aligned}
\tag{10.14}
$$

The vectors y_i defined in this manner can lead to the value of λ_1 and to x_1, the eigenvalue corresponding to λ_1. The method of obtaining the eigenvalue and eigenvector will be illustrated without proof.* In order to provide the illustration, let us first write a remote-terminal program to perform the computations indicated by expression (10-14). A suitable program is

* See, for example, J. G. Herriot, *Methods of Mathematical Analysis and Computation*, John Wiley & Sons, Inc., New York, 1963.

```
 1     DIMENSION  A(10,10),Y(10),YN(10)
 2     PRINT, "INPUT N, TEN OR LESS"
 3     INPUT, N
 4     PRINT, "INPUT  A(1,1)A(1,2),,,A(N,N)"
 5     INPUT, ((A(I,J),J=1,N),I=1,N)
 6     PRINT, "INPUT Y(1),Y(2),,,Y(N)"
 7     INPUT, (Y(I),I=1,N)
 8  1  DO 2 I=1,N
 9     YN(I)=0·
10     DO 2 J=1,N
11  2  YN(I)=YN(I)+A(I,J)*Y(J)
12     PRINT, (YN(I),I=1,N)
13     INPUT, Q
14     DO 3 I=1,N
15  3  Y(I)=YN(I)
16     GO TO 1
17     END
```

In this program, the statements at lines 2 through 7 allow the user to input an initial matrix A and vector y of order up to 10. The statements at lines 8 through 12 compute and print the vector $y_1 = Ay$. At line 13, the user is allowed to specify whether another step of the process is required. If the typed entry is the letter S, the program will terminate. If the entry is any number whatsoever, the program will cause y_1 to replace y, and will repeat lines 8 through 12, thereby computing and printing $y_2 = Ay_1$, and so on.

Example 1. Write all user inputs and machine responses for running the above program with the matrix

$$\begin{pmatrix} 1 & 3 \\ 2 & 2 \end{pmatrix} \quad \text{and the vector} \quad \begin{pmatrix} 1 \\ 0 \end{pmatrix}$$

continuing until vectors through y_8 have been generated.

The inputs and responses are

```
RUN
INPUT N, TEN OR LESS
? 2
    INPUT A(1,1),A(1,2),,,A(N,N)
? 1,3,2,2
    INPUT Y(1),Y(2),,,Y(N)
? 1,0
        1.000000        2.000000
? 0
        7.000000        6.000000
```

? 0

25.00000   26.00000

? 0

103.0000   102.0000

? 0

409.0000   410.0000

? 0

1639.000   1638.000

? 0

6553.000   6554.000

? 0

.2621500E5   .2621400E5

? S
STOP

The above program was Example 1, Section 10.6, which had a largest eigen-value of 4 and corresponding eigenvector of $\binom{1}{1}$. Looking at the vectors printed out above, we see that the vectors generated were

$$\binom{1}{2}, \binom{7}{6}, \binom{25}{26}, \binom{103}{102}, \binom{409}{410}, \binom{1639}{1638}, \binom{6553}{6554}, \binom{26215}{26214}$$

and that after the first few steps the components are always very nearly equal; that is, the vectors themselves are very nearly simple multiples of the vector $\binom{1}{1}$. Since eigenvectors are determined only up to a constant multiple, we can say that the vector y_n of (10-14) is actually approaching the eigenvector x. Now since

$$Ax = \lambda x$$

then if y_n is x, then y_{n+1} will be λx. We note without surprise, then, that in each of the vectors computed in the above example, the components are very nearly four times those of the preceding vector.

It appears, then, that the above program can be used almost directly to find the largest eigenvalue and corresponding eigenvector. Some improve-ment can be made by replacing the statement at line 15 by

15   3 Y(I)=YN(I)/YN(1)

This will serve to keep the components from growing at each stage, and further will cause the first component to approach the actual value of the eigenvalue. If this had been done for Example 1, the printouts would have been

    1.000000   2.000000
? 0
    7.000000   6.000000
? 0
    3.571429   3.714286
? 0
    4.120000   4.080000
? 0
    3.970874   3.980583
? 0
    4.007335   4.004890
? 0
    3.998109   3.998596
? 0
    4.000458   4.000305
? S
STOP

From these results, the eigenvalue 4 and the eigenvector $\begin{pmatrix} 1 \\ 1 \end{pmatrix}$ are apparent.

10.72   Complex Eigenvalues

From Chapter 9 it is known that the characteristic equation may have complex roots, occuring in conjugate pairs. In this case, the eigenvectors are also complex, and the equation

$$(A - \lambda I)x = 0$$

instead of being n equations in n unknowns, is really $2n$ equations in $2n$ unknowns, for both the real and imaginary part of x must satisfy the equation. Let

$$\lambda = \alpha + \beta i$$

be a complex eigenvalue, and let the eigenvector be

$$x = \begin{pmatrix} x_1 + y_1 i \\ x_2 + y_2 i \\ \vdots \\ x_n + y_n i \end{pmatrix}$$

If we substitute these in the above equation and separate real and imaginary parts, the result can be written in the form

$$\begin{pmatrix} a_{11} - \alpha & a_{12} & \cdots & a_{1n} & \beta & 0 & \cdots & 0 \\ a_{21} & a_{22} - \alpha & \cdots & a_{2n} & 0 & \beta & \cdots & 0 \\ \vdots & & & & & & & \\ a_{n1} & a_{n2} & \cdots & a_{nn} - \alpha & 0 & 0 & \cdots & \beta \\ -\beta & 0 & \cdots & 0 & a_{11} - \alpha & a_{12} & \cdots & a_{1n} \\ 0 & -\beta & \cdots & 0 & a_{21} & a_{22} - \alpha & \cdots & a_{2n} \\ \vdots & & & & & & & \\ 0 & 0 & \cdots & -\beta & a_{n1} & a_{n2} & \cdots & a_{nn} - \alpha \end{pmatrix} \begin{pmatrix} x_1 \\ x_2 \\ \vdots \\ x_n \\ y_1 \\ y_2 \\ \vdots \\ y_n \end{pmatrix} = 0$$

These are $2n$ real equations in $2n$ real unknowns which can be solved for the x_i's and y_i's. The eigenvector corresponding to the complex conjugate of λ is the complex conjugate of the eigenvector for λ, so the process needs to be done only once for each pair of complex roots.

Example 1.   Find the eigenvalues and eigenvectors of

$$\begin{pmatrix} -1 & -5 \\ 1 & 3 \end{pmatrix}$$

We set

$$\begin{vmatrix} -1 - \lambda & -5 \\ 1 & 3 - \lambda \end{vmatrix} = 0$$

and obtain the characteristic equation:

$$\lambda^2 - 2\lambda + 2 = 0$$

which has roots:

$$\lambda_1 = 1 + i, \qquad \lambda_2 = 1 - i$$

To find the eigenvector corresponding to λ_1, using the method described above, we write

$$\begin{pmatrix} -2 & -5 & 1 & 0 \\ 1 & 2 & 0 & 1 \\ -1 & 0 & -2 & -5 \\ 0 & -1 & 1 & 2 \end{pmatrix} \begin{pmatrix} x_1 \\ x_2 \\ y_1 \\ y_2 \end{pmatrix} = 0$$

Applying the method of Section 13.34, this reduces to

$$\begin{pmatrix} 1 & 0 & -.2 & .4 \\ 0 & 1 & .4 & .2 \\ 0 & 0 & 0 & 0 \\ 0 & 0 & 0 & 0 \end{pmatrix} \begin{pmatrix} x_2 \\ y_2 \\ y_1 \\ x_1 \end{pmatrix} = 0$$

This has two linearly independent solution vectors

$$\begin{pmatrix} x_1 \\ x_2 \\ y_1 \\ y_2 \end{pmatrix} = \begin{pmatrix} 0 \\ .2 \\ 1 \\ -.4 \end{pmatrix} \quad \text{and} \quad \begin{pmatrix} 1 \\ -.4 \\ 0 \\ -.2 \end{pmatrix}$$

These vectors of themselves are not of interest to us, except to use the numbers in them to construct the complex of vectors

$$\begin{pmatrix} x_1 + y_1 i \\ x_2 + y_2 i \end{pmatrix} = \begin{pmatrix} i \\ .2 - .4i \end{pmatrix} \quad \text{and} \quad \begin{pmatrix} x_1 + y_1 i \\ x_2 + y_2 i \end{pmatrix} = \begin{pmatrix} 1 \\ -.4 - .2i \end{pmatrix}$$

These two vectors are not linearly independent at all, for if we multiply the second by i, we obtain the first. Hence we have really obtained only one eigenvector corresponding to the eigenvalue $1 + i$, and that is

$$\begin{pmatrix} i \\ .2 - .4i \end{pmatrix}$$

The eigenvector corresponding to $1 - i$ is the conjugate of this:

$$\begin{pmatrix} -i \\ .2 + .4i \end{pmatrix}$$

10.73   Determination of All Eigenvalues and Eigenvectors

The method described in Section 10.71 will provide the largest eigenvalue and corresponding eigenvector. Frequently it is necessary to find all eigenvalues and eigenvectors. From the discussions of Section 10.7, it is clear that this can be done by accomplishing the following three steps:

(*1*)   Find the characteristic polynomial.

(*2*)   Solve the characteristic equation for its roots.

(*3*)   Solve sets of linear equations for the eigenvectors.

Chapter 9 gave methods for solving polynomial equations, so we already have computer methods for step (*2*). Section 10.64 gave a computer method for solving systems of linear equations which is satisfactory for step (*3*). Hence the only thing really required is a computer method for generating the characteristic polynomial. In the examples above, we have used very small matrices and found the characteristic polynomial by brute-force expansion of the determinant, but this process is inefficient for large-order matrices. A more efficient method is the Leverrier-Faddeev method, which proceeds as follows:

$$
\begin{aligned}
&\text{Let } A_1 = A && \text{and } p_1 = \text{tr } A \\
&\text{Let } A_2 = A(A_1 - p_1 I), && \text{and } p_2 = (1/2)\, \text{tr } A_2 \\
&\text{Let } A_3 = A(A_2 - p_2 I), && \text{and } p_3 = (1/3)\, \text{tr } A_3 \\
&\quad \vdots \\
&A_n = A(A_{n-1} - p_{n-1} I) && \text{and } p_n = (1/n)\, \text{tr } A_n
\end{aligned}
$$

The numbers $p_1, p_2, \ldots, p_n$ are the required coefficients in the characteristic equation

$$\lambda^n - p_1 \lambda^{n-1} - p_2 \lambda^{n-2} - \cdots - p_n = 0$$

In addition, as a bonus side product of this process, it can be shown that the inverse of A is given by

$$A^{-1} = (1/p_n)(A_{n-1} - p_{n-1} I) \qquad \text{(10-15)}$$

and also, as a sometimes helpful check,

$$A_n - p_n I = 0 \qquad \text{(10-16)}$$

Example 1.   Find the characteristic equation of

$$\begin{pmatrix} 1 & 3 & 2 \\ -2 & 1 & 1 \\ 1 & -2 & -1 \end{pmatrix}$$

Following the above procedure, we have

$$A_1 = \begin{pmatrix} 1 & 3 & 2 \\ -2 & 1 & 1 \\ 1 & -2 & -1 \end{pmatrix}, \qquad p_1 = 1 + 1 - 1 = 1$$

$$A_2 = \begin{pmatrix} 1 & 3 & 2 \\ -2 & 1 & 1 \\ 1 & -2 & -1 \end{pmatrix} \begin{pmatrix} 0 & 3 & 2 \\ -2 & 0 & 1 \\ 1 & -2 & -2 \end{pmatrix} = \begin{pmatrix} -4 & -1 & 1 \\ -1 & -8 & -5 \\ 3 & 5 & 2 \end{pmatrix}$$

$$p_2 = (1/2)(-4 - 8 + 2) = -5$$

$$A_3 = \begin{pmatrix} 1 & 3 & 2 \\ -2 & 1 & 1 \\ 1 & -2 & -1 \end{pmatrix} \begin{pmatrix} 1 & -1 & 1 \\ -1 & -3 & -5 \\ 3 & 5 & 7 \end{pmatrix} = \begin{pmatrix} 4 & 0 & 0 \\ 0 & 4 & 0 \\ 0 & 0 & 4 \end{pmatrix}$$

$$p_3 = (1/3)(4 + 4 + 4) = 4$$

As a check, we see that

$$A_3 - p_3 I = \begin{pmatrix} 4 & 0 & 0 \\ 0 & 4 & 0 \\ 0 & 0 & 4 \end{pmatrix} - \begin{pmatrix} 4 & 0 & 0 \\ 0 & 4 & 0 \\ 0 & 0 & 4 \end{pmatrix} = 0$$

Hence the characteristic equation is

$$\lambda^3 - \lambda^2 + 5\lambda - 4 = 0$$

The flow chart, Figure 10-10, describes this process for an nth-order matrix. According to equation (10-16), the matrix A_n is simply the identity matrix multiplied by p_n, so only the first element of A_n need be calculated to give p_n. The value of p_n is, in fact, the determinant of A, so that if p_n is zero, the matrix is singular. If p_n is not zero, the inverse of A is easily calculated from equation (10-15), and the flow chart includes this calculation. The elements of A^{-1} are the last values obtained for f_{ij}.

The FORTRAN subroutine given below will generate coefficients in accordance with the flow chart, for matrices up to order 20. However, in order that the subscripts will match the notation in the subroutines of Chapter 9, the characteristic equation is written as

$$Q(1)\lambda^N + Q(2)\lambda^{N-1} + \cdots + Q(N+1) = 0$$

The relationship between the p_k of the flow chart and the Q(K) of the subroutine is given by

$$Q(1) = 1, \quad Q(K+1) = -p_k \qquad \text{for } k = 1, \ldots, n$$

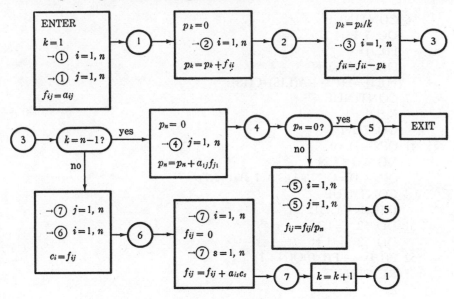

Figure 10-10: Generation of characteristic polynomial

As in the flow chart, the subscripted variable F(I,J) is the inverse matrix unless $Q(N+1)$ happens to be zero.

```
      SUBROUTINE CHAREQ(A,N,Q,F)
      DIMENSION  A(20,20),F(20,20),Q(21),C(20)
      Q(1)=1.
      K=1
      DO 11 I=1,N
      DO 11 J=1,N
   11 F(I,J)=A(I,J)
    1 CONTINUE
      Q(K+1)=0.
      DO 2 I=1,N
      Q(K+1)=Q(K+1)+F(I,I)
    2 CONTINUE
      FK=K
      Q(K+1)=-Q(K+1)/FK
      DO 3 I=1,N
      F(I,I)=F(I,I)+Q(K+1)
    3 CONTINUE
      IF(K-N+1)71,41,71
   71 DO 7 J=1,N
      DO 6 I=1,N
      C(I)=F(I,J)
```

```
  6 CONTINUE
    DO 7 I=1,N
    F(I,J)=0.
    DO 7 IS=1,N
    F(I,J)=F(I,J)+A(I,IS)*C(IS)
  7 CONTINUE
    K=K+1
    GO TO 1
 41 Q(N+1)=0.
    DO 4 J=1,N
    Q(N+1)=Q(N+1)-A(1,J)*F(J,1)
  4 CONTINUE
    IF(Q(N+1))51,5,51
 51 DO 52 I=1,N
    DO 52 J=1,N
 52 F(I,J)=-F(I,J)/Q(N+1)
  5 RETURN
    END
```

With the above subroutine and subroutines of Chapter 9 and Section 10.6, eigenvalues and eigenvectors can be found in a systematic way. There are also methods which, under some conditions, can be used to find all eigenvalues directly from the matrix itself, without generating the characteristic equation first. These methods are available in the literature and will not be reported here.

EXERCISE 28

1. Determine the rank of the following matrices.

   a. $\begin{pmatrix} 1 & 1 \\ 1 & 1 \end{pmatrix}$

   b. $\begin{pmatrix} 1 & 2 & 3 & 4 & 5 \\ 1 & 2 & 3 & 4 & 6 \end{pmatrix}$

   c. $\begin{pmatrix} 1 & 2 & 3 \\ -2 & 1 & -2 \\ -1 & 3 & 1 \end{pmatrix}$

   d. $\begin{pmatrix} 2 & -1 & 3 & 4 \\ 1 & -2 & -2 & -1 \\ 0 & 3 & 7 & 6 \end{pmatrix}$

2. Determine whether the following systems are consistent or inconsistent.

   a. $\begin{aligned} x + 2y + z &= 4 \\ -2x - 4y - 2z &= 3 \end{aligned}$

   b. $\begin{aligned} x + 2y &= 6 \\ x + 3y &= 8 \end{aligned}$

   c. $\begin{aligned} x + 3y &= 7 \\ 2x - y &= 4 \\ 4x + 5y &= 18 \end{aligned}$

   d. $\begin{aligned} x + 2y &= 8 \\ 3x - y &= 2 \\ 2x + y &= 6 \end{aligned}$

3. Solve completely the following systems of equations.

   a. $\begin{aligned} x + 2y &= 0 \\ -2x - 4y &= 0 \end{aligned}$

   b. $\begin{aligned} x + y - z &= 2 \\ x - y + z &= 3 \end{aligned}$

c. $x + 3y - z = 4$ d. $x + 2y + z = 1$
    $2x - y + 2z = 3$ $2x - y + z = 2$
    $3x + 2y + z = 7$ $3x - y + 4z = 3$

4. Find all eigenvalues and eigenvectors for the following matrices.

a. $\begin{pmatrix} 0 & 1 \\ 1 & 0 \end{pmatrix}$ b. $\begin{pmatrix} 1 & 3 \\ -2 & -4 \end{pmatrix}$ c. $\begin{pmatrix} 1 & 2 \\ -2 & -3 \end{pmatrix}$

d. $\begin{pmatrix} 1 & 0 & 1 \\ 0 & 1 & 1 \\ 0 & 0 & 2 \end{pmatrix}$ e. $\begin{pmatrix} 2 & 1 & 2 \\ -1 & 0 & 0 \\ -1 & -1 & -1 \end{pmatrix}$

5. Find the number of multiplications required to find the rank of a 10 by 15 matrix using SUBROUTINE MARANK of Section 10.61, if the rank turns out to be 8.

6. Write a program that will input a system of linear equations up to 20 by 20, call subroutine LINEQ of Section 10.64 to obtain all solutions, and print the result.

7. Write a program that will input a square matrix up to 19 by 19, call subroutine CHAREQ of Section 10.73 to obtain the characteristic equation, call the appropriate subroutine from Chapter 9 to find the largest real root, call subroutine LINEQ of Section 10.64 to find the corresponding eigenvector, and print the result.

Curve Fitting

11.1 INTRODUCTION

One of the fields of elementary mathematics where the digital computer has most to contribute is that of curve fitting. This process is one which is extremely laborious for hand computation, so much so that only quite simple curve-fitting operations were ordinarily done before the day of the digital computer. With the digital computer, however, standard types of programs are available that allow very rapid determination of the best-fitting curve from a wide family of possible choices. No attempt will be made here to catalog the ever-growing variety of standard subroutines available. Instead, we shall concern ourselves with the fundamentals of curve fitting and with some of the basic techniques employed in machine programs for curve fitting, in order to lay a proper foundation upon which the reader can build in areas of his own particular interest.

The basic problem in curve fitting can be described as follows. We have a set of measured values of a quantity x, and an associated set of measured values of another quantity y. We wish to find some functional relation

$$y = f(x)$$

between x and y which is satisfied by our sets of measured values (x_1, y_1), (x_2, y_2), etc., and which will allow us to infer reasonable values of y for other values of x, where we have no measurement. If we were to stop here, our description of the problem would be quite incomplete. What do we mean by "reasonable" values of y? In Figure 11-1 are shown five points (x_1, y_1), ..., (x_5, y_5), and three different curves passing through all five points. The dashed curve is made up of straight-line segments. The dotted curve is a smooth arc. The solid curve is a wildly oscillating function with a singularity. All pass through all the given points. How are we to decide which of these is the best fit? If we have no further information upon which to act, we are intuitively drawn to the smooth, dotted curve as the most reasonable fit. This is the type of curve fitting generally used for interpolation, a process discussed in Chapter 12. Quite frequently, however, we know that for physical reasons the functional relation $y = f(x)$ must have some particular function form, for example,

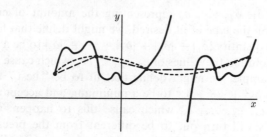

Figure 11-1

$$y = c_1 e^{c_2 x}$$

or perhaps

$$y = c_1 \tan x + c_2 \cos c_3 x$$

where the constants c_1, c_2, c_3 are not known. In such cases, in attempting to find the best fit we are restricted choosing the constants to give the most reasonable fit. We cannot even be sure in this case that our "best" function will pass through all the given points. Ordinarily it will be impossible to choose the constants so that this will happen. The usual practical problem in curve fitting, then, is of the following type. We have some set of values $(x_1, y_1), (x_2, y_2), \ldots, (x_n, y_n)$, which we wish to fit with a curve of the form

$$y = f(x, c_1, c_2, c_3, \ldots, c_m) \tag{11-1}$$

where the function f is completely specified except for the particular values of the constants $c_1, c_2, \ldots, c_m$. We wish to choose these constants so that our measured values will most nearly satisfy the above expression.

Again, we have trouble being sufficiently specific. What do we mean by "most nearly satisfy?" Once we have selected the constants $c_1, \ldots, c_m$, if we substitute in our measured values we will find that the equation

$$f(x_i, c_1, c_2, \ldots, c_m) - y_i = 0$$

is not satisfied exactly, but instead, there is some error in each case:

$$f(x_1, c_1, c_2, \ldots, c_m) - y_1 = \delta_1$$

$$f(x_2, c_1, c_2, \ldots, c_m) - y_2 = \delta_2 \tag{11-2}$$

$$\vdots$$

$$f(x_n, c_1, c_2, \ldots, c_m) - y_n = \delta_n$$

the quantities $\delta_1, \delta_2, \ldots, \delta_n$ representing the amount of error in the fit. Depending upon the type of fit desired, we might define that fit to be "best" that causes the quantity $|\delta_1| + |\delta_2| + |\delta_3| + \cdots + |\delta_n|$ to be a minimum, and accept as our solution the values of $c_1, c_2, \ldots, c_m$ which cause this to happen. As another example, we might define that fit to be "best" that causes the quantity $\delta_1{}^2 + \delta_2{}^2 + \cdots + \delta_n{}^2$ to be a minimum, and accept as our solution the values of $c_1, c_2, \ldots, c_m$ which cause this to happen. Ordinarily, the latter set of c's will turn out to be different from the preceding set. This latter criterion of goodness of fit, the so-called "least-squares" criterion, is by far the most widely used, and is the one we will discuss in the present chapter.

11.2   LINEAR FORMULAS

An extremely wide variety of curve-fitting problems can be grouped into one category, and can be solved with what is basically one machine program. These consist of all curve-fitting problems where the undetermined constants appear only in the linear form; that is, the functional relation between x and y is of the form

$$y = c_1 f_1(x) + c_2 f_2(x) + \cdots + c_m f_m(x) \tag{11-3}$$

where the functions $f_1(x), f_2(x), f_3(x)$, etc., are completely known functions of x. Specific examples of this type will be mentioned shortly. In this case if we have n sets of measured values, $(x_1, y_1), (x_2, y_2), \ldots, (x_n, y_n)$, and we substitute these into the given equation, we *would like* to satisfy the equations

$$c_1 f_1(x_1) + c_2 f_2(x_1) + \cdots + c_m f_m(x_1) = y_1$$

$$c_1 f_1(x_2) + c_2 f_2(x_2) + \cdots + c_m f_m(x_2) = y_2$$

$$\vdots$$

$$c_1 f_1(x_n) + c_2 f_2(x_n) + \cdots + c_m f_m(x_n) = y_n$$

Since the quantities x_1, x_2, etc., are known numbers, the quantities $f_1(x_1)$, $f_2(x_1)$, etc., are known numbers, so the above equations are just a set of n

linear equations for the m unknowns, $c_1, c_2, \ldots, c_m$. If n is equal to m, there is exactly one solution (provided that the determinant of coefficients is not zero), that is, there is one set of values for $c_1, c_2, \ldots, c_m$ that will cause the above equations to be satisfied. If n is less than m, there are fewer equations than unknowns and there are an infinite number of solutions, each of which represents a curve passing through the given points. The problem in this case is not completely determined, and we cannot find a single "best" fit without further information of some sort. If n is greater than m, there are more equations than unknowns, and there is no solution. In this case we look for a solution in the "least-squares" sense. We write the equations in the form

$$c_1 f_1(x_1) + c_2 f_2(x_1) + \cdots + c_m f_m(x_1) - y_1 = \delta_1$$

$$c_1 f_1(x_2) + c_2 f_2(x_2) + \cdots + c_m f_m(x_2) - y_2 = \delta_2 \qquad \textbf{(11-4)}$$

$$\vdots$$

$$c_1 f_1(x_n) + c_2 f_2(x_n) + \cdots + c_m f_m(x_n) - y_n = \delta_n$$

The quantities δ_1, δ_2, etc., are the "residuals." For a perfect fit these residuals would all be zero. Since we cannot make them be zero, we try instead to make them all small by looking for the values of $c_1, c_2, \ldots, c_m$ which will minimize the expression $\delta_1{}^2 + \delta_2{}^2 + \cdots + \delta_n{}^2$. In curve fitting, this last situation, in which there are more equations than unknowns, is the common one. We will discuss its solution shortly, but first we shall describe some particular cases of linear formulas.

11.21   Polynomial Fits

If in the general expression (11.3) of Section 11.2 we take

$$f_1(x) = 1$$

$$f_2(x) = x$$

$$f_3(x) = x^2$$

$$\vdots$$

$$f_m(x) = x^{m-1}$$

we obtain

$$y = c_1 + c_2 x + c_3 x^2 + \cdots + c_m x^{m-1}$$

That is, we are attempting to fit the data with a polynomial of degree $m - 1$. Thus the general method of fitting with a linear formula contains polynomials as a special case. If we have n points we wish to fit with a polynomial and if we have a choice of the degree of the polynomial, the following considerations are of interest.

From Section 11.2 it is clear that we can pass one and only one polynomial

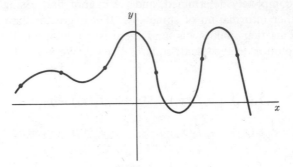

Figure 11-2

of degree $n - 1$ through all n points. As is indicated in Figure 11-2, this may not be a very satisfying fit, since the polynomial may oscillate violently between the points being fitted. A smoother curve is obtained, at the expense of some error in fit at the points themselves, by using a polynomial of lower degree. As an extreme case, Figure 11-3 shows a fit with a polynomial of

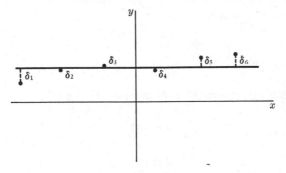

Figure 11-3

degree zero, a horizontal straight line. In most cases, the most satisfying fit is obtained by a polynomial of some intermediate degree, compromising between goodness of fit at the measured values and smoothness of the curve. Figure 11-4 shows the type of fit that might be attainable in the present example with a polynomial of degree 4. Although no completely general rule can be given, it is frequently desirable to conduct polynomial fits using a polynomial whose degree is from 1/2 to 3/4 of the number of points to be fitted.

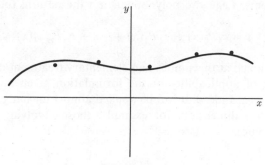

Figure 11-4

11.22   Trigonometric and Other Fits

If in the general expression (11-3) of Section 11.2 we take

$$f_1(x) = \sin bx$$

$$f_2(x) = \sin 2bx$$

$$f_3(x) = \sin 3bx$$

$$\vdots$$

$$f_m(x) = \sin mbx$$

we obtain

$$y = c_1 \sin bx + c_2 \sin 2bx + \cdots + c_m \sin mbx$$

That is, the fitting function is the first m terms of the so-called Fourier sine series. An expansion of this type is frequently the natural one to use for problems in physics or engineering. In applications of this type of fit the constant b would be chosen ahead of time to give the sine functions the desired periodicity.

Instead of sine terms, a series in cosine terms could be used, or a series of mixed sine and cosine terms. These types of series are most useful in problems in which x is constrained to some restricted range of values, and the values outside that region are not of interest.

In passing, it might be mentioned that in some problems it is of interest to express y in terms of some of the more advanced functions of mathematics. In cases in which axial symmetry is present, Bessel functions are frequently the most useful:

$$y = c_1 J_0(\alpha_1 x) + c_2 J_0(\alpha_2 x) + \cdots + c_m J_0(\alpha_m x)$$

In other problems, Legendre polynomials are the natural functions to use:

$$y = c_1 P_0(x) + c_2 P_1(x) + \cdots + c_m P_{m-1}(x)$$

These are mentioned at this point in an elementary treatment only to indicate the wide range of applicability of the formulation given above for fitting by a linear formula. Indeed, it can be applied almost unchanged to problems in more than one dimension, for example, those involving expansions in spherical harmonics.

11.23   Method of Performing the Least-Squares Fit

The method of finding the least-squares solution to equations (11-4) of Section 11.2 will first be derived symbolically and then described by a flow chart. Those equations can be abbreviated

$$\sum_{i=1}^{m} c_i f_i(x_j) - y_j = \delta_j, \qquad j = 1, 2, \ldots, n \qquad \textbf{(11-5)}$$

and we desire to choose the c_i's so that

$$\sum_{j=1}^{n} \delta_j^2 = \text{minimum} \qquad \textbf{(11-6)}$$

This last quantity is to be regarded as a function of the c_i's. A necessary condition for it to be a minimum is that its partial derivatives by each of the c_i's must vanish,

$$\frac{\partial}{\partial c_k} \sum_{j=1}^{n} \delta_j^2 = 0 \qquad \text{for } k = 1, 2, \ldots, m$$

or

$$\sum_{j=1}^{n} \delta_j \frac{\partial \delta_j}{\partial c_k} = 0 \qquad \text{for } k = 1, 2, \ldots, m \qquad \textbf{(11-7)}$$

From (11-5),

$$\frac{\partial \delta_j}{\partial c_k} = f_k(x_j)$$

Substituting this, and the value for δ_j from (11-5) into (11-7), we have

$$\sum_{j=1}^{n} \left[\sum_{i=1}^{m} c_i f_i(x_j) - y_j \right] f_k(x_j) = 0 \qquad \text{for } k = 1, 2, \ldots, m$$

or

$$\sum_{i=1}^{m} c_i \sum_{j=1}^{n} f_i(x_j) f_k(x_j) = \sum_{j=1}^{n} y_j f_k(x_j) \qquad \text{for } k = 1, 2, \ldots, m \quad \textbf{(11-8)}$$

This is a system of m linear equations for the m unknowns $c_1, c_2, \ldots, c_m$, and can be solved by any of the methods of Chapter 10. In order to clarify the structure of the formula, which is perhaps a little confusing because of the double-summation notation, an example will be worked through following the same procedure.

Example 1. Find the formula of the type $y = c_1 e^x + c_2 e^{2x}$ that will give the best least-squares fit to the points $(0, 1)$, $(1, -2)$, $(2, -40)$.

We have

$$c_1 e^0 + c_2 e^0 - 1 = \delta_1$$

$$c_1 e^1 + c_2 e^2 + 2 = \delta_2$$

$$c_1 e^2 + c_2 e^4 + 40 = \delta_3$$

We desire that

$$\delta_1{}^2 + \delta_2{}^2 + \delta_3{}^2 = \text{minimum}$$

so we set

$$\delta_1 \frac{\partial \delta_1}{\partial c_1} + \delta_2 \frac{\partial \delta_2}{\partial c_1} + \delta_3 \frac{\partial \delta_3}{\partial c_1} = 0$$

and

$$\delta_1 \frac{\partial \delta_1}{\partial c_2} + \delta_2 \frac{\partial \delta_2}{\partial c_2} + \delta_3 \frac{\partial \delta_3}{\partial c_2} = 0$$

Now

$$\frac{\partial \delta_1}{\partial c_1} = e^0, \qquad \frac{\partial \delta_2}{\partial c_1} = e^1, \qquad \text{etc.}$$

so we have

$$(c_1 e^0 + c_2 e^0 - 1)e^0 + (c_1 e^1 + c_2 e^2 + 2)e^1 + (c_1 e^2 + c_2 e^4 + 40)e^2 = 0$$

$$(c_1 e^0 + c_2 e^0 - 1)e^0 + (c_1 e^1 + c_2 e^2 + 2)e^2 + (c_1 e^2 + c_2 e^4 + 40)e^4 = 0$$

or, regrouping terms,

$$c_1(e^0 e^0 + e^1 e^1 + e^2 e^2) + c_2(e^0 e^0 + e^2 e^1 + e^4 e^2) = e^0 - 2e^1 - 40e^2$$

$$c_1(e^0 e^0 + e^1 e^2 + e^2 e^4) + c_2(e^0 e^0 + e^2 e^2 + e^4 e^4) = e^0 - 2e^2 - 40e^4$$

or, substituting rough numerical values,

$$63c_1 + 424c_2 = -296$$

$$424c_1 + 3036c_2 = -2198$$

so that

$$c_1 \approx 2, \qquad c_2 \approx -1$$

In order to proceed with the construction of a flow chart, we note that the problem is to solve the set of linear equations

$$\sum_{i=1}^{m} a_{ki} c_i = b_k \qquad \text{for } k = 1, 2, \ldots, m \tag{11-9}$$

where

$$a_{ki} = \sum_{j=1}^{n} f_i(x_j) f_k(x_j) \tag{11-10}$$

$$b_k = \sum_{j=1}^{n} y_j f_k(x_j) \tag{11-11}$$

The process can be written in a particularly nice form in matrix notation. Let

$$F = \begin{pmatrix} f_1(x_1) & f_2(x_1) & \cdots & f_m(x_1) \\ f_1(x_2) & f_2(x_2) & \cdots & f_m(x_2) \\ \vdots & & & \\ f_1(x_n) & f_2(x_n) & \cdots & f_m(x_n) \end{pmatrix} \tag{11-12}$$

and

$$b = \begin{pmatrix} b_1 \\ b_2 \\ b_3 \\ \vdots \\ b_m \end{pmatrix}, \qquad y = \begin{pmatrix} y_1 \\ y_2 \\ \vdots \\ y_m \end{pmatrix}, \qquad c = \begin{pmatrix} c_1 \\ c_2 \\ \vdots \\ c_m \end{pmatrix}$$

Further let $\tilde{F}$ stand for the matrix obtained by interchanging rows and columns in F. (The symbol $\tilde{F}$ is read "F-transpose.") Then

$$\tilde{F} = \begin{pmatrix} f_1(x_1) & f_1(x_2) & \cdots & f_1(x_n) \\ f_2(x_1) & f_2(x_2) & \cdots & f_2(x_n) \\ \vdots & & & \\ f_m(x_1) & f_m(x_2) & \cdots & f_m(x_n) \end{pmatrix}$$

Let

$$A = \begin{pmatrix} a_{11} & a_{12} & \cdots & a_{1m} \\ a_{21} & a_{22} & \cdots & a_{2m} \\ \vdots & & & \\ a_{m1} & a_{m2} & \cdots & a_{mm} \end{pmatrix}$$

With this notation, equations (11-9), (11-10), and (11-11) become

$$Ac = b$$

$$A = \tilde{F}F$$

$$b = \tilde{F}y$$

These equations really say the following.

If we wish the least-squares solution to the system of equations

$$Fc = y \qquad\qquad (11\text{-}13)$$

where there are more equations than unknowns, we multiply both sides by $\tilde{F}$, and then solve the resulting system, which now has the same number of equations and unknowns:

$$(\tilde{F}F)c = \tilde{F}y$$

In order to illustrate these points further, Example 1 will be redone, in matrix notation.

We have

$$f_1(x) = e^x, \qquad f_2(x) = e^{2x}, \qquad x_1 = 0$$

$$y_1 = 1, \qquad x_2 = 1, \qquad y_2 = -2, \qquad x_3 = 2, \qquad y_3 = -40$$

Hence the set of equations we wish to solve is

$$\begin{pmatrix} e^0 & e^0 \\ e^1 & e^2 \\ e^2 & e^4 \end{pmatrix} \begin{pmatrix} c_1 \\ c_2 \end{pmatrix} = \begin{pmatrix} 1 \\ -2 \\ -40 \end{pmatrix}$$

Multiplying both sides by the transpose of the matrix on the left,

$$\begin{pmatrix} e^0 & e^1 & e^2 \\ e^0 & e^2 & e^4 \end{pmatrix} \begin{pmatrix} e^0 & e^0 \\ e^1 & e^2 \\ e^2 & e^4 \end{pmatrix} \begin{pmatrix} c_1 \\ c_2 \end{pmatrix} = \begin{pmatrix} e^0 & e^1 & e^2 \\ e^0 & e^2 & e^4 \end{pmatrix} \begin{pmatrix} 1 \\ -2 \\ -40 \end{pmatrix}$$

Multiplying out the matrices, we have approximately,

$$\begin{pmatrix} 63 & 424 \\ 424 & 3036 \end{pmatrix} \begin{pmatrix} c_1 \\ c_2 \end{pmatrix} = \begin{pmatrix} -296 \\ -2198 \end{pmatrix}$$

These are the same equations for c_1 and c_2 that were obtained in Example 1.

11.24   Flow Chart and Program for Least-Squares Fit

The flow chart, Figure 11-5, describes, in a rather terse form, the performance of a least-squares fit as discussed in Section 11.23. The process described by the flow chart consists of application of a very limited set of routines, as follows:

(1) m different routines for evaluating the functions $f_1(x)$, $f_2(x)$, ..., $f_m(x)$.

(2)   A routine for multiplying a matrix by its transpose.

(3)   A routine for solving an m by m system of linear equations.

The FORTRAN subroutine given below will perform the least-squares fit of a function having M unknown coefficients to N points, where M can be as large as 20 and N as large as 50. The correct functions FU1F(X), FU2F(X), etc., must be inserted, depending on the type of fit wanted. It calls the subroutine GAUSID of Chapter 10 to solve the system of linear equations. Because of the symmetry of the coefficients, the Gauss-Seidel method usually converges for this particular type of problem. (Theoretically it always does, but the use of approximate numbers in the computer sometimes destroys convergence.)

```
SUBROUTINE LSTSQ(X,Y,N,C,M)
DIMENSION  X(50),Y(50),C(20),F(20,50),A(20,20),B(20)
FU1F(X)= ⎱
FU2F(X)= ⎰  insert correct expressions for
 ⋱           f₁(x), ..., fₘ(x)
FUMF(X)= ⎰
```

Where in math notation: insert correct expressions for $f_1(x), \ldots, f_m(x)$

ENTER

$z_{ij} = f_i(x_j)$ $\quad \begin{cases} i = 1, \ldots, m \\ j = 1, \ldots, n \end{cases}$

$a_{ki} = \sum_{j=1}^{n} z_{ij} z_{kj}, \quad \begin{cases} i = 1, \ldots, m \\ k = 1, \ldots, i \end{cases}$

$a_{ik} = a_{ki}, \quad \begin{cases} i = 1, \ldots, m \\ k = i+1, \ldots, m \end{cases}$

$b_k = \sum_{j=1}^{n} y_j z_{kj}, \quad k = 1, \ldots, m$

EXIT to routine for solving system of linear equations

$\sum_{i=1}^{m} a_{ki} c_i = b_k, \quad k = 1, \ldots, m$

Figure 11-5: Least-square fit—expression
linear in coefficients

```
     DO 1 J=1,N
     F(1,J)=FU1F(X(J))
     F(2,J)=FU2F(X(J))
        ⋮
 1   F(M,J)+FUMF(X(J))
     DO 3 I=1,M
     DO 3 K=1,I
     A(K,I)=0.
     DO 2 J=1,N
 2   A(K,I)=A(K,I)+F(I,J)*F(K,J)
 3   A(I,K)=A(K,I)
     DO 4 K=1,M
     B(K)=0
     DO 4 J=1,N
 4   B(K)=B(K)+Y(J)*F(K,J)
     CALL GAUSID(A,M,B,C)
     RETURN
     END
```

11.25   Machine-Time Requirements for the Least-Squares Fit

It is interesting to make a rough estimate of the machine time that will be required to make a fit of an expression having m undetermined coefficients to a set of n points. First m different functions must be evaluated, each a total of n times. The time to evaluate these functions may vary considerably. For a polynomial, only one multiplication is required for each function, since each power of x is obtained by multiplying the preceding power by x. For sines, cosines, exponentials, etc., from 5 to 10 multiplication times might be required for each function. Let us first consider only polynomial fits. For these, a total of nm multiplications will be required to form all the functions $f_i(x_j)$. The next step is to perform the multiplications necessary to obtain the elements a_{ij} and b_j. Each element requires n multiplications, and there are

$$(m + 1) + m + (m - 1) + \cdots + 1 = (m + 1)(m + 2)/2$$

elements, or about $m^2/2$ elements, for a total of about $m^2 n/2$ multiplications. From Chapter 10, the solution of an m by m set of linear equations will require about $m^3/2$ multiplication times. Thus, for a polynomial fit we require about

$$nm + m^2 n/2 + m^3/2 \text{ multiplication times}$$

For example, for a 5th-degree polynomial (6 coefficients) and 8 points, the time is about

$$48 + 144 + 108 = 300 \text{ multiplication times}$$

If $t_m = 10$ milliseconds, this is 3 seconds. If $t_m = 10$ microseconds, it is about 3 milliseconds. For a 19th-degree polynomial (20 undetermined coefficients) and 30 points, the time is about

$$600 + 6000 + 4000 = 10{,}000 \text{ multiplication times}$$

If instead of polynomials, we are making a fit using functions requiring 10 multiplication times to compute, our rough estimate for total time becomes

$$10nm + m^2 n/2 + m^3/2$$

For large values of m and n, this will give times essentially the same as the previous expressions. As has been indicated by the above examples, the computer can quite handily accomplish least-squares fits on systems involving many data points and many undetermined coefficients.

11.3 WEIGHTED LEAST SQUARES

We have said that the most widely used criterion for goodness of fit is that the sum of squares of the residuals,

$$\delta_1{}^2 + \delta_2{}^2 + \delta_3{}^2 + \cdots + \delta_n{}^2$$

be a minimum. A rather common variation of this criterion is to require instead that the expression

$$w_1\delta_1{}^2 + w_2\delta_2{}^2 + \cdots + w_n\delta_n{}^2$$

be a minimum, where $w_1, w_2, \ldots, w_n$ are selected positive numbers. In the upper expression above, we are essentially saying that we are willing to accept as big an error for any one of the points (x_i, y_i) as for any other. If we use the lower expression, we are saying, in effect, that we consider a good fit at one point to have more weight than that at another point. The following two examples will illustrate the effect of weighting.

Example 1.   Find the best least-squares fit of the form $y = c_1 + c_2 x$ for the three points $(0, 0)$, $(1, 1)$, and $(2, 1)$.

Following the procedure of Section 11.23, we write

$$\begin{pmatrix} 1 & 0 \\ 1 & 1 \\ 1 & 2 \end{pmatrix} \begin{pmatrix} c_1 \\ c_2 \end{pmatrix} = \begin{pmatrix} 0 \\ 1 \\ 1 \end{pmatrix}$$

Multiplying by the transpose

$$\begin{pmatrix} 1 & 1 & 1 \\ 0 & 1 & 2 \end{pmatrix} \begin{pmatrix} 1 & 0 \\ 1 & 1 \\ 1 & 2 \end{pmatrix} \begin{pmatrix} c_1 \\ c_2 \end{pmatrix} = \begin{pmatrix} 1 & 1 & 1 \\ 0 & 1 & 2 \end{pmatrix} \begin{pmatrix} 0 \\ 1 \\ 1 \end{pmatrix}$$

So the equations are

$$3c_1 + 3c_2 = 2$$

$$3c_1 + 5c_2 = 3$$

and the solution is $c_1 = 1/6$, $c_2 = 1/2$.

Example 2.   Find the best least-squares fit of the form $y = c_1 + c_2 x$ for the four points $(0, 0)$, $(1, 1)$, $(2, 1)$, and $(2, 1)$.

Here we are weighting the point (2, 1) double by using it twice. A little reflection will show that this is the same as minimizing the expression

$$\delta_1{}^2 + \delta_2{}^2 + 2\delta_3{}^2$$

that is, assigning double weight to the point. For the moment, let us treat the problem as if we had four points, so that we can use the procedures we have already learned. The equations from which we obtain the coefficients are

$$\begin{pmatrix} 1 & 0 \\ 1 & 1 \\ 1 & 2 \\ 1 & 2 \end{pmatrix} \begin{pmatrix} c_1 \\ c_2 \end{pmatrix} = \begin{pmatrix} 0 \\ 1 \\ 1 \\ 1 \end{pmatrix}$$

Upon multiplying by the transpose, we obtain

$$4c_1 + 5c_2 = 3$$

$$5c_1 + 9c_2 = 5$$

and the solution is $c_1 = 2/11$, $c_2 = 5/11$.

Figure 11-6 shows the solutions to both Examples 1 and 2. Clearly the

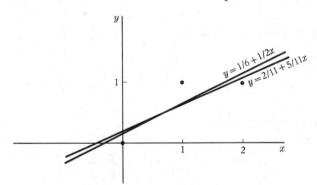

Figure 11-6

weighting in Example 2 has caused the chosen line to be nearer the point (2,1). Weighting is usually required when the experimental data are known to be better for some of the points than for others. There are rules from the field of statistics that say what the weighting factor should be. For example, if the measurement error itself is the same at all data points, then the weighting factor w_i for each point should be proportional to the number of independent measurements available at that point. If the measurements have different intrinsic errors at the different points, the weighting factor assigned should be inversely proportional to the square of the intrinsic error (or standard deviation).

The weighting factor can be absorbed directly into the calculation procedure described in Section 11.23 as follows: If the point (x_j, y_j) is to be assigned a weight w_j, we multiply the jth equation by $\sqrt{w_j}$. This will be demonstrated by reworking Example 2.

Example 3.   Find the best least-squares fit of the form $y = c_1 + c_2 x$ for the three points $(0, 0)$, $(1, 1)$, and $(2, 1)$, giving the last point double weight.

We construct the equations

$$\begin{pmatrix} 1 & 0 \\ 1 & 1 \\ \sqrt{2} & 2\sqrt{2} \end{pmatrix} \begin{pmatrix} c_1 \\ c_2 \end{pmatrix} = \begin{pmatrix} 0 \\ 1 \\ \sqrt{2} \end{pmatrix}$$

Upon multiplying by the transpose, we again obtain the equations

$$4c_1 + 5c_2 = 3$$

$$5c_1 + 9c_2 = 5$$

and the solution is $c_1 - 2/11$, $c_2 = 5/11$.

11.4   NONLINEAR FORMULAS

Occasions arise when it is desirable or necessary to fit an expression with a nonlinear formula. Here the general principles are the same as with linear formulas, but the equations to be solved for the undetermined coefficients in this case will be nonlinear, and their solution presents a more difficult problem. The particular techniques which are of use in this situation will be discussed.

11.41   Formulas Reducible to Linear

If all the undetermined coefficients appear as factors or as exponents in a product, they can be placed in a linear relation by taking logarithms. An example will illustrate this point.

Example 1.   Fit a curve of the form $y = c_1 e^{c_2 x}$ to the points $(0, 1)$, $(1, 2)$, $(2, 6)$.

If we take logarithms of both sides, we obtain

$$\ln y = \ln c_1 + c_2 x$$

Now let

$$k_1 = \ln c_1 \quad \text{and} \quad z = \ln y$$

and we have

$$z = \ln y = k_1 + c_2 x$$

so our problem is now to fit three points of the form (x, z) with this linear expression. The points are $(0, 0)$, $(1, \ln 2)$, $(2, \ln 6)$. We can use the method of Section 11.23 to solve this problem. The equations are

$$\begin{pmatrix} 1 & 0 \\ 1 & 1 \\ 1 & 2 \end{pmatrix} \begin{pmatrix} k_1 \\ c_2 \end{pmatrix} = \begin{pmatrix} 0 \\ \ln 2 \\ \ln 6 \end{pmatrix}$$

Upon multiplying by the transpose, we obtain (roughly)

$$3k_1 + 3c_2 = 2.48$$

$$3k_1 + 5c_2 = 4.28$$

so that

$$k_1 = -.07, \quad c_2 = .9$$

and

$$c_1 = e^{-.07} \approx .9$$

and our fit is

$$y = .9e^{.9x}$$

The solution we have just obtained bears some closer investigation. It is indeed a least-squares fit, but perhaps not the one we really wanted. In taking logarithms of both sides of the initial expression, we have introduced a weighting. We have treated the problem as if it were to minimize

$$\sum \delta_i^2, \quad \text{where } \delta_i = k_i + c_2 x_i - \ln y_i$$

Thus each δ_i is then the error in $\ln y_i$. This is not the same as minimizing

$$\sum \gamma_i^2, \qquad \text{where } \gamma_i = c_1 e^{c_2 x_i} - y_i$$

In this last expression each γ_i is the error in the corresponding y_i. In the former expression, each δ_i is the error in $\ln y_i$. Now if y is in error by an amount Δy, then $\ln y$ is in error by an amount which is approximately $\Delta y / y$. Hence

$$\delta_i \approx \gamma_i / y_i$$

and if we minimize

$$\sum \delta_i^2$$

we are miminizing

$$\sum (\gamma_i / y_i)^2$$

That is, we have weighted each γ_i by the inverse square of y_i. If we wish to weight all the errors in y_i equally, we would have to minimize

$$\sum y_i^2 \delta_i^2$$

Example 2.   Make a least-squares fit of the form $y = c_1 e^{c_2 x}$ to the points $(0, 1)$, $(1, 2)$, $(2, 6)$, giving equal weight to errors in y at all three points.

This is the same as Example 1, except that we now insist that weighting factors be used as described above. We do this by again writing

$$z = \ln y = k_1 + c_2 x$$

and use the method of Section 11.23, multiplying each row by y_i for proper weighting. We obtain

$$\begin{pmatrix} 1 & 0 \\ 2 & 2 \\ 6 & 12 \end{pmatrix} \begin{pmatrix} k_1 \\ c_2 \end{pmatrix} = \begin{pmatrix} 0 \\ 2 \ln 2 \\ 6 \ln 6 \end{pmatrix}$$

so the equations are

$$41 k_1 + 76 c_2 = 67.2760$$

$$76 k_1 + 148 c_2 = 131.7793$$

so that

$$k_1 = -.2000, \qquad c_2 = .9932 \approx 1$$

and

$$c_i = e^{k_1} = .8187 \approx .8$$

Figure 11-7 shows the curves of Examples 1 and 2.

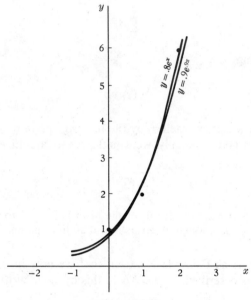

Figure 11-7

Table 1 shows the values of the ordinates of the two curves at the three given points. It will be observed that the curve obtained in Example 2 misses all the points by about the same amount, while the curve obtained in Example 1 misses the point having ordinate 1 by .1, that having ordinate 2 by .2, and

TABLE 1

x	Given y	$y = .9e^{.9x}$ (Example 1)	$y = .8e^x$ (Example 2)
0	1	.9	.8
1	2	2.2	2.2
2	6	5.4	5.9

that having ordinate 6 by .6. That is to say, Example 2 gave us a fit having nearly equal *absolute* errors at the given point, whereas Example 1 gave us a fit having nearly equal *relative* errors.

The situation demonstrated by the above examples is one that applies in general to curve-fitting problems where the coefficients can be put in a linear relation by taking logarithms of both sides of the fitting equation. If the least-squares fit is made directly to the logarithmic expressions, the *relative* errors are being given roughly equal weight. If it is desired to give the *absolute* errors equal weight, then weights proportional to the squares of the ordinates at the measured points must be introduced in performing the fit.

11.42   Formulas Not Reducible to Linear

When it is necessary to make a fit with a formula that cannot be written so as to put the undetermined constants in a linear relationship, then the problem of finding the constants becomes much more difficult. This is the general problem posed at the beginning of the chapter, where the fitting function

$$y = f(x, c_1, c_2, \ldots, c_m)$$

has no special properties we can use to simplify the problem. It is fortunately true that a large percentage of the practical problems fall into the easier types already discussed, but on occasion it is necessary to use this less tractable type to fit. In this case, the problem can be described as follows: We have a set of n relations

$$f(x_i, c_1, c_2, \ldots, c_m) - y_i = \delta_i, \qquad i = 1, \ldots, n \qquad \textbf{(11-14)}$$

and we wish to choose the c's so as to make

$$\sum_{i=1}^{n} \delta_i^2 \text{ a minimum}$$

A necessary condition is that

$$\frac{\partial}{\partial c_j} \sum_{i=1}^{n} \delta_i^2 = 0 \qquad \text{for } j = 1, \ldots, m \qquad \textbf{(11-15)}$$

$$\sum_{i=1}^{n} \delta_i \frac{\partial \delta_i}{\partial c_j} = 0 \qquad \text{for } j = 1, \ldots, m \qquad \textbf{(11-16)}$$

Substituting in the value of δ_i, we have

$$\sum_{i=1}^{n} [f(x_i, c_1, c_2, \ldots, c_m) - y_i] \frac{\partial f(x_i, c_1, c_2, \ldots, c_m)}{\partial c_j} = 0$$

$$\text{for } j = 1, \ldots, m \qquad \textbf{(11-17)}$$

This is a system of m nonlinear equations for the c_j's, equations which may be transcendental, depending on the nature of the function f. In Chapter 8 we discussed methods of solving a transcendental equation in one unknown, and mentioned that these methods could be extended to several equations in several unknowns. The equations may, of course, have more than one root, so the problem of finding a scheme which will converge to the proper root may be a difficult one Another complication is the fact that the expressions for the partial derivative of f by each of the c_j's is required. If f is a very involved function, the writing of these may in itself be quite a difficult task. Nonetheless the problem is by no means a hopeless one for the digital computer, even if one forces the machine to struggle brute force through the calculation as it has just been outlined. A particular iteration scheme, described in Section 11.43, can sometimes be used to obtain the solution rather easily. Before discussing that scheme, in order to make the various aspects of the problem more clear, we will re-do Example 2, Section 11.41, following the procedure outlined above.

Example 1.   Make a least-squares fit of the form

$$y = c_1 e^{c_2 x}$$

to the points $(0, 1)$, $(1, 2)$, and $(2, 6)$, giving equal weights to errors in y at all three points.

Since the formula is reducible to one having a linear relationship among the undetermined constants, the method described in Section 11.41 is the best one for this problem. However, we will use the method described above in order to clarify the above discussion. In our case

$$f(x, c_1, c_2) = c_1 e^{c_2 x}$$

and

$$\frac{\partial f(x, c_1, c_2)}{\partial c_1} = e^{c_2 x}$$

$$\frac{\partial f(x, c_1, c_2)}{\partial c_2} = x c_1 e^{c_2 x}$$

so that, substituting into (11-17), we obtain

$$[c_1 e^{c_2 \cdot 0} - 1]e^{c_2 \cdot 0} + [c_1 e^{c_2 \cdot 1} - 2]e^{c_2 \cdot 1} + [c_1 e^{c_2 \cdot 2} - 6]e^{c_2 \cdot 2} = 0$$

and

$$[c_1 e^{c_2 \cdot 0} - 1].0.c_1 e^{c_2 \cdot 0} + [c_1 e^{c_2 \cdot 1} - 2]1.c_1 e^{c_2 \cdot 1}$$
$$+ [c_1 e^{c_2 \cdot 2} - 6]2.c_1 e^{c_2 \cdot 2} = 0$$

These simplify to

$$c_1 e^{4c_2} + (c_1 - 6)e^{2c_2} - 2e^{c_2} + c_1 - 1 = 0$$

and

$$2c_1 e^{3c_2} + (c_1 - 12)e^{c_2} - 2 = 0$$

These are the equations for c_1 and c_2. In this particular case they are not inordinately difficult to solve, since c_1 appears only to the first power. We will not complete the solution, since it is not very instructive as to the techniques which might be employed on a digital computer. We do point out, however, that these equations are indicative of the complexity of the equations one can become involved in even for rather simple functions $f(x, c_1, c_2, \ldots, c_m)$.

11.43   Iteration Scheme for Formulas Not Reducible to Linear

Let us suppose that we are attempting to fit a set of n points with the function

$$y = f(x, c_1, c_2, \ldots, c_m)$$

and that we already have rough estimates for the constants $c_1, c_2, \ldots, c_m$, for example,

$$c_1 \approx k_1$$

$$c_2 \approx k_2$$

$$\vdots$$

$$c_m \approx k_m$$

where the k_i are known. More precisely, we can say that

$$c_1 = k_1 = E_1$$

$$c_2 = k_2 + E_2$$

$$\vdots$$

$$c_m = k_m + E_m$$

where E_1, E_2, ..., E_m are unknown but small. Then

$$y = f(x, k_1 + E_1, k_2 + E_2, ..., k_m + E_m)$$

or, by Taylor's formula,

$$y = f(x, k_1, k_2, ..., k_m) + E_1 \frac{\partial f(x, k_1, k_2, ..., k_m)}{\partial c_1} + \cdots$$

$$+ E_m \frac{\partial f(x, k_1, k_2, ..., k_m)}{\partial c_m}$$

This is a different formula for y, one in which everything is known but the coefficients E_1, E_2, ..., E_m. Since these appear linearly, they can be determined by the method of Section 11.23. Once these are determined, we have a new estimate for the c_j's. If we desire, we can then use these new c values as k_1, k_2, ..., k_m, and repeat the process. If the initial guess was sufficiently good, the process *may* converge to the desired values $c_1, c_2, ..., c_m$.

Example 1.   Make a least-squares fit of the form

$$y = c_1 e^{c_2 x}$$

to the points $(0, 1)$, $(1, 2)$, and $(2, 6)$, using as an initial estimate $c_1 \approx 1$, $c_2 \approx 1$.

We let $k_1 = 1$, $k_2 = 1$. Then

$$f(x, k_1, k_2) = e^x$$

$$\frac{\partial f}{\partial c_1} = e^{c_2 x}, \qquad \text{so} \qquad \frac{\partial f(x, k_1, k_2)}{\partial c_1} = e^x$$

$$\frac{\partial f}{\partial c_2} = x c_1 e^{c_2 x}, \qquad \text{so} \qquad \frac{\partial f(x, k_1, k_2)}{\partial c_2} = x e^x$$

Hence the formula we desire to use for the first step is

$$y = e^x + E_1 e^x + E_2 x e^x$$

In order to apply the method of Section 11.23, we combine the term e^x with y, giving

$$y - e^x = E_1 e^x + E_2 x e^x$$

Now we can write the equations as

$$\begin{pmatrix} e^0 & 0 \\ e^1 & e^1 \\ e^2 & 2e^2 \end{pmatrix} \begin{pmatrix} E_1 \\ E_2 \end{pmatrix} = \begin{pmatrix} 1 - e^0 \\ 2 - e \\ 6 - e^2 \end{pmatrix}$$

or, approximately,

$$\begin{pmatrix} 1 & 0 \\ 2.718 & 2.718 \\ 7.389 & 14.778 \end{pmatrix} \begin{pmatrix} E_1 \\ E_2 \end{pmatrix} = \begin{pmatrix} 0 \\ -.718 \\ -1.389 \end{pmatrix}$$

Upon multiplying by the transpose, we obtain

$$62.987 E_1 + 116.585 E_2 = -12.216$$

$$116.585 E_1 + 225.781 E_2 = -22.479$$

$$E_1 = -.218, \qquad E_2 = .0132$$

So that our next estimate is

$$c_1 = .782, \qquad c_2 = 1.0132$$

More careful calculation gives for this and succeeding steps:

Step	k_1	k_2	E_1	E_2
1	1	1	$-.218312$	.0131631
2	.781687	1.01316	$-.000921$	.0040536
3	.780766	1.01721	$-.000313$	.0002051
4	.780435	1.01742		

This solution differs slightly from the one obtained in Example 2, Section 11.41. The difference is caused by the fact that the multiplication by y_i in Example 2 gave weighting that was only approximately, not exactly, correct.

The flow chart, Figure 11-8, describes the above iteration scheme. In this flow chart the functions f_1, f_2, etc., are defined by

$$f_1(x, c_1, c_2, \ldots, c_m) = \frac{\partial f(x, c_1, c_2, \ldots, c_m)}{\partial c_1}$$

$$f_2(x, c_1, c_2, \ldots, c_m) = \frac{\partial f(x, c_1, c_2, \ldots, c_m)}{\partial c_2}$$

$$\vdots$$

$$f_m(x, c_1, c_2, \ldots, c_m) = \frac{\partial f(x, c_1, c_2, \ldots, c_m)}{\partial c_m}$$

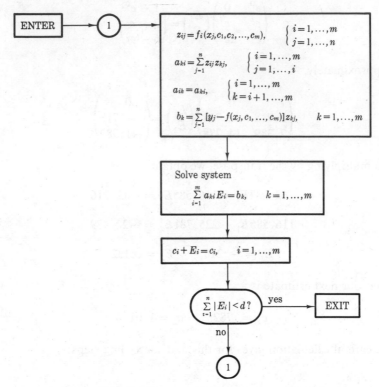

Figure 11-8: Least-square fit—nonlinear expressions

EXERCISE 29

1. Perform the following least-squares fits by hand.

   a. $y = c_1 + c_2 x$,   points $(0, 1)$, $(1, 2)$, $(2, 4)$.
   b. $y = c_1 \sin x + c_2 \cos x$,   points $(0, 0)$, $(\pi/4, 1)$, $(\pi/2, 1)$.
   c. $y = c_1 x + c_2 \ln x$,   points $(1, 1)$, $(2, 2)$, $(3, 3)$.
   d. $y = c_1 + c_2 e^x$,   points $(0, 2)$, $(1, 1)$, $(2, 0)$.

2. Draw a flow chart and write the program for making each of the following fits.

   a. Sixth-degree polynomial, 10 points given.
   b. $y = c_1 + c_2 e^x + c_3 e^{2x} + c_4 e^{3x} + c_5 e^{4x}$,   8 points given.
   c. $y = c_1 + c_2 \cos x + c_3 \cos 2x + c_4 \cos 3x$,   20 points given.

3. Performing the following least-squares fits by hand, giving equal weights to *relative* errors. Draw the curves and compute the errors at the given points.

   a. $y = \dfrac{c_1 e^{c_2 x}}{x^2}$,   points $(1, 1000)$, $(2, 200)$, $(3, 100)$.
   b. $y = c_1 x^{c_2}$,   points $(1, 2)$, $(2, 10)$, $(3, 20)$.

4. Perform the least-squares fits of problem 3, giving equal weights to *absolute* errors. Draw the curves and compute the errors at the given points.

5. Draw a flow chart and write the program for making each of the following fits, using the iteration method of Section 11.43.

   a. $y = c_1 e^{c_2 x} + c_3 e^{c_4 x}$,   10 points given.

   b. $y = c_1 \sin c_2 x + c_3 \sin c_4 x + c_5 \sin c_6 x$,   12 points given.

Interpolation and Differentiation

12.1 INTRODUCTION

We cannot claim to have covered the subject of digital computer use for mathematics through calculus without at least some coverage of the subjects of interpolation and differentiation. This coverage will be brief, compared to that usually given in books on numerical methods. The use of interpolation techniques is not as prevalent in digital computer calculations as in hand calculations, because of the already demonstrated capability of the computer to calculate quite complicated functions quickly. Computer methods for numerical differentiation find their main use in the solution of ordinary or partial differential equations, subjects which will not receive extensive coverage in this text.

It might seem unusual to combine the subjects of interpolation and differentiation in one chapter. As we proceed, however, it will become clear that in numerical work the two subjects are very closely related and do indeed belong as parts of the same discussion.

12.2 INTERPOLATION

The process of interpolation may be regarded as a special case of the general process of curve fitting, discussed in Chapter 11. A function, $y = f(x)$, is known to us only to the extent that we have some set of values (x_1, y_1), $(x_2, y_2), \ldots, (x_m, y_m)$, and we wish to infer reasonable values of y for values of x intermediate between the given ones. The major difference between this problem and the general one of curve-fitting is that we are not interested in having the functional expression $f(x)$ have any particular form, or even having the same form for the entire range of values for x. For interpolation, then, we nearly always use polynomials for $f(x)$, since these are so easy to calculate. We may use different polynomials for different values of x. For example, we might use the first-degree polynomial (straight line) passing through (x_1, y_1) and (x_2, y_2) to interpolate between the first two points, the first-degree polynomial passing through (x_2, y_2) and (x_3, y_3) to interpolate between the second and third points, and so forth. This is the ordinary method of linear interpolation commonly used in obtaining values between those tabulated in trigonometric or other tables. If more accurate interpolation were desired, it would be necessary to fit second-, or third-, or higher-degree polynomials through the nearby points in order to obtain values intermediate between two points. The methods given in Chapter 11 are sufficient to allow us to do this without further ado. However, since we are now restricting ourselves to the case of polynomials, and particularly if we choose to let the given values of x_i be equally spaced, great simplifications occur which lead to formulas much easier to handle. These are the classical interpolation formulas of Newton, Stirling, Bessel, and others, found in standard texts on numerical analysis.

In Chapter 6 we discussed methods of approximating functions by means of power series, and stated that such methods were ordinarily preferable to table look-up for finding values of a function in a computer. For functions whose formulas are extremely complicated in structure, or for functions which are only known at a set of points, a table of values can be stored in the memory, and a program written that will cause the machine to search through this table and then interpolate to obtain a value for the function at any point. Because of the amount of memory required to store a table of values, it is sometimes preferable to store a rather short table of somewhat widely spaced values and use a high-order interpolation scheme, rather than store a long table of closely spaced values and use linear or other low-order interpolation schemes.

12.21   Differences

As mentioned above, all the classic interpolation formulas involve passing a polynomial of degree n through $n + 1$ points and using points from this polynomial to represent the function of interest. For cases in which the values of x are equally spaced, the interpolation formulas are normally stated in terms of differences, so we will present them in this form.

If function $f(x)$ is known at a set of equally spaced values, spaced an amount h apart, the "differences" of $f(x)$ are defined as follows: Let the given values be

$$y_1 = f(x_1)$$

$$y_2 = f(x_2) = f(x_1 + h)$$

$$y_3 = f(x_3) = f(x_1 + 2h)$$

$$y_4 = f(x_4) = f(x_1 + 3h)$$

The first difference of y_1 is defined to be

$$\Delta y_1 = f(x_1 + h) - f(x_1) = y_2 - y_1$$

That of y_2 is

$$\Delta y_2 = f(x_2 + h) - f(x_2) = y_3 - y_2$$

That of y_j is

$$\Delta y_j = y_{j+1} - y_j \qquad\qquad (12\text{-}1)$$

The second difference $\Delta^2 y_1$ is obtained by taking the first difference of Δy_1. Thus

$$\Delta^2 y_1 = \Delta y_2 - \Delta y_1$$

and in like fashion

$$\Delta^2 y_j = \Delta(\Delta y_j) = \Delta y_{j+1} - \Delta y_j$$

Higher-order differences are defined in a similar manner,

$$\Delta^3 y_j = \Delta(\Delta^2 y_j) = \Delta^2 y_{j+1} - \Delta^2 y_j$$

$$\Delta^4 y_j = \Delta(\Delta^3 y_j) = \Delta^3 y_{j+1} - \Delta^3 y_j$$

$$\vdots \qquad\qquad (12\text{-}2)$$

$$\Delta^k y_j = \Delta(\Delta^{k-1} y_j) = \Delta^{k-1} y_{j+1} - \Delta^{k-1} y_j$$

Table 1 demonstrates an easy format for taking differences by hand. In this table each difference is obtained by subtracting the two neighboring differences in the next column to the left, taking in each case the lower value minus the upper.

TABLE 1

x_1	y_1				
		Δy_1			
x_2	y_2		$\Delta^2 y_1$		
		Δy_2		$\Delta^3 y_1$	
x_3	y_3		$\Delta^2 y_2$		$\Delta^4 y_1$
		Δy_3		$\Delta^3 y_2$	
x_4	y_4		$\Delta^2 y_3$		
$\vdots$					
x_{m-3}	y_{m-3}				
		Δy_{m-3}			
x_{m-2}	y_{m-2}		$\Delta^2 y_{m-3}$		
		Δy_{m-2}		$\Delta^3 y_{m-3}$	
x_{m-1}	y_{m-1}		$\Delta^2 y_{m-2}$		
		Δy_{m-1}			
x_m	y_m				

Example 1. Write the difference table for $y = x^4$, where x takes on the integer values 0, 1, 2, ..., 6.

x	y	Δy	$\Delta^2 y$	$\Delta^3 y$	$\Delta^4 y$	$\Delta^5 y$
0	0					
		1				
1	1		14			
		15		36		
2	16		50		24	
		65		60		0
3	81		110		24	
		175		84		0
4	256		194		24	
		369		108		
5	625		302			
		671				
6	1296					

If the first entry in the table is denoted by (x_1, y_1), then, according to the definitions given above, each of the table entries has a unique designation, for example, $\Delta^2 y_1 = 14$, $\Delta^3 y_3 = 84$, etc.

Note that in the above table the fourth differences are constant and so the fifth and all higher differences are zero. This is because $y = x^4$ is a polynomial of fourth degree. It can be proved in general that, for a polynomial of nth degree, all nth differences are constant and all higher differences are zero. In this respect the differences are somewhat analogous to derivatives.

Note also that the successive differences are obtained by repeated subtractions, and that in accordance with the accuracy discussions in Chapter 3, the relative error will have a tendency to grow as one proceeds to higher-order differences. This fact is often used to assist in the detection of errors in tabulated values. A small and unnoticeable error in a tabulated value can produce a large and easily detectable discontinuity in a list of third or fourth differences. The high-order differences are not to be trusted unless the table entries themselves are known to high accuracy.

12.22   Machine Construction of Difference Tables

In order to apply the classical interpolation formulas, it is frequently desirable to construct a table of differences and store it in the memory so that the differences need not be recalculated each time they are used. This can easily be done by repeated applications of equations (12-1) and (12-2). The very simple flow chart in Figure 12-1 describes this operation, for constructing a

START

$$\Delta y_j = y_{j+1} - y_j, \quad j = 1, \ldots, m-1$$

$$\Delta^k y_j = \Delta^{k-1} y_{j+1} - \Delta^{k-1} y_j, \quad j = 1, \ldots, m-k$$

STOP

Figure 12-1: Formation of a difference table

table of up to nth-order differences from a table having m stored values $y_1, y_2, \ldots, y_m$, corresponding to equally spaced values of x.

A FORTRAN program for constructing such a difference table is quite straightforward. If we denote the quantity y_j by the FORTRAN variable Y(J) and the quantity $\Delta^k y_j$ by the FORTRAN variable DY(K,J), then the FORTRAN equivalents of the equations in the flow chart are

$$DY(1,J) = Y(J+1) - Y(J)$$
$$DY(K,J) = DY(K-1,J+1) - DY(K-1,J)$$

In writing a program following the flow chart it is only necessary to use these statements in DO loops, with a DIMENSION statement which will set aside the required amount of storage for the table.

Example 1.   Write a FORTRAN program that will input M values of y, where M may be as large as 1000, and construct a difference table for the first five differences.

A suitable program is

```
      DIMENSION  Y(1000),DY(5,1000)
      READ 100,M
      READ 101,Y(1)
      MM = M − 1
      DO 2 J = 1,MM
      READ 101,Y(J+1)
    2 DY(1,J) = Y(J+1) − Y(J)
      DO 3 K = 2,5
      L = M − K
      DO 3 J = 1,L
    3 DY(K,J) = DY(K−1,J+1) − DY(K−1,J)
      STOP
  100 FORMAT(I4)
  101 FORMAT(E12.4)
      END
```

12.3   CLASSICAL INTERPOLATION FORMULAS

If a polynomial of degree n,

$$y = c_1 + c_2 x + c_3 x^2 + \cdots + c_{n+1} x^n$$

is to pass through $n + 1$ given points, and if we substitute the points in this equation, we obtain $n + 1$ equations in the $n + 1$ unknowns $c_1, c_2, \ldots, c_{n+1}$. Once we solve these equations and determine these coefficients, the polynomial is completely determined. There are still different ways of writing the polynomial, however; the terms do not have to be written in ascending, distinct powers of x, but can be grouped into different factors or terms. For example, the polynomial

$$y = -7 + 12x - 6x^2 + x^3$$

can be written as

$$y = (x - 1) + (x - 1)(x - 2)(x - 3)$$

and it is still precisely the same polynomial, although the numbers which appear explicitly therein are different. The interpolation formulas write an nth-degree polynomial in such a way that the differences as defined above are used directly, and the coefficients c_1, c_2, etc., never need be actually calculated. The different interpolation formulas essentially represent just different groupings of terms in the same basic polynomial.

12.31   Newton's Forward Interpolation Formula

An nth-degree polynomial through the $n + 1$ points (x_0, y_0), (x_1, y_1), ..., (x_n, y_n), where the values of x are equally spaced an amount h apart, can be written

$$y = y_0 + \frac{\Delta y_0}{h}(x - x_0) + \frac{\Delta^2 y_0}{2h^2}(x - x_0)(x - x_1)$$

$$+ \frac{\Delta^3 y_0}{3! h^3}(x - x_0)(x - x_1)(x - x_2)$$

$$+ \cdots + \frac{\Delta^n y_0}{n! h^n}(x - x_0)(x - x_1) \cdots (x - x_{n-1}) \qquad \textbf{(12-3)}$$

This is known as Newton's forward interpolation formula. If we let

$$u = (x - x_0)/h$$

then it can be written

$$y = y_0 + \Delta y_0 u + \frac{\Delta^2 y_0}{2!} u(u - 1) + \frac{\Delta^3 y_0}{3!} u(u - 1)(u - 2)$$

$$+ \cdots + \frac{\Delta^n y_0}{n!} u(u - 1)(u - 2) \cdots (u - n + 1) \qquad \textbf{(12-4)}$$

Example 1.   Given the table of values

x	1.0	1.1	1.2	1.3	1.4	1.5
y	2.0	2.1	2.3	2.7	3.5	4.7

Find the value of y at $x = 1.05$ using Newton's forward interpolation formula with $n = 4$.

The difference table, including up to fourth differences, is

x	y	Δy	$\Delta^2 y$	$\Delta^3 y$	$\Delta^4 y$
1.0	2.0				
		.1			
1.1	2.1		.1		
		.2		.1	
1.2	2.3		.2		.1
		.4		.2	
1.3	2.7		.4		-.2
		.8		0	
1.4	3.5		.4		
		1.2			
1.5	4.7				

We have also

$$h = .1, \qquad x_1 = 1.0, \qquad u = \frac{1.05 - 1.0}{.1} = .5$$

Hence by (12-4),

$$y = 2.0 + (.1)(.5) + \frac{(.1)}{2!}(.5)(-.5) + \frac{.1}{3!}(.5)(-.5)(-1.5)$$

$$+ \frac{.1}{4!}(.5)(-.5)(-1.5)(-2.5)$$

$$= 2.0 + .05 - .0125 + .00625 - .00390625$$

$$= 2.03984375$$

Remembering the discussions of accuracy in Chapter 3, this calculation is of little value unless the values of x and y are known more accurately than is implied by the above table. If the value of y at $x = 1.0$ is only as good as is given in the table, then the number 2.0 represents a value whose absolute error is .05. Hence, the absolute error in y at $x = 1.05$ as determined above is at least .05, so the best that can be said for the answer is that $y = 2.04 \pm .05$. Linear interpolation would have given a result that was just as accurate. The higher-order differences, which contributed to the less significant figures in the answer, can be used with confidence *only* if the original table entries were known with high accuracy. If the formula (12-3) is to be used on a computer, the calculation can be made more efficient by a better grouping of terms, as

$$y = (((\cdots ((\Delta^n y_0/n!)(u - n + 1) + \Delta^{n-1} y_0/(n - 1)!)(u - n + 2)$$

$$+ \cdots + \Delta^3 y/3!)(u - 2) + \Delta^2 y_0/2!)(u - 1) + \Delta y_0)u + y_0 \qquad \textbf{(12-5)}$$

In this formula, each difference $\Delta^k y_0$ appears in the form $\Delta^k y_0/k!$. Let $\delta^k y_0 = \Delta^k y_0/k!$. If stored differences are to be used, the quantities $\delta^k y_0$ are more efficient for use than the quantities $\Delta^k y_0$. With these "adjusted differences," equation (12-5) becomes

$$y = (((\cdots ((\delta^n y_0)(u - n + 1) + \delta^{n-1} y_0)(u - n + 2)$$

$$+ \cdots + \delta^3 y_0)(u - 2) + \delta^2 y_0)(u - 1) + \delta y_0)u + y_0 \qquad \textbf{(12-6)}$$

Example 2.   Given M values of Y already stored, where M $\leqslant$ 1000, for M equally spaced values of x beginning with XBEG and spaced DX apart, and a stored table of adjusted differences DNY(K, J) $= \delta^k y_J$, through third differences, write a FORTRAN subroutine that will perform third-order interpolation using formula (12-6).

A suitable subroutine is

```
      SUBROUTINE FNEWS3(Y,M,DNY,XBEG,DX,XINT,YINT)
      DIMENSION Y(1000),DNY(3,1000)
      MO=(XINT−XBEG+DX)/DX
    1 IF(MO)20,20,2
    2 IF(MO−M)3,3,20
    3 IF(M−3−MO)4,5,5
    4 MO=M−3
    5 FMO=MO−1
      XO=XBEG+FMO*DX
      U=(XINT−XO)/DX
      YINT=(((U−2.)*DNY(3,MO)+DNY(2,MO))*(U−1.)
     ·+DNY(1,MO))*U+Y(MO)
      RETURN
   20 PRINT 101
      STOP
  101 FORMAT(31H OUT OF RANGE FOR INTERPOLATION)
      END
```

In this subroutine the machine first computes MO, the subscript that should correspond to the tabulated x value just before XINT, the value at which the interpolation is to be performed. Statements 1 and 2 then check to see if MO is less than 1 or greater than M, in which case the words "OUT OF RANGE FOR INTERPOLATION" are printed and the problem stopped. If MO is in range, statement 3 then checks to see if MO is greater than $M - 3$, for if it is, third differences will not exist in the difference table. In this case, M is set equal to $M - 3$, and then XO, the value of x at the tabulated value Y(MO) is computed, and equation (12-6) used to obtain YINT, the interpolated value of Y. In this subroutine, six multiplications or divisions are performed to find an interpolated value. In addition, 3M divisions must have been done to initially store the difference table; 3000 memory locations are required to store the difference table. Three subtractions and three divisions were saved by using the stored differences. If the subroutine is to be called at least M times in the problem, and if memory space is readily available the stored differences are worthwhile. Otherwise, differences should be computed as needed.

It is possible to arrange equation (12-4) so that the contributions of the higher differences are determined recursively, and the process can be carried to any desired order and stopped. If the coefficient of $\Delta^k y_0$ in (12-4) is denoted by a_k, then a_k is seen to be given by

$$a_k = a_{k-1}(u - k + 1)/k$$

The flow chart in Figure 12-2 shows how the calculation can be done using this relation. The flow chart assumes that the differences $\Delta^k y_0$ and the values of x_0, h, and x are available and describes the computation of y by nth-order interpolation.

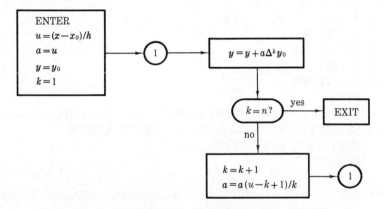

Figure 12-2: Interpolation by Newton's forward formula

Example 3.   Given M values of Y already stored, where M ⩽ 1000, for M equally spaced values of X beginning with XBEG and spaced DX apart, write a FORTRAN subroutine which will perform Nth-order interpolation by Newton's forward formula, where N is an input number to the subroutine.

A suitable program is

```
      SUBROUTINE  FNEW(Y,M,XBEG,DX,N,XINT,YINT)
      DIMENSION  Y(1000),D(1000)
      MO=(XINT−XBEG+DX)/DX
      IF(M−N−1)21,21,1
    1 IF(MO)20,20,2
```

```
   2 IF(MO−M)3,3,20
   3 IF(M−MO−N)4,5,5
   4 MO=M−N
   5 FMO=MO−1
     XO=XBEG+FMO*DX
     U=(XINT−XO)/DX
     A=U
     YINT=Y(MO)
   6 DO 7 I=1,N
     J=MO+I−1
   7 D(I)=Y(J+1)−Y(J)
     IF(N−1)21,10,8
   8 DO 9 K=2,N
     DO 9 I=K,N
     L=N+K−I
   9 D(L)=D(L)−D(L−1)
  10 DO 11 K=1,N
     YINT=YINT+A*D(K)
     FK=K+1
  11 A=A*(U−FK+1.)/FK
     RETURN
  20 PRINT 101
     STOP
  21 PRINT 102
     STOP
 101 FORMAT(31H OUT OF RANGE FOR INTERPOLATION)
 102 FORMAT(45H ORDER OF INTERPOLATION
     INCORRECTLY SPECIFIED)
     END
```

In this subroutine statements 6 through 9 compute the differences needed, in a somewhat devious way to conserve storage. After these statements are completely executed, we have $D(1) = \Delta y_{mo}$, $D(2) = \Delta^2 y_{mo}$, etc., $D(K) = \Delta^k y_{mo}$. The reader may find it instructive to confirm this by following through that part of the program, say with $M = 500$ and $N = 3$. Statements 5 through 6 calculate the quantities in the first box of the flow chart, Figure 12-2, and statements 10 through 11 actually calculate the interpolated value of y in accordance with the flow chart.

12.32   Newton's Backward Interpolation Formula

The polynomial on the right side of (12-3) can be written

$$y = y_n + \frac{\Delta y_{n-1}}{h}(x - x_n) + \frac{\Delta^2 y_{n-2}}{2! h^2}(x - x_n)(x - x_{n-1}) + \cdots$$

$$+ \frac{\Delta^n y_0}{n! h^n}(x - x_n) \cdots (x - x_1) \tag{12-7}$$

In this form it is known as Newton's backward interpolation formula. If we let

$$u = (x - x_n)/h$$

this formula takes the form

$$y = y_n + \Delta y_{n-1} u + \frac{\Delta^2 y_{n-2}}{2!} u(u + 1) + \cdots + \frac{\Delta^n y_0}{n!} u(u + 1) \cdots (u + n - 1) \tag{12-8}$$

Examples similar to those of Section 12.31 could be given for this formula but will be omitted here. If the differences $\Delta^k y_j$, and the values x_n, h, and x are available, the flow chart, Figure 12-3, describes calculation of y by equation (12-8).

12.33   Stirling's Interpolation Formula

This formula is a so-called central difference formula. It is written as a polynomial of degree $2n$, passing through the $2n + 1$ points (x_{-n}, y_{-n}), (x_{-n+1}, y_{-n+1}), ..., (x_{-1}, y_{-1}), (x_0, y_0), (x_1, y_1), ..., (x_n, y_n). If we let

$$u = (x - x_0)/h$$

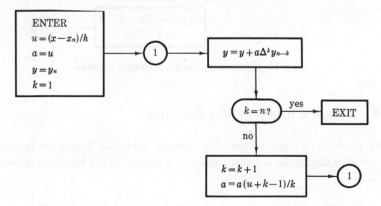

Figure 12-3: Interpolation by Newton's backward formula

it can be written

$$y = y_0 + u \frac{\Delta y_{-1} + \Delta y_0}{2} + \frac{u^2}{2} \Delta^2 y_{-1}$$

$$+ \frac{u(u^2 - 1^2)}{3!} \frac{\Delta^3 y_{-2} + \Delta^3 y_{-1}}{2} + \frac{u^2(u^2 - 1^2)}{4!} \Delta^4 y_{-2}$$

$$+ \cdots$$

$$+ \frac{u(u^2 - 1^2) \cdots [u^2 - (n-1)^2]}{(2n-1)!} \frac{\Delta^{2n-1} y_{-n} + \Delta^{2n-1} y_{-n+1}}{2}$$

$$+ \frac{u^2(u^2 - 1^2) \cdots [u^2 - (n-1)^2]}{(2n)!} \Delta^{2n} y_{-n} \qquad \textbf{(12-9)}$$

In this formula, each time n is increased by one, two new terms are added. The flow chart, Figure 12-4, shows how the formula can be used, assuming the values y_j, the differences $\Delta^k y_j$, and the values of x, x_0, and h are already stored.

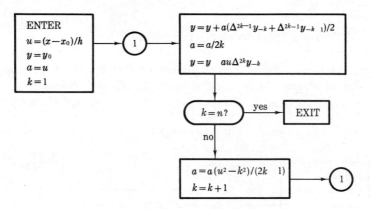

Figure 12-4: Interpolation by Stirling's formula

12.34   Bessel's Interpolation Formula

Bessel's formula is also a central difference formula, based on passing a polynomial of degree $2n + 1$ through $2n + 2$ points. In the notation of Section 12.34 it is

$$y = y_0 + u\Delta y_0$$

$$+ \frac{u(u-1)}{2!} \frac{\Delta^2 y_{-1} + \Delta^2 y_0}{2} + \frac{(u-1/2)u(u-1)}{3!} \Delta^3 y_{-1}$$

$$+ \frac{u(u^2-1^2)(u-2)}{4!} \frac{\Delta^4 y_{-2} + \Delta^4 y_{-1}}{2}$$

$$+ \frac{(u-1/2)u(u^2-1^2)(u-2)}{5!} \Delta^5 y_{-2}$$

$$+ \cdots$$

$$+ \frac{u(u^2-1^2)\cdots[u^2-(n-1)^2](u-n)}{(2n)!} \frac{\Delta^{2n} y_{-n} + \Delta^{2n} y_{-n+1}}{2}$$

$$+ \frac{(u-1/2)u(u^2-1^2)\cdots(u-n)}{(2n+1)!} \Delta^{2n+1} y_{-n} \qquad \textbf{(12-10)}$$

In this formula, as in Stirling's formula, each time n is increased by one, two terms are added to the expression. The flow chart, Figure 12-5, shows how the formula can be used, assuming the values y_j, the differences $\Delta^k y_j$, and the values of x, x_0, and h are already stored.

12.35   Lagrange's Interpolation Formula

This formula is the equation for an nth-degree polynomial through $n + 1$ points (x_0, y_0), (x_1, y_1), ..., (x_n, y_n) which are not necessarily equally spaced. It is

$$y = \frac{(x-x_1)(x-x_2)\cdots(x-x_n)}{(x_0-x_1)(x_0-x_2)\cdots(x_0-x_n)} y_0$$

$$+ \frac{(x-x_0)(x-x_2)\cdots(x-x_n)}{(x_1-x_0)(x_1-x_2)\cdots(x_1-x_n)} y_1$$

$$+ \cdots$$

$$+ \frac{(x-x_0)(x-x_1)\cdots(x-x_{n-1})}{(x_n-x_0)(x_n-x_1)\cdots(x_n-x_{n-1})} y_n \qquad \textbf{(12-11)}$$

It can be seen that this formula involves large numbers of multiplications and hence becomes quite slow if n is large. Since the other classical interpola-

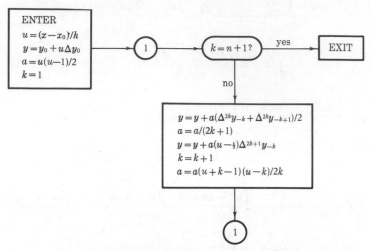

Figure 12-5: Interpolation by Bessel's formula

tion formulas as given in the preceding sections do not work for unequal intervals, this one is used, perhaps more often than it should be. Spline fits, to be discussed in Section 12-4, offer advantages over Lagrange interpolation for unequally spaced intervals. Also, the formulas of the preceding sections can be adapted to unequal intervals by the use of so-called "divided differences," a process that will not be described here.

12.36   Remainder Terms

The above formulas represent ways of approximating a function by a polynomial of degree n (or of degree $2n$ or $2n + 1$ for the central difference formulas). In Chapter 6 we discussed the method of approximating a function by a polynomial of degree n, using the Taylor series. In that case we found that the difference between the given function and the polynomial could be expressed as a remainder term, which was then the error made when the polynomial was used to represent the function. Remainder terms exist for the above formulas also. They are:

For Newton's forward interpolation formula:

$$R_n = \frac{f^{(n+1)}(\xi)}{(n+1)!}(x - x_0)(x - x_1)\cdots(x - x_n)$$

$$= \frac{h^{n+1}f^{(n+1)}(\xi)}{(n+1)!}u(u-1)\cdots(u-n) \qquad (12\text{-}12)$$

where ξ is some value intermediate between x_0 and x_n.

For Newton's backward interpolation formula:

$$R_n = \frac{h^{n+1}f^{(n+1)}(\xi)}{(n+1)!} u(u+1)(u+2)\cdots(u+n) \qquad \text{(12-13)}$$

For Stirling's formula:

$$R_n = \frac{h^{2n+1}f^{(2n+1)}(\xi)}{(2n+1)!} u(u^2-1)(u^2-2^2)\cdots(u^2-n^n) \qquad \text{(12-14)}$$

For Bessel's formula:

$$R_n = \frac{h^{2n+2}f^{(2n+2)}(\xi)}{(2n+2)!} u(u^2-1)(u^2-2^2)\cdots(u^2-n^2)(u-n-1) \quad \text{(12-15)}$$

For Lagrange's formula:

$$R_n = \frac{f^{(n+1)}(\xi)}{(n+1)!}(x-x_0)(x-x_1)\cdots(x-x_n) \qquad \text{(12-16)}$$

In these remainder terms, as in the Taylor remainder, our problem in estimating the remainder comes in estimating the derivative $f^{(n+1)}(\xi)$ [or $f^{(2n+1)}(\xi)$, or $f^{(2n+2)}(\xi)$] at the unknown point ξ. If we know the function f, we can make the estimate in the same manner as we did for the Taylor formula, by actually taking the $(n+1)$st derivative and then determining an upper limit on its value. If we do not know the function f, the best we can do is to assume that the $(n+1)$st derivative is approximately given by

$$f^{(n+1)}(\xi) \approx \frac{\Delta^{n+1}y_0}{h^{n+1}}$$

or for the central difference formulas

$$f^{(2n+1)}(\xi) \approx \frac{\Delta^{2n+1}y_{-n} + \Delta^{2n+1}y_{-n+1}}{2h^{2n+1}}$$

and

$$f^{(2n+2)}(\xi) \approx \frac{\Delta^{2n+2}y_{-n-1} + \Delta^{2n+2}y_{-n}}{2h^{2n+1}}$$

If these estimates of the derivatives are substituted back in (12-12) to (12-15) and the results compared to the corresponding interpolation formulas, it is seen that the error as estimated in this fashion is just equal to the next term in the interpolation formula beyond those being used. It is possible to use this type of error estimate to choose the order of interpolation automatically. The flow chart, Figure 12-6, illustrates how this might be done for Stirling's formula. The flow chart assumes that the values y_j, the differences $\Delta^k y_j$, the values x, x_0, h, and a specified relative error E are available at the beginning of the calculation.

The flow chart, Figure 12-6, shows an automatic exit after $k = 5$, even if the specified accuracy has not been achieved. Some arrangement such as this is advisable. At this point differences up through the tenth difference have been used, and practical problems where the tabulated information is sufficiently accurate to warrant such high-order interpolation are extremely rare.

Example 1.   Given M values of Y already stored, where M $\leqslant$ 100, for M equally spaced values of X beginning with XBEG and spaced DX apart and difference tables up through eleventh differences, write a FORTRAN subroutine that will do interpolation by Stirling's formula, using that order of interpolation which will make the relative error less than ERR.

A suitable program is

```
     SUBROUTINE STIR(Y,M,DY,XBEG,DX,XINT,ERR,YINT)
     DIMENSION Y(100),DY(11,100),E(6)
     N=1
     MO=(XINT-XBEG+DX)/DX
   1 IF(MO)20,20,2
   2 IF(MO-M)3,3,20
   3 IF(MO-N)4,4,5
   4 MO=N+1
   5 IF(M-MO-N)6,7,7
   6 MO=M-N
   7 FMO=MO-1
     XO=XBEG+FMO*DX
     U=(XINT-XO)/DX
     A=U
     YINT=Y(MO)
   8 K=1
   9 KK=2*K
     MM=MO-K
     E(K)=A*(DY(KK-1,MM)+DY(KK-1,MM+1))/2.
     IF(K-5)10,10,13
  10 IF(ABSF(E(K))-ABSF(YINT*ERR))14,14,11
```

```
   11 YINT=YINT+E(K)
      FKK=KK
      A=A/FKK
      YINT=YINT+A*U*DY(KK,MM)
      FK=K
      A=A*(U+FK)*(U-FK)/(FKK+1.)
      K=K+1
      GO TO 9
   13 PRINT 101, YINT,E(K)
   14 RETURN
   20 PRINT 102
      STOP
  101 FORMAT(33H DESIRED ACCURACY NOT ACHIEVED,
                        Y=,E12.4,4HERR=,E12.4)
  102 FORMAT(31H OUT OF RANGE FOR INTERPOLATION)
      END
```

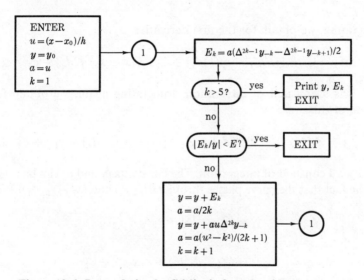

ENTER
$u=(x-x_0)/h$
$y=y_0$
$a=u$
$k=1$

$E_k = a(\Delta^{2k-1}y_{-k} - \Delta^{2k-1}y_{-k+1})/2$

$k>5$? →yes→ Print y, E_k EXIT

no

$|E_k/y| < E$? →yes→ EXIT

no

$y = y + E_k$
$a = a/2k$
$y = y + au\Delta^{2k}y_{-k}$
$a = a(u^2 - k^2)/(2k+1)$
$k = k+1$

Figure 12-6: Interpolation by Stirling's formula with error control

12.4   SPLINE FITS

A well-known way of interpolating by hand is to plot points and draw a smooth curve through these points using a draftsman's spline. The expert at drawing such curves can do so in such a way that not only the curve itself, but also its slope and its curvature, are continuous functions. Interpolation is done by merely reading points off the smooth curve thus generated. This type of interpolation is not at all difficult on a computer, and offers several advantages over the classical interpolation methods discussed above.

12.41   Derivation of the Equations

Let the given set of points be $(x_1, y_1), (x_2, y_2), \ldots, (x_m, y_m)$, arranged in order of increasing values of x. We accomplish the spline fit by connecting each pair of adjacent points with a section of a third-degree polynomial, matching up the sections so that the first and second derivatives are continuous at each point. Let $Z_1, Z_2, \ldots, Z_m$ be the values of the second derivative at the points. Then between points (x_k, y_k) and (x_{k+1}, y_{k+1}), the second derivative has the value

$$y'' = Z_k \frac{x_{k+1} - x}{d_k} + Z_{k+1} \frac{x - x_k}{d_k} \tag{12-17}$$

where $d_k = x_{k+1} - x_k$.

Integrating, we obtain for the first derivative

$$y' = -Z_k[(x_{k+1} - x)^2/2d_k] + Z_{k+1}[(x - x_k)^2/2d_k] + c_1 \tag{12-18}$$

where c_1 is a constant of integration. Integrating again, we obtain for the equation of the curve

$$y = Z_k[(x_{k-1} - x)^3/6d_k] + Z_{k+1}[(x - x_k)^3/6d_k] + c_1 x + c_2 \tag{12-19}$$

where c_2 is a constant of integration. The constants c_1 and c_2 can be evaluated from the fact that the curve passes through (x_k, y_k) and (x_{k+1}, y_{k+1}). We have

$$y_k = (Z_k d_k^2/6) + c_1 x_k + c_2$$

$$y_{k+1} = (Z_{k+1} d_k^2/6) + c_1 x_{k+1} + c_2$$

from which

$$c_1 = [(y_{k+1} - y_k)/d_k] - [(Z_{k+1} - Z_k)d_k/6] \tag{12-20}$$

$$c_2 = [(y_k x_{k+1} - y_{k+1} x_k)/d_k] - [(Z_k x_{k+1} - Z_{k+1} x_k)d_k/6] \tag{12-21}$$

Substituting these values in (12-19), we have for the equation of the curve,

$$y = [Z_k(x_{k+1} - x)^3/6d_k] + [Z_{k+1}(x - x_k)^3/6d_k]$$

$$+ [(x_{k+1} - x)(y_k/d_k - Z_k d_k/6)]$$

$$+ [(x - x_k)(y_{k+1}/d_k - Z_{k+1} d_k/6)] \tag{12-22}$$

In this equation all quantities are known except Z_k and Z_{k+1}, the values of the second derivative at the end points of the interval. One condition which will help determine these values is that the slope at (x_k, y_k) as determined from equation (12-18) must be the same as that determined by the corresponding formula for the interval (x_{k-1}, y_{k-1}) to (x_k, y_k). When the value of c_1 from (12-20) is used in (12-18) the equation becomes

$$y' = -[Z_k(x_{k+1} - x)^2/2d_k] + [Z_{k+1}(x - x_k)^2/2d_k]$$

$$+ [(y_{k+1} - y_k)/d_k] - [(Z_{k+1} - Z_k)d_k/6] \tag{12-23}$$

The corresponding relation for the preceding interval is

$$y' = -[Z_{k-1}(x_k - x)^2/2d_{k-1}] + [Z_k(x - x_{k-1})^2/2d_{k-1}]$$

$$+ [(y_k - y_{k-1})/d_{k-1}] - [(Z_k - Z_{k-1})d_{k-1}/6] \tag{12-24}$$

At the point (x_k, y_k), these relations give

$$y_k' = (-Z_k d_k/2) + [(y_{k+1} - y_k)/d_k] - [(Z_{k+1} - Z_k)d_k/6]$$

$$= (Z_k d_{k-1}/2) + [(y_k - y_{k-1})/d_{k-1}] - [(Z_k - Z_{k-1})d_{k-1}/6] \tag{12-25}$$

or, collecting the unknowns Z_{k-1}, Z_k, and Z_{k+1} on one side of the equation,

$$Z_{k-1}(d_{k-1}/6) + Z_k[(d_{k-1} + d_k)/3] + Z_{k+1}(d_k/6)$$

$$= [(y_{k+1} - y_k)/d_k] - [(y_k - y_{k-1})/d_{k-1}] \tag{12-26}$$

We have an equation like this for each of the internal points, that is, $k = 2, 3, \ldots, m - 1$. There are $m - 2$ equations in the m unknowns $Z_1, Z_2, \ldots, Z_m$. Two more conditions may be specified in order to determine these quantities completely. It is customary to place some additional condition on Z_1 and Z_m, the values of the second derivative at the end points. There are several reasonable choices for these values, and the particular choice will influence the shape of the fit, especially near the end points. We will demonstrate the fit for the use where the second derivative at each end is a linear extrapolation of the value at the two adjacent points. Stated another way, we will require the third derivative to be continuous at (x_2, y_2) and at (x_{m-1}, y_{m-1}). From equation (12-17) the third derivative is

$$y''' = -Z_k/d_k + Z_{k+1}/d_k \qquad\qquad \text{(12-27)}$$

Equating values for $k = 1$ and $k = 2$,

$$-Z_1/d_1 + Z_2/d_1 = -Z_2/d_2 + Z_3/d_2 \qquad\qquad \text{(12-28)}$$

or

$$-Z_1/d_1 + Z_2(1/d_1 + 1/d_2) - Z_3/d_2 = 0 \qquad\qquad \text{(12-29)}$$

In like manner, equating values for $k = m - 2$ and $k = m - 1$,

$$-Z_{m-2}/d_{m-2} + Z_{m-1}/d_{m-2} = Z_{m-1}/d_{m-1} + Z_m/d_{m-1}$$

or

$$-Z_{m-2}/d_{m-2} + Z_{m-1}(1/d_{m-2} + 1/d_{m-1}) - Z_m/d_{m-1} = 0 \qquad \text{(12-30)}$$

Equations (12-29) and (12-30), along with equations (12-26), constitute m equations in m unknowns for the quantities $Z_1, Z_2, \ldots, Z_m$. The eaqutions are particularly simple, since only three of the unknowns appear in each. Once these equations are solved and the Z_k determined, equation (12-22) can be used directly for finding y for any value of x between x_1 and x_m.

12.42   Computer Use of the Spline Fit

Spline-fit interpolation in the table of values $(x_1, y_1), (x_2, y_2), \ldots, (x_m, y_m)$ consists of determining which two points (x_k, y_k) and (x_{k+1}, y_{k+1}) the given value of x lies between and then finding y by the formula

$$y = c_{1,k}(x_{k+1} - x)^3 + c_{2,k}(x - x_k)^3 + c_{3,k}(x_{k+1} - x) + c_{4,k}(x - x_k) \quad \text{(12-31)}$$

where the constants $c_{1,k}, c_{2,k}, c_{3,k}, c_{4,k}$ have been previously computed and stored. A flow chart for this operation is shown in Figure 12-7.

Example 1.   Write a FORTRAN subroutine for spline-fit interpolation in the table of values (x_1, y_1) to (x_m, y_m), where m may be as large as 100, and where the constants $C(1, k)$, $C(2, k)$, $C(3, k)$, and $C(4, k)$ are already computed and stored.

A suitable subroutine is

```
SUBROUTINE SPLINE(X,Y,M,C,XINT,YINT)
DIMENSION X(100),Y(100),C(4,100)
IF(XINT−X(1))7,1,2
```

1 YINT = Y(1)
   RETURN
2 K = 1
3 IF(XINT − X(K + 1))6,4,5
4 YINT = Y(K + 1)
   RETURN
5 K = K + 1
   IF(M − K)7,7,3
6 YINT = (X(K + 1) − XINT)*(C(1,K)*(X(K + 1) − XINT)**2
   + C(3,K))
   YINT = YINT + (XINT − X(K))*(C,(2,K)*(XINT − X(K))**
   2 + C(4,K))
   RETURN
7 PRINT 101
   STOP
101 FORMAT(31H OUT OF RANGE FOR INTERPOLATION)
     END

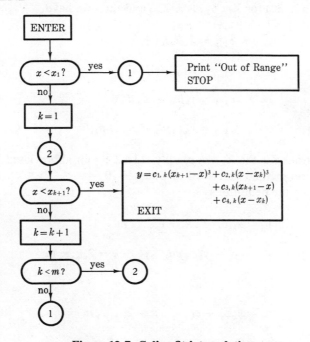

Figure 12-7: Spline-fit interpolation

In the above subroutine, the location of XINT in the table is found by comparing it with X(K + 1) for K = 1, 2, etc., until a value X(K + 1) greater than XINT is found. If the values of XINT for which this subroutine is to be used are spread uniformly through the table, an average of M/2 tests will be required to find the proper place in the table. A more efficient way of performing this search can be used, a method analogous to the bisection method of finding a root of an equation discussed in Section 8.3. To use this method, we would first compare XINT with X(M/2). If XINT is larger, we would next compare with X(3M/4), or if smaller, with X(M/4), etc. By this process, the number of steps required to find the correct interval for interpolation is about $\log_2(M)$. For M = 100, then, about 7 comparisons are required, as opposed to an average of 50 for the program as shown. For larger tables, the effect becomes more dramatic. The preparation of a flow chart and program for the improved method is left as an exercise for the reader.

12.43   Determination of Constants

In Section 12.42 it was assumed that the constants $c_{1,k}$, $c_{2,k}$, $c_{3,k}$, $c_{4,k}$ were available. These constants must be determined by solving equations (12-26), (12-29), and (12-30) for $Z_1, Z_2, \ldots, Z_m$, and then we have

$$c_{1,k} = Z_k/6d_k \tag{12-32}$$

$$c_{2,k} = Z_{k+1}/6d_k \tag{12-33}$$

$$c_{3,k} = y_k/d_k - Z_k\, d_k/6 \tag{12-34}$$

$$c_{4,k} = y_{k+1}/d_k - Z_{k+1}\, d_k/6 \tag{12-35}$$

The equations for the Z_k are readily solved by an abbreviated version of the elimination method described in Chapter 10.

Let us make the definitions

$$p_k = d_k/6, \qquad k = 1, \ldots, m = 1$$

$$e_k = (y_k - y_{k-1})/d_{k-1}, \qquad k = 2, \ldots, m$$

$$b_1 = 0$$

$$b_k = e_{k+1} - e_k, \qquad k = 2, \ldots, m = 1$$

$$b_m = 0$$

Then we can write the system of equations (12-26), (12-29), and (12-30) in matrix form as

$$AZ = b$$

where

$$A =$$

$$\begin{pmatrix} -1/d_1 & 1/d_1 + 1/d_2 & -1/d_2 & 0 & \cdots & 0 & 0 \\ p_1 & 2(p_1 + p_2) & p_2 & 0 & \cdots & 0 & 0 \\ 0 & p_2 & 2(p_2 + p_3) & p_3 & \cdots & 0 & 0 \\ 0 & & & & & & \\ \vdots & \vdots & \vdots & \vdots & & \vdots & \vdots \\ 0 & 0 & \cdots & & p_{m-2} & 2(p_{m-2} + p_{m-1}) & p_{m-1} \\ 0 & 0 & \cdots & & -1/d_{m-2} & 1/d_{m-2} + 1/d_{m-1} & -1/d_{m-1} \end{pmatrix}$$

Making use of the special nature of the above matrix, the Z_i can be found and then the c_{ij} determined by using the flow chart shown in Figure 12-8.

START

Input $x_k, y_k, \quad k = 1, \ldots, m$

$$\left.\begin{aligned} d_k &= x_{k+1} - x_k \\ p_k &= d_k/6 \\ e_k &= (y_{k+1} - y_k)/d_k \end{aligned}\right\} \quad k = 1, \ldots, m-1$$

$b_k = e_k - e_{k-1}, \quad k = 2, \ldots, m-1$

$a_{1,2} = -1 - d_1/d_2$

$a_{1,3} = d_1/d_2$

$a_{2,3} = p_2 - p_1 a_{1,3}$

$a_{2,2} = 2(p_1 + p_2) - p_1 a_{1,2}$

$a_{2,3} = a_{2,3}/a_{2,2}$

$b_2 = b_2/a_{2,2}$

$$\left.\begin{aligned} a_{k,k} &= 2(p_{k-1} + p_k) - p_{k-1} a_{k-1,k} \\ b_k &= b_k - p_{k-1} b_{k-1} \\ a_{k,k+1} &= p_k/a_{k,k} \\ b_k &= b_k/a_{k,k} \end{aligned}\right\} \quad k = 3, \ldots, m-1$$

$a_{m,m-1} = 1 + d_{m-2}/d_{m-1} + a_{m-2,m-1}$

$a_{m,m} = -d_{m-2}/d_{m-1} - a_{m,m-1} a_{m-1,m}$

$b_m = b_{m-2} - a_{m,m-1} b_{m-1}$

$z_m = b_m/a_{m,m}$

$z_k = b_k - a_{k,k+1} z_{k+1}, \quad k = m-1, \ldots, 2$

$$\left.\begin{aligned} z_1 &= -a_{1,2} z_2 - a_{1,3} z_3 \\ c_{1,k} &= z_k/6d_k \\ c_{2,k} &= z_{k+1}/6d_k \\ c_{3,k} &= y_k/d_k - z_k p_k \\ c_{4,k} &= y_{k+1}/d_k - z_{k+1} p_k \end{aligned}\right\} \quad k = 1, \ldots, m-1$$

STOP

Figure 12-8: Determination of coefficients for a spline fit

THE following FORTRAN subroutine will accomplish the calculation for up to 100 points:

```
        SUBROUTINE SPLICO(X,Y,M,C)
        DIMENSION X(100),Y(100),D(100),P(100),E(100),C(4,100),
        A(100,3),B(100),Z(100)
        MM=M-1
        DO 2 K=1,MM
        D(K)=X(K+1)-X(K)
        P(K)=D(K)/6.
    2   E(K)=(Y(K+1)-Y(K))/D(K)
        DO 3 K=2,MM
    3   B(K)=E(K)-E(K-1)
        A(1,2)=-1.-D(1)/D(2)
        A(1,3)=D(1)/D(2)
        A(2,3)=P(2)-P(1)*A(1,3)
        A(2,2)=2.*(P(1)+P(2))-P(1)*A(1,2)
        A(2,3)=A(2,3)/A(2,2)
        B(2)=B(2)/A(2,2)
        DO 4 K=3,MM
        A(K,2)=2.*(P(K-1)+P(K))-P(K-1)*A(K-1,3)
        B(K)=B(K)-P(K-1)*B(K-1)
        A(K,3)=P(K)/A(K,2)
    4   B(K)=B(K)/A(K,2)
        Q=D(M-2)/D(M-1)
        A(M,1)=1.+Q+A(M-2,3)
        A(M,2)=-Q-A(M,1)*A(M-1,3)
        B(M)=B(M-2)-A(M,1)*B(M-1)
        Z(M)=B(M)/A(M,2)
        MN=M-2
        DO 6 I=1,MN
        K=M-I
    6   Z(K)=B(K)-A(K,3)*Z(K+1)
        Z(1)=-A(1,2)*Z(2)-A(1,3)*Z(3)
        DO 7 K=1,MM
        Q=1./(6.*D(K))
        C(1,K)=Z(K)*Q
        C(2,K)=Z(K+1)*Q
        C(3,K)=Y(K)/D(K)-Z(K)*P(K)
    7   C(4,K)=Y(K+1)/D(K)-Z(K+1)*P(K)
        RETURN
  101   FORMAT(2E12.4)
        END
```

12.5 NUMERICAL DIFFERENTIATION

When dealing with a function which is defined by an analytic expression, it is usually possible to find an analytic expression for its derivative by the methods of elementary calculus. If the function is extremely complicated, or if the function is known as a table of values, it may be necessary to resort to numerical differentiation. To perform numerical differentiation, we represent the function by one of the interpolation formulas and then differentiate the formula as many times as desired. For example, for Newton's forward interpolation formula we have

$$\frac{dy}{dx} = \frac{dy}{du}\frac{du}{dx} = \frac{1}{h}\left\{ \Delta y_0 + \frac{\Delta^2 y_0}{2!}\left[(u-1)+u\right] \right.$$

$$+ \frac{\Delta^3 y_0}{3!}\left[(u-1)(u-2)+(u)(u-2)+u(u-1)\right] + \cdots$$

$$+ \frac{\Delta^n y_0}{n!}\left[(u-1)(u-2)\cdots(u-n+1)+u(u-2)\cdots\right.$$

$$\left.\left.(u-n+1)+\cdots+u(u-1)\cdots(u-n+2)\right]\right\}$$

$$\frac{dy^2}{dx^2} = \frac{1}{h^2}\left\{ \Delta^2 y_0 + \frac{\Delta^3 y_0}{3!}\left[2(u-2)+2(u-1)+2u\right] + \cdots \right.$$

$$+ \frac{\Delta^n y_0}{n!}\left[2(u-2)(u-3)\cdots(u-n+1)+2u(u-3)\cdots\right.$$

$$\left.\left.(u-n+1)+\cdots+2u(u-1)\cdots(u-n+3)\right]\right\}$$

etc. It is frequently of interest to have the value just at $x = x_0$, or $u = 0$. There we have

$$(dy/dx)_{x_0} = (1/h)(\Delta y_0 - \Delta^2 y_0/2 + \Delta^3 y_0/3 + \cdots + (-1)^{n-1}\Delta^n y_0/n)$$

and

$$(d^2 y/dx^2)_{x_0} = (1/h^2)\{\Delta^2 y_0 - \Delta^3 y_0 + \cdots$$

$$+ (-1)^n 2\Delta^n y_0 \times [1 + 1/2 + \cdots + 1/(n-1)]/n\}$$

Similar expressions can be developed for the other interpolation formulas. For the spline fit, the relations are simply

$$dy/dx = -3c_{1,k}(x_{k+1} - x)^2 + 3c_{2,k}(x - x_k)^2 - c_{3,k} + c_{4,k}$$

$$d^2y/dx^2 = 6c_{1,k}(x_{k+1} - x) + 6c_{2,k}(x - x_k)$$

Derivatives obtained from the interpolation formulas tend to be less accurate than function values themselves. Since accuracy is quite difficult to estimate, it will not be discussed here, but the reader is warned that it can be a serious problem.

12.51   Flow Charts

Direct differentiation of the classical interpolation formulas is somewhat messy because of the many products involved. It is easier to modify the flow charts for the classical interpolation formulas to give values of the derivatives. One merely differentiates all quantities in the flow chart which involve x. The flow charts given in Figures 12-9, 12-10, and 12-11 describe the calculation of the function and its first two derivatives using the classical interpolation formulas. Prime denotes derivative by x.

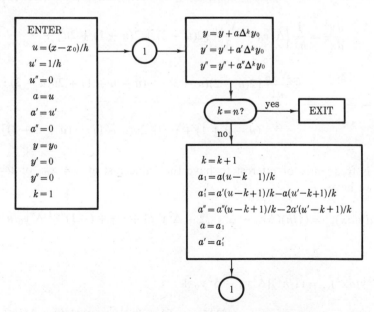

Figure 12-9: Newton's forward interpolation for a function, first and second derivatives

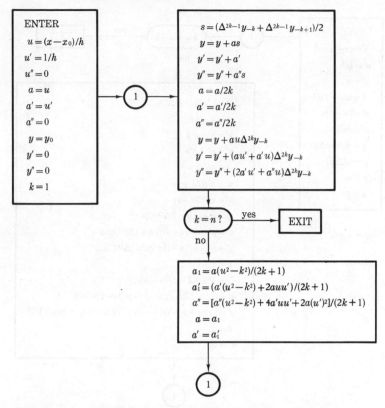

Figure 12-10: Stirling's interpolation for a function, first and second derivatives

EXERCISE 30

1. Given the table of values

x	1	1.1	1.2	1.3	1.4	1.5	1.6	1.7	1.8	1.9	2.0
y	43.4	47.7	52.1	56.4	60.8	65.1	69.5	73.8	78.2	82.5	86.9

   a. Write a table of differences.
   b. Find the value at 1.05, using Newton's forward interpolation formula, $n = 3$.
   c. Find the value at 1.95, using Newton's backward interpolation formula, $n = 3$.
   d. Find the value at 1.55, using Stirling's formula, $n = 2$.
   e. Find the value at 1.55, using Bessel's formula, $n = 2$.

2. a. Write a FORTRAN program that will read in 10 values of y, $y_1, \ldots, y_{10}$, an initial value x_1 and spacing h, then read x and call subroutine STIR of Section 12.36, and print y, for as many input values of x as may be given.
   b. Check your program and subroutine STIR by using $x_1 = 0$, $h = 1$, and 10 points from the curve $y = x^2 - 5x$. Try ERR = .1, .001, .0000001.

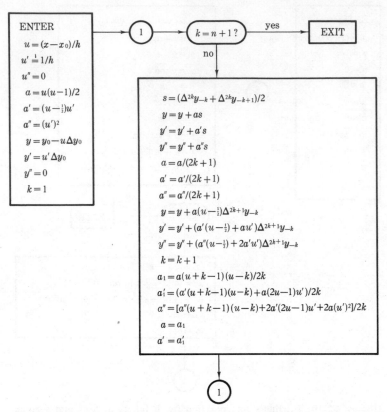

Figure 12-11: Bessel's interpolation for a function, first and second derivatives

   c. For $x = 1, 2, \ldots, 10$, a function is defined by the expression $y = \ln(x!)$. Use your program, with ERR $= .01$, to find y for $x = 1.5$ and $x = 7.5$.

3. a. Write a FORTRAN program that will read in 10 pairs of values (x_1, y_1), $\ldots$, (x_{10}, y_{10}), call subroutine SPLICO of Section 12.43 to determine constants for a spline fit, then read x and call subroutine SPLINE of Section 12.42, and print y, for as many input values of x as may be available.

   b. Check your program by running problem 2b. (You cannot specify the error, of course.)

   c. Use your program to perform problem 2c.

4. Given M values of y already stored, where M $\leq$ 100, for M equally spaced values of x beginning with XBEG and spaced DX apart, and difference tables through eleventh differences, write a FORTRAN subroutine BESS that will do interpolation by Bessel's formula, using the order of interpolation that will make the relative error less than ERR. (See Example 1, Section 12.36.)

5. Prepare a new flow chart and FORTRAN subroutine for the spline fit, based on the improved search method described in the last paragraph of Section 12.42.

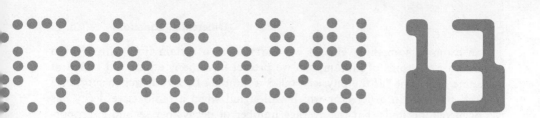

Differential Equations

13.1   INTRODUCTION

In problems in science and engineering, we are usually interested in studying the properties of a physical system and predicting its behavior. We do this by constructing a mathematical model of the physical system. We assign names to the things we can measure about the system and then represent them by symbols x, y, t, etc. The basic laws of physics tell us what relationships should exist between these quantities, and enable us to write equations involving the mathematical symbols which represent them. If these equations can be solved, so that we can obtain numerical values for the various quantities, then we can say that we understand the mathematical model, and hopefully we have gained some useful information or insight concerning the behavior of the physical system it represents.

The physical laws which give us the equations quite frequently involve not only the quantities of interest but also rates of change of these quantities. Mathematically these rates of change are represented as derivatives; the equations to be solved are called differential equations.

As is indicated by the generality of the above paragraphs, the field of differential equations is a broad one, having far-reaching applications in science and engineering. In this chapter we will give only a brief introduction to this important subject.

13.2   AN ELEMENTARY EXAMPLE

In a radioactive substance—say, radium—each atom is in an unstable condition, and sooner or later will cast off some of its excess energy as radiation. When it has done this it is no longer an atom of radium but has become an atom of something else. If we start out at a certain time with a given number of atoms of radium, can we predict how many atoms will be left at some later time? If the physicist takes a sample of the radioactive substance and places it in a detector which will signal when a radioactive decay has occurred, he finds that the average number of decays per second is proportional to the amount of material present. This appears to be the basic law governing radioactive decay. In order to state this law in mathematical terms, let

$$y = \text{number of atoms of radium present in a sample}$$

Since each decay signifies the disappearance of an atom of radium, the number of decays per second is

$$-dy/dt$$

The above law then states that

$$-dy/dt = \lambda y \tag{13-1}$$

where λ is a constant of proportionality determined experimentally.

Equation (13-1) is the differential equation that describes the problem. We would say that it is solved if given any particular time, t, we can determine the value of y.

We can write equation (13-1) as

$$dy/y = -\lambda \, dt \tag{13-2}$$

and, integrating both sides we have

$$\ln y = -\lambda t + c \tag{13-3}$$

where c is a constant of integration. Suppose that when we started, at time 0, there were exactly y_0 atoms. Substituting $y = y_0$ and $t = 0$ in (13-3), we find that

$$\ln y_0 = c \tag{13-4}$$

which fixes the value of c. Substituting this back into (13-3),

$$\ln y = -\lambda t + \ln y_0 \tag{13-5}$$

and taking exponentials of both sides,

$$y = y_0 \, e^{-\lambda t} \tag{13-6}$$

This equation represents the solution. Given any value of t, the corresponding value of y can be determined directly.

13.3 SYSTEMS OF LINEAR EQUATIONS WITH CONSTANT COEFFICIENTS

A very important class of differential equations is that in which there are several dependent variables, $y_1, y_2, \ldots, y_n$, and one independent variable, t, and the basic physical laws lead to the equations

$$dy_1/dt = a_{11}y_1 + a_{12}y_2 + \cdots + a_{1n}y_n$$

$$dy_2/dt = a_{21}y_1 + a_{22}y_2 + \cdots + a_{2n}y_n \qquad \textbf{(13-7)}$$

$$\vdots$$

$$dy_n/dt = a_{n1}y_1 + a_{n2}y_2 + \cdots + a_{nn}y_n$$

where the a_{ij} are constants.

Such a set of equations can be written for nearly any mechanical system of masses, springs, and shock absorbers, and its solution will give the motions of the masses as a function of time if they are given some initial motion. Such a set also can be written for an electrical circuit made up of resistors, capacitors, and inductors, and its solution will give the currents and voltages in the various parts of the circuit as a function of time for any set of initial currents or voltages. Such a system of equations also arises in the study of the stability of nearly any mechanical, electrical, chemical, or biological system. Because the equations can represent voltages in an electrical system, analog computers are frequently used to solve equations of the above type. There are many occasions, however, when digital computer solutions are needed.

Higher-order equations of the form

$$a_1 \frac{d^n y}{dt^n} + a_2 \frac{d^{n-1}y}{dt^{n-1}} + \cdots + a_n y = 0$$

can also be written as a system of equations such as (13-7) by letting

$$y = y_1$$

$$y_1' = y_2$$

$$\vdots$$

$$\frac{d^{n-1}y}{dt^{n-1}} = y_n$$

Matrix notation can be used for the system of equations (13-7) if we define the derivative of a matrix to be the matrix made up of the derivatives of the individual elements; that is,

$$
\frac{d}{dt}\begin{pmatrix} y_1 \\ y_2 \\ \vdots \\ y_n \end{pmatrix} = \begin{pmatrix} \dfrac{dy_1}{dt} \\ \dfrac{dy_2}{dt} \\ \vdots \\ \dfrac{dy_n}{dt} \end{pmatrix} \tag{13-8}
$$

Then if we let

$$
y = \begin{pmatrix} y_1 \\ y_2 \\ \vdots \\ y_n \end{pmatrix}, \qquad A = \begin{pmatrix} a_{11} & a_{12} & \cdots & a_{1n} \\ a_{21} & a_{22} & \cdots & a_{2n} \\ \vdots & & & \\ a_{n1} & a_{n2} & \cdots & a_{nn} \end{pmatrix} \tag{13-9}
$$

the system of equations (13-7) can be written

$$
dy/dt = Ay \tag{13-10}
$$

Because of the resemblance of equation (13-10) to equation (13-1) of Section 13.2, let us try a solution of the form

$$
y = \begin{pmatrix} c_1 e^{\lambda t} \\ c_2 e^{\lambda t} \\ \vdots \\ c_n e^{\lambda t} \end{pmatrix} = c e^{\lambda t}, \qquad \text{where } c = \begin{pmatrix} c_1 \\ c_2 \\ \vdots \\ c_n \end{pmatrix} \tag{13-11}
$$

Then

$$
dy/dt = \lambda c e^{\lambda t} \tag{13-12}
$$

and substituting in (13-10), we obtain

$$
\lambda c e^{\lambda t} = A c e^{\lambda t}
$$

or

$$
\lambda c = A c
$$

or

$$
(A - \lambda I)c = 0 \tag{13-13}
$$

This is precisely the equation of Section 10.7 for determining the eigenvalues and eigenvectors of the matrix A. We found in Chapter 10 that there are n values of λ for which this system has a solution, and these are the eigenvalues of the matrix A. If these values are discrete, there is one eigen-vector, u_i, corresponding to each eigenvalue λ_i. Thus we have found not one, but n different solutions:

$$u_1 e^{\lambda_1 t}, \, u_2 e^{\lambda_2 t}, \, \ldots, \, u_n e^{\lambda_n t}$$

It can be proved that, for a system of linear differential equations such as (13-7), any linear combination of solutions is also a solution, so the most general solution we have found is

$$y = k_1 u_1 e^{\lambda_1 t} + k_2 u_2 e^{\lambda_2 t} + \cdots + k_n u_n e^{\lambda_n t} \qquad \text{(13-14)}$$

where $\lambda_1, \ldots, \lambda_n$ are the eigenvalues of A, $u_1, \ldots, u_n$ are the corresponding eigenvectors, and $k_1, \ldots, k_n$ are arbitrary constants. Ordinarily in the physical problem we will have side conditions that determine the value of these constants. For example, we may know the values of $y_1, y_2, \ldots, y_n$ for $t = 0$, that is, initial conditions for the problem of interest, and these will determine the k_i.

Example 1.   Solve the system $y' = Ay$, where

$$A = \begin{pmatrix} -1 & -2 & 2 \\ 3 & 4 & -2 \\ -1 & -1 & 3 \end{pmatrix}$$

subject to the initial condition

$$y = \begin{pmatrix} 1 \\ 2 \\ 2 \end{pmatrix} \qquad \text{when } t = 0$$

The eigenvalues for A are obtained by setting

$$\begin{vmatrix} -1 - \lambda & -2 & 2 \\ 3 & 4 - \lambda & -2 \\ -1 & -1 & 3 - \lambda \end{vmatrix} = 0$$

If we expand this determinant we obtain the characteristic equation

$$-\lambda^3 + 6\lambda^2 - 11\lambda + 6 = 0$$

which has the roots

$$\lambda_1 = 1, \qquad \lambda_2 = 2, \qquad \lambda_3 = 3$$

The eigenvector corresponding to λ_1 is obtained by setting

$$(A - \lambda_1 I)c = 0$$

or

$$\begin{pmatrix} -2 & -2 & 2 \\ 3 & 3 & -2 \\ -1 & -1 & 2 \end{pmatrix} \begin{pmatrix} c_1 \\ c_2 \\ c_3 \end{pmatrix} = 0$$

This set of equations is satisfied if $c_1 = 1$, $c_2 = -1$, and $c_3 = 0$. Thus an eigenvector corresponding to λ_1 is

$$u_1 = \begin{pmatrix} 1 \\ -1 \\ 0 \end{pmatrix}$$

In like manner we can find that eigenvectors corresponding to λ_2 and λ_3 are

$$\begin{pmatrix} 0 \\ 1 \\ 1 \end{pmatrix} \quad \text{and} \quad \begin{pmatrix} 1 \\ -1 \\ 1 \end{pmatrix}$$

Hence the general solution is

$$y = \begin{pmatrix} 1 \\ -1 \\ 0 \end{pmatrix} k_1 e^t + \begin{pmatrix} 0 \\ 1 \\ 1 \end{pmatrix} k_2 e^{2t} + \begin{pmatrix} 1 \\ -1 \\ 1 \end{pmatrix} k_3 e^{3t}$$

This can be written

$$y = \begin{pmatrix} 1 & 0 & 1 \\ -1 & 1 & -1 \\ 0 & 1 & 1 \end{pmatrix} \begin{pmatrix} k_1 e^t \\ k_2 e^{2t} \\ k_3 e^{3t} \end{pmatrix}$$

When $t = 0$, we have

$$\begin{pmatrix} 1 \\ 2 \\ 2 \end{pmatrix} = \begin{pmatrix} 1 & 0 & 1 \\ -1 & 1 & -1 \\ 0 & 1 & 1 \end{pmatrix} \begin{pmatrix} k_1 \\ k_2 \\ k_3 \end{pmatrix}$$

so that

$$k_1 = 2, \quad k_2 = 3, \quad k_3 = -1$$

and the solution is

$$y = \begin{pmatrix} 1 & 0 & 1 \\ -1 & 1 & -1 \\ 0 & 1 & 1 \end{pmatrix} \begin{pmatrix} 2e^t \\ 3e^{2t} \\ -e^{3t} \end{pmatrix}$$

or

$$y_1 = 2e^t - e^{3t}$$
$$y_2 = -2e^t + 3e^{2t} + e^{3t}$$
$$y_3 = 3e^{2t} - e^{3t}$$

13.31   Linear Independence and Multiple Eigenvalues

In the above discussions it has been assumed that the eigenvalues λ_i are all discrete; that is, the characteristic equation has no multiple roots. In this case we obtain n different solutions of the form $u_i e^{\lambda_i t}$. These solutions are linearly independent; that is, no one solution is a linear combination of the other solutions. In Example 1, Section 13-3, for instance, it is impossible to add terms containing e^{2t} and e^{3t} and obtain a term containing e^t. In the theory of differential equations, it is shown that a system of n first-order differential equations has exactly n linearly independent solutions, so that the solution given by (13-14) is indeed the most general solution for the system.

If the characteristic equation has multiple roots, however, the above procedure may or may not give n linearly independent solutions.

Example 1. Find the general solution of

$$y' = \begin{pmatrix} 1 & 0 \\ 0 & 1 \end{pmatrix} y$$

The eigenvalues are $\lambda_1 = 1$, $\lambda_2 = 1$, and the eigenvectors are solutions of the equations

$$\begin{pmatrix} 0 & 0 \\ 0 & 0 \end{pmatrix} \begin{pmatrix} x_1 \\ x_2 \end{pmatrix} = 0$$

Since there are two unknowns and the rank of the coefficient matrix is zero, there are two linearly independent solutions. By inspection, these may be taken as

$$\begin{pmatrix} 1 \\ 0 \end{pmatrix} \quad \text{and} \quad \begin{pmatrix} 0 \\ 1 \end{pmatrix}$$

With these two eigenvectors, we can write the general solution

$$y = k_1 \begin{pmatrix} 1 \\ 0 \end{pmatrix} e^t + k_2 \begin{pmatrix} 0 \\ 1 \end{pmatrix} e^t$$

Even though the functions e^t and e^t are not linearly independent, we still have two linearly independent solutions because there were two independent eigenvectors corresponding to the double eigenvalue. This does not happen in every case.

Example 2. Find the general solution of

$$y' = \begin{pmatrix} 0 & 1 \\ -1 & 2 \end{pmatrix} y$$

To find the eigenvalues, we set

$$\begin{vmatrix} -\lambda & 1 \\ -1 & 2 - \lambda \end{vmatrix} = 0$$

or

$$\lambda^2 - 2\lambda + 1 = 0$$

which has the solutions

$$\lambda_1 = 1, \qquad \lambda_2 = 1$$

To find the eigenvector corresponding to λ_1, set

$$\begin{pmatrix} -1 & 1 \\ -1 & 1 \end{pmatrix} \begin{pmatrix} x_1 \\ x_2 \end{pmatrix} = 0$$

These equations are satisfied if $x_1 = x_2 = 1$. Hence an eigenvector is

$$\begin{pmatrix} 1 \\ 1 \end{pmatrix}$$

Clearly if we repeat the process with λ_2, we get nothing new, just the same thing over again. There must be another linearly independent solution, however, and some other method is needed to find it. One method is that known as variation of parameters. Since $y = u_1 e^t$ is a solution, we try making the substitution

$$y = v(t)e^t, \qquad \text{where } v(t) = \begin{pmatrix} v_1(t) \\ v_2(t) \end{pmatrix}$$

is as yet unspecified. Then

$$y' = v'e^t + ve^t$$

Substituting in the original equation,

$$v'e^t + ve^t = A v e^t$$

or

$$v' = (A - I)v = \begin{pmatrix} -1 & 1 \\ -1 & 1 \end{pmatrix} v$$

Now try

$$v = \begin{pmatrix} 1 \\ 1 \end{pmatrix} t + \begin{pmatrix} h_1 \\ h_2 \end{pmatrix}$$

Then

$$v' = \begin{pmatrix} 1 \\ 1 \end{pmatrix} = \begin{pmatrix} -1 & 1 \\ -1 & 1 \end{pmatrix} \left[\begin{pmatrix} 1 \\ 1 \end{pmatrix} t + \begin{pmatrix} h_1 \\ h_2 \end{pmatrix} \right]$$

Simplifying

$$\begin{pmatrix} 1 \\ 1 \end{pmatrix} = \begin{pmatrix} 0 \\ 0 \end{pmatrix} t + \begin{pmatrix} -h_1 + h_2 \\ -h_1 + h_2 \end{pmatrix}$$

So the equation is satisfied if $h_1 = 0$ and $h_2 = 1$. Hence

$$y = ve^t$$

is a solution if

$$v = \begin{pmatrix} 1 \\ 1 \end{pmatrix} t + \begin{pmatrix} 0 \\ 1 \end{pmatrix} = \begin{pmatrix} t \\ t + 1 \end{pmatrix}$$

Hence two linearly independent solutions are

$$y = \begin{pmatrix} 1 \\ 1 \end{pmatrix} e^t \quad \text{and} \quad y = \begin{pmatrix} t \\ t + 1 \end{pmatrix} e^t$$

and the general solution is

$$y = \begin{pmatrix} 1 \\ 1 \end{pmatrix} k_1 e^t + \begin{pmatrix} t \\ t + 1 \end{pmatrix} k_2 e^t$$

The method used in Example 1 can be generalized to apply to any case with multiple roots. Suppose λ_i is a root of multiplicity r for the equation

$$y' = Ay \tag{13-15}$$

and let q_1 be an eigenvector corresponding to λ_i. Making the substitution

$$y = ve^{\lambda_i t}$$

we have

$$y' = v'e^{\lambda_i t} + \lambda_i ve^{\lambda_i t} = Ave^{\lambda_i t}$$

or

$$v' + \lambda_i v = Av$$

or

$$v' = (A - \lambda_i I)v \tag{13-16}$$

Now let

$$v = q_1 t + q_2$$

Then equation (13-16) becomes

$$q_1 = (A - \lambda_i I)q_1 t + (A - \lambda_i I)q_2$$

Since q_1 is an eigenvector corresponding to λ_i,

$$(A - \lambda_i I)q_1 = 0 \tag{13-17}$$

and the equation becomes

$$q_1 = (A - \lambda_i I)q_2 \tag{13-18}$$

Hence equation (13-16) is satisfied if we can choose a vector q_2 to satisfy (13-18). [Whether or not there is a solution depends, according to Theorem 1 of Section 10.62, on the rank of the matrix $A - \lambda_i I$ and the rank of the augmented matrix consisting of the elements of $A - \lambda_i I$ and one additional column consisting of the elements of q_1. If the system is inconsistent, then either all the linearly independent solutions corresponding to λ_i have already been found, or q_1 on the right-hand side of (13-18) needs to be replaced by some other eigenvector or linear combination of eigenvectors corresponding to λ_i.] If we can solve (13-18), then, we have another solution to the differential equation

$$y = (q_1 t + q_2)e^{\lambda_i t}$$

If λ_i is a root of multiplicity higher than 2, let

$$v(t) = q_1 t^2 + 2q_2 t + q_3$$

where q_1 and q_2 are the vectors determined above; when this is substituted into equation (13-16), we obtain

$$2q_1 t + 2q_2 = (A - \lambda_i I)q_1 t^2 + (A - \lambda_i I)2q_2 t + (A - \lambda_i I)q_3$$

Using (13-17) and (13-18) in this relation, it simplifies to

$$2q_2 = (A - \lambda_i I)q_3 \tag{13-19}$$

and if we choose q_3 to satisfy this relation, we have another solution to the original equation:

$$y = (q_1 t^2 + 2q_2 t + q_3)e^{\lambda_i t}$$

Continuing in this manner, we can usually generate the other linearly independent solutions up to a number equal to the multiplicity of the root.

Example 3.   Find the general solution of

$$y' = \begin{pmatrix} 0 & 1 & 0 \\ 0 & 0 & 1 \\ -8 & -12 & -6 \end{pmatrix} y$$

To find the eigenvalues we set

$$\begin{vmatrix} -\lambda & 1 & 0 \\ 0 & -\lambda & 1 \\ -8 & -12 & -6-\lambda \end{vmatrix} = 0$$

and obtain the characteristic equation

$$-\lambda^3 - 6\lambda^2 - 12\lambda - 8 = 0$$

which has the roots

$$\lambda_1 = -2, \qquad \lambda_2 = -2, \qquad \lambda_3 = -2$$

To find the eigenvector, we set

$$\begin{pmatrix} 2 & 1 & 0 \\ 0 & 2 & 1 \\ -8 & -12 & -4 \end{pmatrix} q_1 = 0$$

and find that this is satisfied by

$$q_1 = \begin{pmatrix} 1 \\ -2 \\ 4 \end{pmatrix}$$

and so one solution is

$$y = \begin{pmatrix} 1 \\ -2 \\ 4 \end{pmatrix} e^{-2t}$$

To find the vector q_2, we set

$$\begin{pmatrix} 2 & 1 & 0 \\ 0 & 2 & 1 \\ -8 & -12 & -4 \end{pmatrix} q_2 = \begin{pmatrix} 1 \\ -2 \\ 4 \end{pmatrix}$$

This set of equations is satisfied by

$$q_2 = \begin{pmatrix} 1 \\ -1 \\ 0 \end{pmatrix}$$

and so another solution is

$$y = \begin{pmatrix} t+1 \\ -2t-1 \\ 4t \end{pmatrix} e^{-2t}$$

To find the vector q_3 we set

$$\begin{pmatrix} 2 & 1 & 0 \\ 0 & 2 & 1 \\ -8 & -12 & -4 \end{pmatrix} q_3 = \begin{pmatrix} 2 \\ -2 \\ 0 \end{pmatrix}$$

This set of equations is satisfied by

$$q_3 = \begin{pmatrix} 1 \\ 0 \\ -2 \end{pmatrix}$$

and so another solution is

$$y = \begin{pmatrix} t^2 + 2t + 1 \\ -2t^2 - 2t \\ 4t^2 - 2 \end{pmatrix} e^{-2t}$$

and so the general solution can be written

$$y = \begin{pmatrix} 1 \\ -2 \\ 4 \end{pmatrix} k_1 e^{-2t} + \begin{pmatrix} t+1 \\ -2t-1 \\ 4t \end{pmatrix} k_2 e^{-2t} + \begin{pmatrix} t^2 + 2t + 1 \\ -2t^2 - 2t \\ 4t^2 - 2 \end{pmatrix} k_3 e^{-2t}$$

13.32 Computer Methods

From the examples of the preceding sections it is clear that all the computer techniques needed to solve systems of differential equations of the type illustrated above has already been given in Chapters 9 and 10. We must be able to find the eigenvalues and eigenvectors of a matrix, a subject covered in Section 10.7. Occasionally, in the case of multiple eigenvalues, we must solve a set of nonhomogeneous linear equations of the type

$$(A - \lambda_i I)x = q_1$$

The method of Section 10.64 will suffice for this. It might be noted that the methods of Sections 10.2 and 10.3 do not work for this system, since

$$\det(A - \lambda_i I) = 0$$

EXERCISE 31

1. Find the complete solution of the following sets of linear equations:

a. $y' = \begin{pmatrix} 2 & 2 \\ 4 & -3 \end{pmatrix} y$

b. $y' = \begin{pmatrix} 1 & 1 & -3 \\ 0 & 2 & -3 \\ 0 & 0 & -1 \end{pmatrix} y$

c. $y' = \begin{pmatrix} 1 & 0 & 0 \\ 0 & 2 & 2 \\ 0 & 4 & -3 \end{pmatrix} y$

d. $y' = \begin{pmatrix} 0 & 1 & 0 \\ 0 & 0 & 1 \\ 1 & 0 & -3 \end{pmatrix} y$

2. Write a FORTRAN program that will find the complete solution of a 2 by 2 system of homogeneous linear differential equations with constant coefficients.

3. A symmetric matrix is one whose corresponding elements across the main diagonal are equal, $a_{ij} = a_{ji}$. All the eigenvalues of a real symmetric matrix are real. Write a FORTRAN program, calling upon the subroutines of Chapters 9 and 10, that will find all linearly independent solutions of $y' = Ay$, for A a real symmetric matrix of order up to 20.

13.4   THE GENERAL FIRST-ORDER EQUATION

So far, the differential equations we have looked at have been amenable to analytic solution. The use of the computer for such equations has been as an aid in evaluating various constants involved in the solution. For most types of differential equations, however, an analytical solution cannot be obtained, and then the computer has the very different function of finding numerical values which trace out the solution. We consider this problem only for a first-order equation in one dependent and one independent variable, which we may write

$$dy/dx = f(x, y) \qquad\qquad \textbf{(13-20)}$$

If we consider x and y as coordinates of a point in an (x, y) plane, then this equation defines a slope at each point. The curve

$$y = g(x) \qquad\qquad \textbf{(13-21)}$$

is said to be a solution of the differential equation if at every point on the curve its slope is equal to the slope defined by equation (13-20). Except under very unusual conditions, there is one and only one solution curve through each point in the (x, y) plane. In a physical problem we usually know one point of interest, and wish to trace the solution curve through that point. The situation is as shown in Figure 13-1. There is some point (x_0, y_0) at which we are to start, and some unknown solution curve through that point, whose slope at every point is given by (13-20).

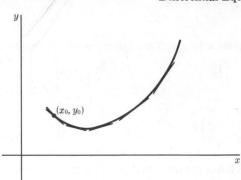

Figure 13-1

The basic process we must follow to trace this solution curve is quite clear. Since the slope at each point tells us the direction the curve is going, we must start at the initial point, move in the direction of the slope as defined by equation (13-20) to a new point, find a new slope, and move again in a new direction to a next point, and so forth.

The fundamental difficulty with this process is also quite clear, and is shown in Figure 13-2. The slope changes from point to point along the curve with

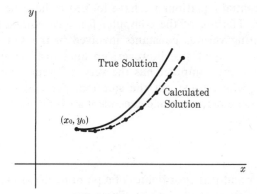

Figure 13-2

each infinitesimal change in position, while in a numerical process we must take finite, even if small, steps. Thus at the points we skip over in calculating, we will be using the wrong slope, and the solution we obtain may tend to wander away from the true solution.

It will be shown shortly that once an error occurs at a step in the calculation, it may tend to grow through later steps, even if the steps themselves are not introducing further errors. The way in which an error tends to propagate from step to step depends on the particular solution method in use, on the equation being solved, and in many cases also on the step size being used.

The way in which the error propagates is a measure of the stability of the solution method. One may concern oneself with either absolute or relative error in considering stability. If the absolute error tends to grow without bound, the method is said to be absolutely unstable, whereas if the absolute error tends to approach zero, the method is said to be absolutely stable. The intermediate case, where the error neither grows without bound nor approaches zero, is sometimes termed orbital stability. If relative error is being considered, one may speak of a method as being relatively stable or relatively unstable.

13.41 Methods of Numerical Solution

In the selection of a method for numerical solution of a differential equation, one must concern himself with the question of accuracy, that is, how accurately does the method give the next point on the solution curve, given some earlier points, and with the question of stability, that is, how do errors, once made, propagate.

No one method of obtaining solutions for differential equations like equation (13-20) turns out to be the best in all cases. There are many methods in use, each one representing a compromise among several competing considerations. In each of these methods, information about the solution curve at one or more points is used to obtain an approximation to a new point on the solution curve. Those methods that make use of information at only one point on the solution curve to compute a next point are normally termed Runge-Kutta methods. These methods have the advantage that they are easy to apply, since the only information required is a starting point and the step size to be used. They have the disadvantage that it is somewhat difficult to estimate the error.

Those methods that make use of information at more than one point on the solution curve to compute a next point tend to give more accuracy for a given amount of computation, since they make more use of available information. The most popular methods of this type are the so-called predictor-corrector methods, in which a new point is first computed based on some previous points, and then its value is adjusted using a correction formula to reduce the error. Such methods have the disadvantage that since they require several points on the solution curve, they are not self-starting. Given only the starting point, we must first generate some additional points on the curve by some other method to have enough points to apply a predictor-corrector method. Also, whenever the step size being used is reduced to enhance the accuracy of the solution or increased to reduce the amount of computation, some other method must be used to obtain a set of points suitable for restarting the predictor-corrector method. Thus, when predictor-corrector methods are used, a one-step method is ordinarily used for starting the solution or changing step size.

13.42 Runge-Kutta Methods

Suppose we have been computing the solution of the equation

$$y' = f(x, y) \tag{13-22}$$

and that the last point calculated was the point (x_n, y_n). We wish now to find a next point (x_{n+1}, y_{n+1}), where $x_{n+1} = x_n + h$; that is, we wish to take a step of size h in the independent variable x. By Taylor's formula, equation (6-1), we may write

$$y_{n+1} = y_n + hy_n' + \frac{h^2}{2!} y_n'' + \frac{h^3}{3!} y_n''' + \cdots + R_k \tag{13-23}$$

where y_n, y_n', etc., are values of y and its derivatives at the point x_n, and R_k is the remainder term if the expansion is terminated after k terms. We have assumed that y_n is known. The value of y_n' is also available to us from equation (13-22), for we have that

$$y_n' = f(x_n, y_n) \tag{13-24}$$

However, the values of y_n'', y_n''', etc., are not directly available to us.

The simplest approximation for y_{n+1} is to neglect all these higher derivatives and approximate (13-23) by

$$y_{n+1} = y_n + hf(x_n, y_n) \tag{13-25}$$

This formula can be used to generate a value for y_{n+1}, and then used recursively to generate the values for y_{n+2}, y_{n+3}, etc. It is known as Euler's formula. Historically it probably ranks as the first of the methods for numerical solution of differential equations, but it is not widely used because its qualifications in the departments of both accuracy and stability are not impressive. Because of its simplicity, however, we will use it to demonstrate the accuracy and stability considerations which apply to all numerical solution methods.

In the accuracy department, the error involved in a single step of the calculation is represented by the difference between equations (13-25) and (13-23). If h is small, the difference is roughly $(h^2/2)y_n''$. By differentiating (13-24), we find that

$$y_n'' = f_x(x_n, y_n) + f_y(x_n, y_n)f(x_n, y_n) \tag{13-26}$$

so the error introduced in a single step is

$$E = (h^2/2)[f_x(x_n, y_n) + f_y(x_n, y_n)f(x_n, y_n)] \tag{13-27}$$

Example 1. The equation $y' = x + y$ is being integrated by Euler's method. If $x_n = 1$, $y_n = 1$, and $h = .1$, what is the value of y_{n+1}? What is the approximate error in this value?

$$\text{By (13-25),} \quad y_{n+1} = 1 + .1(2) = 1.2$$

$$\text{By (13-27),} \quad E = .005(1 + 2) = .015$$

The actual solution to the differential equation is $y = 3e^{x-1} - x - 1$, which has the value $y = 1.2156$, which is consistent with the values of y_{n+1} and E just obtained.

Since the error is proportional to h^2, changing the step size will change the error in proportion to the square of the step size.

The Euler method is said to be a first-order formula. Most of the integration methods commonly used have a higher order of accuracy.

To investigate the stability question, let the value y_n be in error by an amount ϵ_n; that is, let the true value of y for $x = x_n$ be $y_n + \epsilon_n$. Then the value which should have been obtained for y_{n+1} from (13-25) is

$$y_{n+1} + \epsilon_{n+1} = y_n + \epsilon_n + hf(x_n, y_n + \epsilon_n)$$

$$\approx y_n + \epsilon_n + hf(x_n, y_n) + h\epsilon_n f_y(x_n, y_n)$$

or, subtracting (13-25) from this expression,

$$\epsilon_{n+1} = \epsilon_n(1 + hf_y(x_n, y_n)) \tag{13-28}$$

As long as f_y is positive, the absolute error will grow at each step, essentially in an exponential fashion. Thus the method is absolutely unstable if f_y is positive. If f_y is negative (and h is less than $1/f_y$) the absolute error will be reduced at each step, and the method is absolutely stable.

It should be noted that the error ϵ_{n+1} was computed assuming that the Euler formula (13-25) is exact and does not include any additional error caused by the fact that (13-25) neglects terms contained in the full Taylor's formula, (13-23). It is a measure of the growth of some previous error, without any new contribution in the current step. As such, it is a measure of the stability of the method, without reference to the accuracy of the approximation for individual steps. The question of stability is not directly related to the question of accuracy but is a separate and distinct consideration.

To investigate relative stability, divide both sides of (13-28) by y_{n+1}, giving

$$\frac{\epsilon_{n+1}}{y_{n+1}} = \frac{\epsilon_n}{y_{n+1}} (1 + hf_y(x_n, y_n))$$

Using (13-25),

$$\frac{\epsilon_{n+1}}{y_{n+1}} = \frac{\epsilon_n}{y_n} \left[\frac{1 + hf_y(x_n, y_n)}{1 + hf(x_n, y_n)/y_n} \right] \tag{13-29}$$

Thus the relative error will increase if the quantity in brackets is greater than one and decrease if it is less than one.

Example 2. The equation $y' = x + y$ is to be integrated, starting at $x_0 = 1$, $y_0 = 1$, and taking steps of size $h = .1$. Investigate the absolute and relative stability.

Using (13-28) we find that

$$\epsilon_{n+1} = 1.1\epsilon_n$$

If this relation is applied successively starting at step zero, we find that

$$\epsilon_n = (1.1)^n \epsilon_0$$

Thus as n increases, ϵ_n increases without bound, and so the method is absolutely unstable. Using (13-29), we find that

$$\frac{\epsilon_{n+1}}{y_{n+1}} = \frac{\epsilon_n}{y_n} \left[\frac{1.1}{1.1 + .1x_n/y_n} \right]$$

As long as y_n is positive, we have

$$\frac{\epsilon_{n+1}}{y_{n+1}} < \frac{\epsilon_n}{y_n}$$

and so the method is relatively stable.

As was mentioned earlier, Euler's method is not much used in practice, and the preference is for methods which use a more accurate approximation to equation (13-23), that is, methods which represent (13-23) correctly to higher powers in h. As already indicated, the values of y_n'', y_n''', etc., called for in (13-23) are not directly available. However, they can be obtained by differentiating (13-22),

$$y'' = f_x + f_y y' = f_x + f_y f$$

$$y''' = f_{xx} + f x_y y' + f_{xy}f + f_{yy}f y'+ f_y f_y y' \qquad \textbf{(13-30)}$$

$$= f_{xx} + 2f_{xy}f + f_{yy}f^2 + f_y^2 f \qquad \text{etc.}$$

An approximation to (13-23) accurate through h^2 is thus

$$y_{n+1} = y_n + hf(x_n, y_n) + \frac{h^2}{2} [f_x(x_n, y_n) + f_y(x_n, y_n)f(x_n, y_n)] \qquad \textbf{(13-31)}$$

This formula could be used as a more accurate one-step integration formula. However, its use would require the evaluation of the partial derivatives of $f(x, y)$ at each step as well as evaluation of the function itself. For many functions, these partial derivatives are difficult to find directly. Thus it is preferable to write (13-31) in a form in which they do not appear. This can be done at the expense of finding $f(x, y)$ for an additional pair of values of x and y. Let

$$k_1 = hf(x_n, y_n) \tag{13-32}$$

and

$$k_2 = hf(x_n + h/2, y_n + k_1/2) \tag{13-33}$$

If this last expression is expanded in a Taylor series, we have

$$k_2 = hf(x_n, y_n) + h^2[f_x(x_n, y_n)(h/2) + f_y(x_n, y_n)(k_1/2)] + O(h^3)$$

where $O(h^3)$ indicates terms involving third or higher power of h. Hence (13-30) can be written, to terms accurate through h^2, as

$$y_{n+1} = y_n + k_2 \tag{13-34}$$

Equations (13-32), (13-33), and (13-34) thus form a scheme for obtaining y_{n+1}, given x_n and y_n, to accuracy h^2, by evaluating $f(x, y)$ at two points, (x_n, y_n) and $(x_n + h/2, y_n + k_1/2)$. The equations describe a second-order Runge-Kutta process.

Figure 13-3 illustrates the process. Starting from point P, with coordinates (x_n, y_n), we proceed along line L, whose slope is $f(x_n, y_n)$. The point

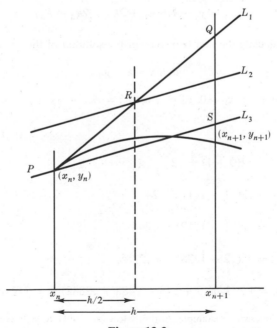

Figure 13-3

$(x_n + h, y_n + k)$, labeled Q in the figure, is the point which would be produced by the original Euler method. The point R, halfway between P and Q, is used to find a new slope, $f(x_n + h/2, y_n + k/2)$. In the figure, the line L_2 represents this slope. A line through P having this slope is used to find the point S, the location for the point (x_{n+1}, y_{n+1}).

The method just described is known as the modified Euler method and is one of a number of possible second-order Runge-Kutta methods. Other such methods can be generated by eliminating the f_x and f_y terms from (13-31) in different ways.

By far the most commonly used Runge-Kutta method is the fourth-order method, one which corresponds to retaining terms through h^4 in equation (13-23). It will be described without derivation and developed into a FORTRAN subroutine.* Given a step size h and a solution point (x_n, y_n), we obtain the next point (x_{n+1}, y_{n+1}) by the following steps:

$$k_1 = hf(x_n, y_n)$$

$$k_2 = hf(x_n + h/2, y_n + k_1/2)$$

$$k_3 = hf(x_n + h/2, y_n + k_2/2)$$

$$k_4 = hf(x_n + h, y_n + k_3)$$

$$y_{n+1} = y_n + 1/6(k_1 + 2k_2 + 2k_3 + k_4)$$

Example 3.   Calculate the first two steps of the solution of the equation

$$dy/dx = x + y$$

starting from the point $x_0 = 0$, $y_0 = 1$, with $h = .2$.

We have

$$k_1 = .2(0 + 1) = .2$$

$$k_2 = .2(.1 + 1.1) = .24$$

$$k_3 = .2(.1 + 1.12) = .244$$

$$k_4 = .2(.2 + 1.244) = .2888$$

$$y_1 = 1 + 1/6(.2 + .48 + .488 + .2888) = 1.24280$$

* For a derivation, see, for example, Anthony Ralston and H. S. Wilf, eds., *Mathematical Methods for Digital Computers*, Vol. 1, John Wiley & Sons, Inc., New York, 1967.

For the next interval,

$$k_1 = .2(.2 + 1.24280) = .288560$$

$$k_2 = .2(.3 + (1.24280 + .14428)) = .337416$$

$$k_3 = .2(.3 + (1.24280 + .16871)) = .342302$$

$$k_4 = .2(.4 + (1.24280 + .34230)) = .397020$$

$$y_2 = 1.24280 + 1/6(.288560 + .674832 + .684604 + .397020) = 1.58364$$

Example 4. Calculate the first step of the solution of

$$dy/dx = x + y$$

starting from the point $x_0 = 0$, $y_0 = 1$, with $h = .4$.

We have

$$k_1 = .4(0 + 1) = .4$$

$$k_2 = .4(.2 + 1.2) = .56$$

$$k_3 = .4(.2 + 1.28) = .592$$

$$k_4 = .4(.4 + 1.592) = .7968$$

$$y_1 = 1 + 1/6(.4 + 1.12 + 1.184 + .7968) = 1.58345$$

In the above two examples, we have presumably reached the same point on the solution curve, the first time by two smaller steps, the second time by one larger one. The values obtained for y at $x = .4$ are seen to differ slightly. For the above examples, an analytical solution can be found. It is $y = 2e^x - x - 1$, which for $x = .4$ has the value $y = 1.58364$. The error in the first case is below the level of accuracy to which the numbers were carried. The error in the second case was about .00019. In the fourth-order Runge-Kutta method, the error over an interval is proportional to h^5, so we would expect that reducing the step size by a factor of 2 should reduce this error by a factor of 32.

A flow chart for the Runge-Kutta method might appear as in Figure 13-4. This flow chart will take a starting set of values, (x_n, y_n), and compute new values, increasing x_n by an amount h each time, until the specified final value of x is reached. Normally the slowest part of the calculation is computing $f(x, y)$. For each point, the function $f(x, y)$ must be calculated four

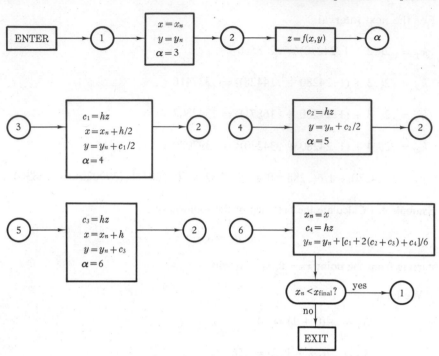

Figure 13-4: Integration by the Runge–Kutta method

times, so the total execution time is determined by the size chosen for h and the time required for a computation of $f(x, y)$. If h is too large, the accuracy will be poor. If h is too small, the time requirement will be excessively large. The FORTRAN subroutine given below does the calculation following the flow chart in Figure 13-4. The floating-point quantity D is used for the interval size.

```
      SUBROUTINE  RUNG(XN,YN,D,XEND,YEND)
      GIVF(X,Y)=   [insert correct expression for f(x, y)]
      DD=.5*D
    1 X=XN
      Y=YN
      J=1
    2 Z=GIVF(X,Y)
      GO  TO(3,4,5,6),J
    3 C1=D*Z
      X=XN+DD
      Y=YN+C1/2.
      J=2
      GO  TO 2
```

```
  4 C2=D*Z
    Y=YN+C2/2.
    J=3
    GO TO 2
  5 C3=D*Z
    X=XN+D
    Y=YN+C3
    J=4
    GO TO 2
  6 XN=X
    C4=D*Z
    YN=YN+(C1+2.*(C2+C3)+C4)/6.
    IF(XN-XEND)1,7,7
  7 XEND=XN
    YEND=YN
    RETURN
    END
```

13.43   The Euler-Romberg Method

One of the drawbacks of the Runge-Kutta methods is the estimation of the error incurred in each step. As was indicated, one way of coping with this problem is to build a program which will run each step with two step sizes, compare the results, and alter the basic step size as required to control the error. It will be recalled that a similar process was used in Chapter 7 as a means of adjusting the step sizes to give the desired accuracy in performing quadrature. It will also be recalled that one method of quadrature gained considerably by combining results for various step sizes, to obtain a result of improved accuracy. That was the Romberg method. The same approach can be applied to the Runge-Kutta integration methods and will be demonstrated here for the simplest of these, the Euler method. Suppose we have reached the point (x_n, y_n) and wish to take another step of size h. Let

$$Y_1 = y_n + hf(x_n, y_n) \tag{13-35}$$

be the result of applying the Euler formula (13-25) with step size h. Let Y_2 be the result obtained by applying the formula twice with step size $h/2$; that is, let

$$Z_1 = y_n + (h/2)f(x_n, y_n)$$

$$Y_2 = Z_1 + (h/2)f(x_n + h/2, Z_1)$$

The true value of y_{n+1} is given by Taylor's formula, (13-23). Using the relation (13-26), Taylor's formula can be written

$$y_{n+1} = y_n + hf(x_n, y_n) + (h^2/2)[f_x(x_n, y_n) + f_y(x_n, y_n)f(x_n, y_n)] + O(h^3)$$

$$(13\text{-}36)$$

Comparing this with (13-35), we see that the error in Y_1 can be written as

$$y_{n+1} - Y_1 = (h^2/2)[f_x(x_n, y_n) + f_y(x_n, y_n)f(x_n, y_n)] + O(h^2) \quad (13\text{-}37)$$

To find the error in Y_2, we write

$$Y_2 = y_n + (h/2)f(x_n, y_n) + (h/2)f(x_n + h/2, y_n + (h/2)f(x_n, y_n))$$

$$= y_n + hf(x_n, y_n) + (h^2/4)[f_x(x_n, y_n) + f_y(x_n, y_n)f(x_n, y_n)]$$

Subtracting this from (13-36), we have

$$y_{n+1} - Y_2 = (h^2/4)[f_x(x_n, y_n) + f_y(x_n, y_n)f(x_n, y_n)] + O(h^3) \quad (13\text{-}38)$$

Comparing this with (13-37), we see that the h^2 term is easily eliminated by doubling (13-38) and subtracting (13-37) from it, giving

$$y_{n+1} - 2Y_2 + Y_1 = O(h^3)$$

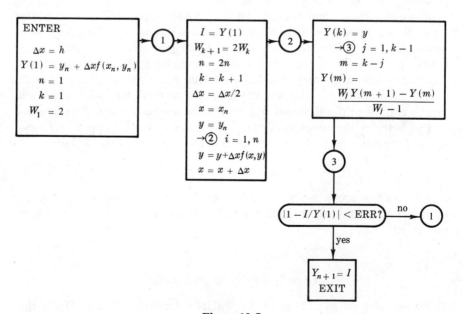

Figure 13-5

Thus Y_1 and Y_2 each approximate y_{n+1} to order h^2, but the combination $2Y_2 - Y_1$ approximates y_{n+1} to order h^3. If we were to compute a Y_3 by applying Euler's formula with step size $h/4$, and then combine it properly with Y_1 and Y_2, a still higher order of approximation could be obtained.

It is seen that the process is analogous to that used in Romberg integration in Chapter 7. The main difference is that in the case of Romberg integration the series in x had only even powers, whereas the present series in h contains all powers. In fact, the process is so similar that only two types of modification are required to change the flow chart for Romberg integration, Figure 7-9, into one for Euler-Romberg integration. First, the process of computing the trapezoidal sums, $TR(k)$, of Figure 7-9, must be replaced by the process of computing Y_k, obtained by applying Euler's formula 2^k times with step size $h/2^k$. Second, the relations $W_1 = 4$ and $W_{k+1} = 4W_k$ must be replaced by $W_1 = 2$ and $W_{k+1} = 2W_k$. This has been done to obtain the flow chart, Figure 13-5. A subroutine based on this flow chart is given below.

```
      SUBROUTINE OILROM(XN,YN,DX,YP,ERR,FAIL)
      DIMENSION Y(20),W(20)
      D=DX
      Y(1)=YN+D*DERIV(XN,YN)
      N=1
      K=1
      W(1)=2.
    1 FI=Y(1)
      W(K+1)=2.*W(K)
      N=2*N
      K=K+1
      D=D/2.
      X=XN
      T=YN
      DO 2 I=1,N
      T=T+D*DERIV(X,T)
    2 X=X+D
      Y(K)=T
      KK=K-1
      DO 3 J=1,KK
      M=K-J
    3 Y(M)=(W(J)*Y(M+1)-Y(M))/(W(J)-1.)
      FAIL=ABS(1.-FI/Y(1))
      IF(FAIL-ERR)6,6,4
    4 IF(K-20)1,6,6
    6 YP=Y(1)
      RETURN
      END
```

The subroutine performs one integration step. Input quantities are XN and YN, the step size DX, and the allowed error ERR. Outputs are YP, which stands for y_{n+1}, and a quantity FAIL. This is the value of the relative error at the time of exit. If an accuracy corresponding to ERR is not achieved by the time the interval has been subdivided into 2^{20} parts, the exit occurs anyhow, and the value of FAIL can be used to detect this condition. The subroutine requires a driver routine to specify the starting point and step size and to call the subroutine repeatedly if several steps are to be taken.

The above subroutine provides automatic error control only to the extent that it controls the amount of new error introduced in each step. It does not prevent the growth of errors previously introduced. The question of stability is a difficult one and will not be treated here, but the reader is cautioned that the method will be unstable for some differential equations.*

13.44   Predictor-Corrector Methods

As mentioned earlier, there are many integration schemes that depend on values at more than one previous point in order to estimate the next point on the solution curve. Among these are the predictor-corrector methods.

Suppose that for equally spaced values $x_1, x_2, \ldots, x_n$ we have obtained solution values $y_1, y_2, \ldots y_n$ by some method, and now want to obtain y_{n+1}. We can write

$$y_{n+1} - y_n = \int_{x_n}^{x_{n+1}} y' \, dx \tag{13-39}$$

Now y' is a function of x, $y'(x)$, which we know at the points x_n, x_{n-1}, etc., because from the differential equation $y_n' = f(x_n, y_n)$, $y_{n-1}' = f(x_{n-1}, y_{n-1})$ and so forth. If we approximate $y'(x)$ by a polynomial passing through the points (x_n, y_n'), (x_{n-1}, y_{n-1}'), etc., we can integrate (13-39) directly to obtain a formula for y_{n+1}. For example, if we use only the two preceding points, x_n and x_{n-1}, the polynomial would be the straight line

$$y'(x) = \frac{x_n - x}{h} y_{n-1}' + \frac{x - x_{n-1}}{h} y_n' \tag{13-40}$$

where $h = x_n - x_{n-1}$. If this is integrated from x_n to x_{n+1} we have

$$\int_{x_n}^{x_{n+1}} y'(x) \, dx = -\frac{(x_n - x)^2}{2h} y_{n-1}' + \frac{(x - x_{n-1})^2}{2h} y_n' \bigg|_{x_n}^{x_{n+1}}$$

$$= -\frac{h}{2} y_{n-1}' + 2hy_n' - \frac{h}{2} y_n'$$

---

* For some discussion of the stability problem, see, e.g. Thomas McCalla, *Introduction to Numerical Methods and FORTRAN Programming*, John Wiley & Sons, Inc., New York, 1967.

or, substituting into (13-39),

$$y_{n+1} = y_n + (h/2)(3y_n' - y_{n-1}') \tag{13-41}$$

This formula will allow a prediction of y_{n+1}. Note, however, that the formula was based on (13-40), an expression for linear interpolation between x_{n-1} and x_n, and we have used it to integrate from x_n to x_{n+1}. Using an interpolation formula to extrapolate outside the range can be expected to produce inaccurate results. Thus it would appear advisable to attempt to improve on this estimate of y_{n+1}. Now that we have a value for y_{n+1} we can compute $y_{n-1}' = f(x_{n+1}, y_{n+1})$. Then (13-40) can be rewritten in terms of the points x_n and x_{n+1} as

$$y'(x) = \frac{x_{n+1} - x}{h} y_n' + \frac{x - x_n}{h} y_{n+1}' \tag{13-42}$$

If this is integrated from x_n to x_{n+1} we have

$$\int_{x_n}^{x_{n+1}} y'(x)\,dx = -\frac{(x_{n+1} - x)^2}{2h} y_n' + \frac{(x - x_n)^2}{2h} y_{n+1}' \bigg|_{x_n}^{x_{n+1}}$$

$$= (h/2)(y_n' + y_{n+1}')$$

Substituting into (13-39), we have

$$y_{n+1} = y_n + (h/2)(y_n' + y_{n+1}') \tag{13-43}$$

This time we have used an interpolation formula based on x_n and x_{n+1} and have integrated between the same two points, so the result should be more accurate than the previous one. Formula (13-43) can be used to obtain another, hopefully better, estimate of y_{n+1}. In fact, it could be used a number of times, to correct the estimate of y_{n+1} still further. Ordinarily an efficient way to proceed is to use the correction formula (13-43) just once, and compare the value of y_{n+1} obtained with that obtained from the prediction formula (13-41). If the difference is too large, reduce the step size h and restart with a smaller step size.

The formulas just derived belong to a class known as Adams-Moulton formulas.* Other predictor-corrector formulas of this class can be derived by using a higher order polynomial to approximate $y'(x)$ in (13-39). For this

---

* For a more complete discussion of these and other predictor-corrector formulas, see, e.g., R. W. Hamming, *Numerical Methods for Scientists and Engineers*, McGraw-Hill, New York, 1962.

purpose it is convenient to use Newton's backward interpolation formula, (12-7), to represent $y'(x)$. We can write

$$y'(x) = y_n' + \Delta y_{n-1}' u + \frac{\Delta^2 y_{n-2}'}{2!} u(u+1) + \frac{\Delta^3 y_{n-3}'}{3!} u(u+1)(u+2) + \cdots$$

$$(13\text{-}44)$$

In this formula, $u = (x - x_n)/h$ so $dx = h\,du$. To obtain the predictor formula, we integrate the above from $x = x_n$ to $x = x_{n+1}$ (or $u = 0$ to $u = 1$).

$$y_{n+1} - y_n = \int_{x_n}^{x_{n+1}} y'\,dx = \int_0^1 \left[y_n' + \Delta y_{n-1}' u + \frac{\Delta^2 y_{n-2}'}{2!} u(u+1) \right.$$

$$\left. + \frac{\Delta^3 y_{n-3}'}{3!} u(u+1)(u+2) + \cdots \right] h\,du$$

$$= h[y_n' + (1/2)\Delta y_{n-1}' + (5/12)\Delta^2 y_{n-2}'$$

$$+ (3/8)\Delta^3 y_{n-3}' + \cdots] \quad (13\text{-}45)$$

If the first difference is replaced by its value in terms of y_n' and y_{n-1}', we have

$$y_{n+1} - y_n = (h/2)(3y_n' - y_{n-1}') + (5/12)h\,\Delta^2 y_{n-2}' + \cdots, \quad (13\text{-}46)$$

which is the predictor formula (13-41) with remainder terms. If the second difference is also replaced by its value in terms of y_n', y_{n-1}', and y_{n-2}' we have another predictor formula

$$y_{n+1} = y_n + (h/12)(23y_n' - 16y_{n-1}' + 5y_{n-2}') + (3/8)h\,\Delta^3 y_{n-3}' + \cdots \quad (13\text{-}47)$$

The first remainder term in this case depends on a third difference rather than on the second difference as does (13-46). Other, higher order predictor formulas can be obtained in an analogous manner.

To obtain the corrector formulas we rewrite (13-44) in terms beginning at x_{n+1} instead of x_n. Thus

$$y'(x) = y_{n+1}' + \Delta y_n' u + \frac{\Delta^2 y_{n-1}'}{2!} u(u+1) + \frac{\Delta^3 y_{n-2}'}{3!} u(u+1)(u+2) + \cdots$$

$$(13\text{-}48)$$

In this formula $u = (x - x_{n+1})/h$.

Integrating from $x = x_n$ to $x = x_{n+1}$, or $u = -1$ to $u = 0$ we have

$$y_{n+1} - y_n = \int_{x_n}^{x_{n+1}} y' \, dx$$

$$= \int_{-1}^{0} \left[y'_{n+1} + \Delta y_n' u + \frac{\Delta^2 y'_{n-1}}{2!} u(u+1) \right.$$

$$\left. + \frac{\Delta^3 y'_{n-2}}{3!} u(u+1)(u+2) + \cdots \right] h \, du$$

$$= -h[-y'_{n+1} + (1/2)\Delta y_n' + (1/12)\Delta^2 y'_{n-1}$$

$$+ (1/24)\Delta^3 y'_{n-2} + \cdots] \quad \textbf{(13-49)}$$

If the first difference is replaced by its values in terms of y'_{n+1} and y_n', we have

$$y_{n+1} = y_n + (h/2)(y_n' + y'_{n+1}) - (h/12) \, \Delta^2 y'_{n-1} + \cdots \quad \textbf{(13-50)}$$

which is the corrector formula (13-43) with remainder terms. If the second difference is also replaced by its value in terms of y'_{n+1}, y_n', and y'_{n-1}, we have a higher order corrector formula

$$y_{n+1} = y_n + (h/12)(5y'_{n+1} + 8y_n' - y'_{n-1}) + (h/24)\Delta^3 y_{n-2} + \dots \quad \textbf{(13-51)}$$

Again, higher order corrector formulas can be obtained in an analogous manner.

13.45   Step Size in Predictor-Corrector Methods

It was pointed out in Section 13.44 that the difference in the values of y_{n+1} as given by the predictor formula and the corrector formula could be used to obtain an accuracy estimate. To see how this can be done, consider the predictor-corrector formulas (13-47) and (13-51). It is seen that the error in y_{n+1} as determined from the predictor formula is about

$$E_1 = (3/8)h \, \Delta^3 y'_{n-3} \quad \textbf{(13-52)}$$

and the error in y_{n+1} as determined from the corrector formula is about

$$E_2 = (1/24)h \, \Delta^3 y'_{n-2} \quad \textbf{(13-53)}$$

Let y be the true value at $x = x_{n+1}$, y^p_{n+1} the value given by the predictor formula, and y^c_{n+1} the value given by the corrector formula. Then

$$y - y^p_{n+1} \approx E_1$$

and

$$y - y^c_{n+1} \approx E_2$$

Subtracting the second of these relations from the first, we have

$$y^c_{n+1} - y^p_{n+1} \approx E_1 - E_2$$

If $\Delta^3 y'_{n-3}$ and $\Delta^3 y'_{n-2}$ are of the same magnitude, then $E_1 \approx 9E_2$ and

$$y^c_{n+1} - y^p_{n+1} \approx 8E_2$$

or

$$E_2 \approx (1/8)(y^c_{n+1} - y^p_{n+1}) \qquad \textbf{(13-54)}$$

Of course we have no guarantee that $\Delta^3 y'_{n-3}$ and $\Delta^3 y'_{n-2}$ are of the same magnitude, nor indeed that they even correctly represent the errors in y^p_{n+1} and y^c_{n+1}. To obtain an upper bound estimate we would have had to make a more vigorous derivation of the predictor-corrector formulas in Section 13.44. In using Newton's backward interpolation formula we would have had to include the remainder term given by (12-13). As pointed out in Section 12.36 however, that approach would still leave us with the problem of determining an upper bound for some derivative of $y^{(4)}(x)$ at some unknown point ξ, a situation similar to that which occurred in the quadrature formulas of Chapter 7. As in that case, we dismiss the process as being generally too difficult to accomplish analytically and content ourselves with the observations that if the step size h is small the quantities $\Delta^3 y'_{n-3}$ and $\Delta^3 y'_{n-2}$ should be good approximations of the errors for their respective formulas and that furthermore they should be near each other in value. Thus with some reservation we accept (13-54) as the estimate of the error in y^c_{n+1}.

Having accepted (13-54) as a basis for an error estimate we can now write a FORTRAN subroutine to apply the Adams-Moulton formulas with automatic step size control. If we use the formulas (13-47) and (13-51) as the predictor-corrector formulas, we need to store three consecutive values of y and y' in order to compute the next value. To change step size at any point we must restart the solution at that point with three new starting values corresponding to the new step size. It is easiest to double or halve the step size, since in that way we can use values already available. To double the step size, we need to have saved five consecutive values at the old size. By dropping out two of these, we will have three equally spaced values at the new step size. Two new values must then be computed at this new step size before we will have enough points to double the step size again. To halve the step size, it is necessary to interpolate values between some of those already available. For accuracy, it is wise to move back slightly from the point at which the step size became too large and to use a central interpolation formula for this purpose. Bessel's interpolation formula, Section 13-34, is frequently used for this purpose. It takes on a particularly simple form when used for the mid-point of an interval. Taking $u = 1/2$ and retaining terms only up to the fourth difference, we have

$$y_{1/2} = y_0 + (1/2)\Delta y_0 - (1/16)(\Delta^2 y_{-1} + \Delta^2 y_0) + \cdots$$

Substituting the values of these differences as obtained from Section 12.21

$$y_{1/2} = (1/2)(y_0 + y_1) - (1/16)(y_2 - y_1 - y_0 + y_{-1})$$

If the last accurate value of y we have obtained is y_n, then we may use this formula to obtain

$$y_{n-3/2} = (1/2)(y_{n-2} + y_{n-1}) - (1/16)(y_n - y_{n-1} - y_{n-2} + y_{n-3}) \quad \textbf{(13-55)}$$

In order to retain a set of five values, so that the step size can again be reduced if required, it is well to have the machine also compute

$$y_{n-5/2} = (1/2)(y_{n-3} + y_{n-2}) - (1/16)(y_{n-1} - y_{n-2} - y_{n-3} + y_{n-4}) \quad \textbf{(13-56)}$$

The derivatives may be interpolated in like manner, and then the solution restarted with the three values $y_{n-2}, y_{n-5/2}, y_{n-1}$.

In order to incorporate the above information into a program let us assume we have five points Y(1), Y(2), Y(3), Y(4), and Y(5) and the corresponding derivatives YP(1), YP(2), YP(3), YP(4), and YP(5). At the next step we are to compute Y(6) and YP(6). With this notation the predictor formula (13-47) becomes

$$Y(6) = Y(5) + (H/12)*(23.*YP(5) - 16.*YP(4) + 5.*YP(3)) \quad \textbf{(13-57)}$$

and the corrector formula (13-51) becomes

$$Y(6) = Y(5) + (H/12.)*(5.*YP(6) + 8.*YP(5) - YP(4)) \quad \textbf{(13-58)}$$

If the accuracy check shows that the two results are satisfactorily close together we want to proceed with another step the same size, first making the redefinitions

$$\begin{array}{ll}
Y(5) = Y(6) & YP(5) = YP(6) \\
Y(4) = Y(5) & YP(4) = YP(5) \\
Y(3) = Y(4) & YP(3) = YP(4) \\
Y(2) = Y(3) & YP(2) = YP(3) \\
Y(1) = Y(2) & YP(1) = YP(2)
\end{array}$$

If the accuracy check shows that the two results are unnecessarily close together we want to double the step size by setting $H = 2.*H$ and making the redefinitions

$$\begin{array}{ll}
Y(5) = Y(6) & YP(5) = YP(6) \\
Y(3) =: Y(2) & YP(3) = YP(2)
\end{array}$$

The values of Y(4) and YP(4) remain unchanged. We cannot double the step size again until we have taken two successful steps at this size, to obtain new Y(1), Y(2), YP(1), and YP(2).

If the accuracy check shows that the two results of (13-57) and (13-58) are not sufficiently close together we want to back up and try again with a smaller step size. To do so we use the interpolation formulas (13-55) and (13-56) to obtain intermediate values of Y and YP. For the Y's the relations can be written

$$Q1 = .5*(Y(2) + Y(3)) - .0625*(Y(4) - Y(3) - Y(2) + Y(1))$$
$$Q2 = .5*(Y(3) + Y(4)) - .0625*(Y(5) - Y(4) - Y(3) + Y(2))$$

and then

$$\begin{array}{l}
Y(5) = Y(4) \\
Y(1) = Y(2) \\
Y(4) = Q2 \\
Y(2) = Q1
\end{array}$$

with a similar set of relations for the YP's. Figure 13-6 summarizes the redefinitions and interpolations used in the step size control scheme just described. The FORTRAN subroutine below incorporates the scheme.

```
SUBROUTINE ADMO(XO,Y,YP,H,ERR,XEND)
DIMENSION Y(6),YP(6)
GIVF(X,Z) =   [insert correct expression for f(x, z)]
```

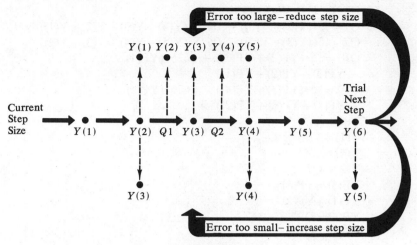

Figure 13-6

```
    X=XO
    ERP=16.*ERR
    ERG=.01*ERP
10  A=H/12.
    K=1
20  X=X+H
    YS=Y(5)+A*(23.*YP(5)-16.*YP(4)+5.*YP(3))
    YP(6)=GIVF(X,YS)
    Y(6)=Y(5)+A*(5.*YP(6)+8.*YP(5)-YP(4))
    YP(6)=GIVF(X,Y(6))
    E=ABS(Y(6)-YS)
    IF(E-ERP*Y(6))30,30,60
30  IF(X-XEND)35,100,100
35  IF(E-ERG*Y(6))50,50,40
40  K=K+1
    DO 45 I=1,5
    Y=Y(I+1)
45  YP(I)=YP(I+1)
    GO TO 20
50  IF(K-3)40,55,55
55  Y(5)=Y(6)
    Y(3)=Y(2)
    YP(5)=YP(6)
    YP(3)=YP(2)
    K=1
    H=H+H
    GO TO 10
```

```
   60 IF(H−1.E−7*X)65,999,999
   65 Q1 = .5*(Y(2)+Y(3))−.0625*(Y(4)−Y(3)−Y(2)+Y(1))
      Q2 = .5*(Y(3)+Y(4))−.0625*(Y(5)−Y(4)−Y(3)+Y(2))
      QP1 = .5*(YP(2)+YP(3))−.0625*(YP(4)
      −YP(3)−YP(2)+YP(1))
      QP2 = .5*(YP(3)+YP(4))−.0625*(YP(5)
      −YP(4)−YP(3)+YP(2))
      Y(5) = Y(4)
      Y(1) = Y(2)
      Y(4) = Q2
      Y(2) = Q1
      YP(5) = YP(4)
      YP(1) = YP(2)
      YP(4) = QP2
      YP(2) = QP1
      X = X−2.*H
      H = .5*H
      GO TO 10
  100 XEND = X
      RETURN
  999 PRINT 1000
      STOP
 1000 FORMAT(33H ROUNDOFF ERROR PREVENTS
      SOLUTION)
      END
```

   This subroutine requires as inputs five consecutive values of Y and YP,
Y(1) through Y(5) and YP(1) through YP(5); XO, the X value corresponding
to the last of these; H the step size by which these values are separated; ERR,
the acceptable relative error; and XEND, the value of X to which the integration is to be performed. It returns when this value is exceeded, with XEND
reset to the last X value actually used, and the corresponding Y value stored
as Y(6). The statements following 20 perform a time step, using the formulas
(13-57) and (13-58). The series of IF tests determines the conditions for the
next step. If E, the difference between predicted and corrected values for
Y(6), is greater than the allowed value, the routing is to statement 60, where
the step size is to be reduced. If the step size is already so small that roundoff
error will dominate, an error exit to statement 999 is taken. Otherwise the
step size is reduced and the program branches back to statement 10 to take a
new step. At statement 35, the program determines if the step size is too small,
that is if the calculated error is less than one hundredth of the required error.
If this is the case the program branches to statement 50, where the step size
is to be doubled. First a check is made, and the step size is not doubled unless

three consecutive successful steps have been taken at the current size. This fact is determined from the quantity K, which is increased by one each time a successful step is taken at the current size and reset to one each time the step size is changed.

13.46   Stability for the Predictor-Corrector Methods

As has been the case with the other methods discussed, the accuracy control imposed in SUBROUTINE ADMO of Section 13.45 in the form of automatic step size control does not insure an accurate solution to the differential equation. There is still the stability problem, and for some differential equations the solution may be instable.*

13.5   SYSTEMS OF DIFFERENTIAL EQUATIONS

The methods of Section 13.4 are all readily extensible to systems of ordinary differential equations, and even the FORTRAN subroutines given there can be modified in a straightforward manner to be applicable to systems of equations. To solve the system of k equations

$$y_1' = f_1(x, y_1, y_2, \ldots, y_k)$$

$$y_2' = f_2(x, y_1, y_2, \ldots, y_k) \tag{13-59}$$

$$\vdots$$

$$y_k' = f_k(x, y_1, y_2, \ldots, y_k)$$

using the Runge-Kutta method one simply applies the method of Section 13.42 to all equations simultaneously. In subroutine RUNG, the variables Y, Z, and YN are changed to dimensioned variables, and the function GIV(X,Y) replaced by the whole set of functions corresponding to (13.59), and appropriate DO loops introduced to perform the operations for each subscript value. An analogous situation applies for the other methods in Section 13.4. Equations involving higher derivatives can be transformed into the form represented by the system (13-59) by the addition of new variables. For example, the equation

$$x^2 y'' + yy' + xy^3 = 0$$

* For a discussion of stability in predictor-corrector methods, see, e.g. Daniel D. McCracken and William S. Dorn, *Numerical Methods and FORTRAN Programming*, John Wiley and Sons, Inc., New York, 1964.

can be written as

$$y' = u$$

$$u' = -(yy' + xy^3)/x^2$$

The concepts regarding accuracy and stability also carry over from a single equation to systems of equations, compounded in difficulty as one might expect.

EXERCISE 32

1. The equation $y' = y$ is to be solved by the Euler method, starting with $x = 0$, $y = 1$.
   a. Find an expression for the absolute error at step $n+1$, ϵ_{n+1}, in terms of the absolute error at step n, ϵ_n, and the step size h.
   b. Find an expression for the error at step n in terms of the error at step zero, ϵ_0.
   c. If the initial error is .001, find the error after 10 steps at step size .1; 100 steps at step size .01; 1000 steps at step size .001.

2. For the modified Euler method, equations (13-32), (13-33), and (13-34), derive a formula that will relate the absolute error in y at step $n+1$ to that at step n.

3. Use the formula derived in problem 2 to determine the stability of the modified Euler method in solving the equation $y' = x + y$ with starting values $x = 1$, $y = 1$.

4. Prepare a remote-terminal routine that will serve as a driver routine for SUBROUTINE OILROM of Section 13.43, one which will ask for XN, YN, DX, and ERR as inputs, call OILROM, print out YP and FAIL, then ask for new inputs. For the function DERIV(X,Y), use DERIV=X+Y. Run the program with the following inputs:
   a. X = 1, Y = 1, DX = .1, ERR = .01
   b. X = 1, Y = 1, DX = 2, ERR = .01
   c. X = 1, Y = 1, DX = 2, ERR = .0001

5. Rerun the program of problem 4c with the following PRINT statements added in SUBROUTINE OILROM.

   PRINT, D   after D = .5*D
   PRINT, M,Y(M)   after M = K − J
   PRINT, FAIL   after the statement that computes FAIL.

   On the printout, note that each printed value of FAIL corresponds to a halving of the step size. From the changes in size of FAIL, what is the apparent order of accuracy of the Euler-Romberg method when four subdivisions are used? Eight? Sixteen? From the discussions following equation (13-38), what would you expect the order of accuracy to be for these cases?

6. Using the Runge-Kutta method and a step size of .1, calculate four steps of the solution of

   a. $y' = xy,$ $y(0) = 1.$
   b. $y' = y + e^{-x},$ $y(0) = 2.$
   c. $y' = x + 2y,$ $y(0) = .5$

7. Modify subroutine RUNG of Section 13.42 to print out $x, y,$ and y'

   a. At each step.
   b. Every 10 steps.
   c. Every time x has increased by .5.

8. From the relations (13-45) and (13-49), derive a form of the Adams-Moulton formulas accurate through terms representing third differences.

9. Write a version of SUBROUTINE ADMO of Section 13.45 that uses the predictor-corrector equations derived in problem 8.

10. Revise SUBROUTINE ADMO of Section 13.45 to make two applications of the corrector formula at each step instead of one.

11. Write a version of SUBROUTINE RUNG of Section 13.42 that can be used to solve a system of 3 differential equations.

12. Write a version of SUBROUTINE RUNG of Section 13.42 that can be used to solve a system of N differential equations, where N is an input number from 1 to 5.

Index

A
O